P. Webster.

Refractories

Fourth Edition

Refractories

F. H. NORTON

Professor Emeritus
Massachusetts Institute of Technology
Consultant

McGRAW-HILL BOOK COMPANY

New York   San Francisco   Toronto   London   Sydney

Preface

During the last twenty years, since the Third Edition of "Refractories" was published, many changes have taken place—both in the use and production of refractories. New raw materials, such as sea-water magnesite, have come into use; and others, like diaspore, have been exhausted. New products, such as the direct-bonded magnesite-chrome brick, the fusion-cast basic brick, the high-fired super-duty brick, high-use-limit insulating firebrick, and improved castables, all have made possible more efficient production in many industries. There have been many changes in consumer demands; the steel industry is going over rapidly to the basic oxygen converter which uses fewer and less costly refractories than the rapidly disappearing open-hearth furnace, and the steam-power generator now uses very little in the way of refractories.

It seems appropriate at this time to issue an up-to-date revision of this work. However, the field has become so broad that it has seemed unwise to try to combine both the heavy refractories and the technical ceramics, as in the Third Edition. Therefore, here we shall only concern ourselves with the heavy refractories and leave for another volume the pure oxides, porcelains, special electrical bodies, and nuclear fuels. This decision has made it advisable to make considerable rearrangements in the text. The various subjects have been approached not from the point of view of the scientist, but rather as an interpretation of the scientific advances in the field that will be understandable to the producer and user of refractories, many of whom cannot be experts in all of the many fields of science now applied to the development of new refractories. It is interesting to remember that every major development of use to the industry or the public was known in the scientific literature twenty or thirty years earlier. The only important exception is the transistor, which came into use in a few years. Therefore, anything that can be done to bring out new ideas where practical minds can make use of them will be well worth while to our industry.

The author wishes to thank those who have so kindly helped in this revision. He is particularly indebted to Bruce S. Olds, Leonard Gallagher, and Bruce M. Putnam, all of Arthur D. Little, Inc. Louise Halm of the Institut de Recherches de la Sidérurgie and M. Y. Letort of Société Française de Céramique have been most helpful in bringing him up to date on Continental refractory practices.   J. H. Chesters and J. White have aided in keeping up to date on British use of refractories.   R. E. Birch of the Harbison-Walker Refractories Company has been very helpful in providing information concerning the United States refractories industry.   E. C. Leiberg of the Corhart Refractories Company has been most generous in providing information on fused-cast refractories.

It is also desired to thank all those individuals connected with the refractories industry who have been so helpful in supplying data and illustrations.

F. H. Norton

Contents

Part 4 PROPERTIES

Part 5   USE

Refractories

PART ONE

Introduction

CHAPTER ONE

Scope of
the Refractories Industry
in the United States

1.1 Introduction. Since the publication of the Third Edition of "Refractories" there have been many changes in the industry. It is hoped that in this volume the important ones can be pointed out. One of the more concrete indicators is the production statistics given in the next section.

1.2 Statistics of the Refractories Industry. The annual value of refractories produced in the period from 1947 to 1965 has shown a fairly steady climb of about 16 million dollars a year, as shown in Fig. 1.1. A

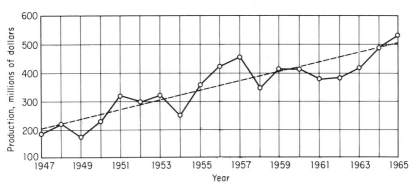

FIG. 1.1 Total production of refractories in the United States.

TABLE 1.1 Refractories Industry Statistics—Clay Refractories Production*

Year	Fireclay bricks: Bricks, blocks, or tiles for locomotive and other firebox lining, etc., 9-in. equivalents including superduty and insulating firebrick			High-alumina bricks: Bricks, blocks, or tiles containing over 40 % alumina, 9-in. equivalents			Insulating firebrick			Total clay firebricks		
	Thousands	Value	Unit value	Thousands	Value	Unit value	Thousands	Value	Unit value	Thousands	Value	Unit value
1947	791,307	$62,777,000	$79.33	25,184	$5,347,000	$215.89	44,805	$6,027,000	$134.52	861,296	$74,151,000	$86.09
1948	721,986	67,269,000	93.17	22,498	4,896,000	217.62	43,668	5,997,000	137.33	788,152	78,162,000	99.17
1949	580,976	56,499,000	97.25	16,459	4,392,000	266.84	29,239	5,840,000	199.73	626,674	66,731,000	106.48
1950	623,597	67,597,000	108.40	17,620	4,757,000	269.98	40,986	7,635,000	186.28	682,203	79,989,000	117.25
1951	820,575	94,720,000	115.43	23,643	6,825,000	288.67	56,052	11,076,000	197.60	900,270	112,621,000	125.10
1952	724,757	87,800,000	121.14	22,251	6,655,000	299.09	60,343	11,510,000	190.74	807,351	105,965,000	131.25
1953	613,259	79,805,000	130.13	17,694	5,337,000	301.63	53,781	9,925,000	184.54	684,734	95,067,000	138.84
1954	481,654	64,066,000	133.01	16,763	5,547,000	330.91	39,568	8,024,000	202.79	537,985	77,637,000	144.31
1955	562,859	78,359,000	139.22	21,132	7,138,000	337.78	54,178	11,196,000	206.65	638,169	96,693,000	151.52
1956	604,528	92,070,000	152.30	23,593	8,631,000	365.83	62,490	13,698,000	219.20	690,611	114,399,000	165.65
1957	561,162	94,412,000	168.24	23,101	9,460,000	409.51	57,498	13,583,000	236.23	641,761	117,455,000	183.02
1958	380,197	71,901,000	189.12	17,395	7,464,000	429.09	38,600	9,386,000	243.16	436,192	88,751,000	203.47
1959	403,829	70,135,000	173.67	20,031	8,919,000	445.26	44,596	10,672,000	239.30	468,456	89,726,000	191.54
1960	411,487	73,846,000	179.46	21,962	9,978,000	454.33	47,250	11,681,000	247.22	480,699	95,505,000	198.68
1961	341,200	62,519,000	183.23	23,346	10,851,000	464.79	39,412	9,834,000	249.52	403,958	83,204,000	205.97
1962	341,227	61,356,000	179.81	28,539	12,998,000	455.45	42,852	10,823,000	252.57	412,618	85,177,000	206.43
1963	344,196	63,346,000	184.04	32,596	15,373,000	471.62	44,302	11,337,000	255.90	421,094	90,056,000	213.86
1964	362,262	67,023,000	185.01	38,443	19,150,000	498.14	54,330	13,443,000	247.43	455,035	99,616,000	218.92
1965	393,390	75,237,000	191.25	45,955	23,894,000	519.94	62,203	16,249,000	261.23	501,548	115,380,000	230.05

*"Facts for Industry," U.S. Department of Commerce.

4

TABLE 1.2 Refractories Industry Statistics—Nonclay Refractories Production*

Year	Silica			Magnesite and chrome			Other nonclay, includes those of alumina and silicon carbide			Total nonclay refractories		
	Thousands	Value	Unit value	Thousands	Value	Unit value	Thousands	Value	Unit value	Thousands	Value	Unit value
1947	313,001	$29,407,000	$ 93.95	61,223	$ 23,329,000	$381.05	13,323	$18,571,000	$1,393.91	387,547	$ 71,307,000	$184.00
1948	321,443	34,118,000	106.14	66,170	27,790,000	419.98	12,752	20,533,000	1,610.18	400,365	82,441,000	205.91
1949	266,596	30,320,000	113.73	54,355	23,591,000	434.02	8,026	16,975,000	2,115.00	328,977	70,886,000	215.47
1950	312,334	38,222,000	122.38	74,663	34,087,000	456.54	11,385	24,147,000	2,120.95	398,382	96,456,000	242.12
1951	368,653	51,686,000	140.20	96,979	49,143,000	506.78	15,280	33,415,000	2,186.85	480,912	134,244,000	279.14
1952	336,579	46,797,000	139.04	87,128	45,607,000	523.45	11,168	25,139,000	2,250.98	434,875	117,543,000	270.29
1953	343,827	54,033,000	157.15	97,775	53,657,000	548.78	13,853	30,664,000	2,213.53	455,455	138,354,000	303.77
1954	228,768	37,875,000	165.56	66,696	37,712,000	565.43	10,025	27,830,000	2,776.06	305,489	103,417,000	338.53
1955	328,414	55,563,000	169.19	107,254	60,864,000	567.48	24,555	33,985,000	1,384.04	460,223	150,412,000	326.82
1956	316,653	58,979,000	186.26	111,304	69,467,000	624.12	27,400	41,303,000	1,507.41	455,357	169,749,000	372.78
1957	304,210	62,002,000	203.81	1,687,480	106,825,000	63.30	24,322	40,391,000	1,660.68	2,016,012	209,218,000	103.78
1958	202,685	42,190,000	208.16	259,492	69,841,000	269.15	23,934	39,070,000	1,632.41	486,111	151,101,000	310.84
1959	200,566	40,905,000	203.95	257,001	88,358,000	343.80	33,172	52,655,000	1,587.33	490,739	181,918,000	370.70
1960	183,297	39,069,000	213.15	235,288	89,602,000	380.82	25,194	57,606,000	2,286.50	443,779	186,277,000	419.56
1961	145,027	29,643,000	204.40	258,214	94,072,000	364.32	24,427	54,065,000	2,213.33	427,668	177,780,000	415.70
1962	119,161	23,497,000	197.19	272,680	93,122,000	341.51	25,884	59,250,000	2,289.06	417,725	175,869,000	421.02
1963	119,290	22,499,000	188.61	312,750	110,805,000	354.29	26,033	66,529,000	2,555.56	458,073	199,833,000	436.25
1964	127,781	23,981,000	187.67	394,043	139,810,000	354.81	28,752	71,885,000	2,500.17	550,576	235,676,000	428.05
1965	109,586	21,109,000	192.62	335,092	137,096,000	409.13	32,464	85,268,000	2,626.54	477,142	243,473,000	510.27

*"Facts for Industry," U.S. Department of Commerce.

TABLE 1.3 Refractories Industry Statistics—Plastics, Castables, and Mortars Production*

Year	Refractory mortars (clay), plastics, and castables			Refractory mortars (nonclay), plastics, and castables			Total refractory mortars, plastics, and castables			Fireclay (raw and prepared)		
	Tons	Value	Unit value	Tons	Value	Unit value	Tons	Value	Unit value	Tons	Value	Unit value
1947	204,965	$10,986,000	$53.60	109,596	$ 7,859,000	$ 71.71	314,561	$18,845,000	$ 59.91	406,695	$ 8,334,000	$20.49
1948	232,022	12,197,000	52.57	125,390	9,327,000	74.38	357,412	21,524,000	60.22	350,378	5,807,000	16.57
1949	181,285	11,277,000	62.21	103,293	8,266,000	80.02	284,578	19,543,000	68.67	329,470	5,818,000	17.66
1950	213,576	14,401,000	67.43	136,822	11,520,000	84.20	350,398	25,921,000	73.98	380,385	7,606,000	20.00
1951	252,650	18,034,000	71.38	180,744	16,166,000	89.44	433,394	34,200,000	78.91	446,426	10,223,000	22.90
1952	256,369	18,703,000	72.95	187,581	16,528,000	88.11	443,950	35,231,000	79.36	397,144	9,464,000	23.83
1953	274,921	19,904,000	72.40	217,492	19,527,000	71.31	492,413	39,431,000	89.78	376,601	11,775,000	31.27
1954	254,432	20,188,000	79.35	168,982	15,596,000	92.29	423,414	35,784,000	84.51	. . . . . .	No data	
1955	319,150	26,513,000	83.07	303,521	28,117,000	92.64	622,671	54,630,000	87.73	776,966	22,669,000	29.18
1956	367,428	31,220,000	84.97	291,046	28,585,000	98.21	658,474	59,805,000	90.82	621,504	26,324,000	42.36
1957	343,831	29,381,000	87.20	339,155	34,333,000	101.23	682,986	64,314,000	94.17	757,965	21,763,000	28.71
1958	295,817	28,029,000	94.75	311,345	32,444,000	104.21	607,162	60,473,000	99.60	645,807	18,874,000	29.23
1959	317,972	31,138,000	97.93	302,668	33,748,000	111.50	620,640	64,886,000	104.55	819,175	23,633,000	28.85
1960	307,929	29,361,000	95.35	278,494	32,150,000	115.44	586,423	61,511,000	104.89	721,308	27,377,000	37.95
1961	322,977	31,266,000	96.81	371,082	40,006,000	107.81	694,059	71,272,000	102.69	586,534	23,054,000	39.31
1962	322,579	30,981,000	96.04	350,751	37,847,000	107.90	673,330	68,828,000	102.22	569,666	28,290,000	49.66
1963	367,567	35,224,000	95.83	419,949	43,694,000	104.05	787,516	78,918,000	100.21	558,047	31,969,000	57.29
1964	416,332	41,395,000	99.43	464,238	51,159,000	110.20	880,570	92,554,000	105.11	628,175	36,092,000	57.46
1965	452,466	46,010,000	101.69	493,052	56,363,000	114.31	945,518	102,373,000	108.27	709,098	37,871,000	53.41

*"Facts for Industry," U.S. Department of Commerce.

TABLE 1.4 Refractories Industry Statistics—Other Refractories Production*

Year	Glasshouse refractories			Crucibles, ladle-brick and casting-pit refractories			Other fireclay products			Total, all refractories		
	Tons	Value	Unit value	Tons	Value	Unit value	Tons	Value	Unit value	Tons	Value	Unit value
1947	30,963	$3,689,000	$119.14	247,946	$15,317,000	$ 61.78	......	$3,435,000	......	2,249,008	$195,078,000	$ 86.74
1948	25,505	3,404,000	133.46	289,501	19,948,000	68.90	......	2,457,000	......	2,211,313	213,743,000	96.66
1949	17,218	2,539,000	147.46	238,959	16,620,000	69.55	......	2,034,000	......	1,825,876	184,171,000	100.87
1950	19,919	3,177,000	159.50	291,792	23,459,000	80.40	......	2,049,000	......	2,123,079	238,657,000	112.41
1951	25,601	3,819,000	149.17	351,550	29,048,000	82.63	......	2,687,000	......	2,638,153	326,842,000	123.89
1952	20,513	3,373,000	164.43	313,744	26,321,000	83.89	......	2,680,000	......	2,417,577	300,577,000	124.33
1953	14,207	3,262,000	229.61	360,227	31,928,000	88.63	......	4,550,000	......	2,383,637	324,367,000	136.08
1954	14,060	2,519,000	179.16	234,604	23,602,000	100.60	......	8,079,000	......	1,515,552	258,379,000	170.49
1955	18,389	3,602,000	195.88	343,813	37,119,000	107.96	......	4,227,000	......	2,860,231	369,352,000	129.13
1956	21,770	4,443,000	218.79	342,142	39,267,000	114.77	......	4,723,000	......	2,789,858	418,710,000	150.08
1957	17,837	3,962,000	222.12	306,704	38,911,000	126.87	340,413	5,685,000	$16.70	4,763,678	461,308,000	96.84
1958	14,534	3,598,000	247.56	289,051	28,447,000	98.42	211,856	4,608,000	21.75	2,690,713	355,852,000	132.25
1959	15,098	4,065,000	269.24	359,171	37,692,000	104.94	232,811	5,147,000	22.11	3,006,090	407,067,000	135.41
1960	18,792	5,254,000	279.59	297,856	32,550,000	109.28	241,396	5,012,000	20.76	2,790,253	413,486,000	148.19
1961	15,681	4,943,000	315.22	293,651	32,881,000	111.97	176,339	4,096,000	23.23	2,597,890	382,115,000	147.09
1962	13,402	4,151,000	309.73	277,356	31,724,000	114.38	170,567	4,012,000	23.52	2,534,664	383,958,000	151.48
1963	13,516	4,391,000	324.87	289,039	34,495,000	119.34	202,936	4,606,000	22.70	2,730,221	428,089,000	156.80
1964	12,667	4,291,000	338.75	346,180	42,636,000	123.16	206,608	5,327,000	25.78	3,079,811	497,559,000	161.56
1965	15,518	5,273,000	339.80	338,748	44,034,000	129.99	195,763	5,322,000	27.19	3,183,335	536,281,000	168.47

*"Facts for Industry," U.S. Department of Commerce.

more detailed picture is given in Tables 1.1 to 1.4, from which some interesting facts may be gleaned over this period.

The annual production of fireclay bricks has been reduced to one-half, while the unit price has a little more than doubled, thus giving a small increase in the total dollar volume. The number of high-alumina bricks has nearly doubled and the unit price has more than doubled, giving a total value with a fivefold increase. The insulating firebricks show a 50 percent increase in production and 100 percent increase in unit price.

The number of silica brick produced over this period has dropped down to one-third, largely because of the obsolescence of the open-hearth furnace. However, the unit price has doubled. The basic refractories have made an amazing sixfold increase in number of units produced, but there is still little change in unit price.

Plastics, mortars, and castables with a clay base have doubled in both tonnage and price, while the nonclay materials have increased five times in tonnage and two times in price. Monolithic, gunned linings account for much of this gain.

Glasshouse refractories have decreased in tonnage to about one-half, while the unit value has gone up three times. At present better refractories are giving longer life in the tank.

Crucible and pouring-pit refractories have increased in volume about 50 percent, but the unit price has doubled.

Perhaps the most striking facts from these tables are first the tremendous increase in basic refractories, due largely to their use in the basic oxygen converter, and second the enormous growth of the high-grade plastics and castables.

1.3 Recent Developments in the Industry. In regard to materials, the tendency seems to be along the lines of higher purity. Ganister is being scrubbed to remove a portion of alumina, titania, and iron oxide; chromite is being beneficiated to remove the silicate minerals; and seawater magnesite is now being produced with a higher purity than before. New deposits of refractory materials have been discovered, especially in under-developed countries.

High firing temperatures have been employed for superduty fireclay brick and basic brick with distinct improvements in their properties. Along with the high firing temperature, efforts have been made to produce brick with lower porosity.

Fusion-cast refractories, because of their good performance under severe conditions, have seen increasing use, especially in the steel industry.

Steam-power generators, once one of the larger users of refractories, now employ very little, since waterwalls take up most of the interior area. The trend toward nuclear power has been increasing, so that now a number of large, efficient power generators are in operation. These require almost no heavy refractories.

The steel industry has had profound changes, with the basic oxygen converter, continuous casting, and vacuum treatment making numerous changes in refractory use.

Monolithic furnace walls are being used to a greater extent because of the high cost of bricklaying. Improvements in castables and gunning methods have made possible major repairs which prolong the life of the lining. Prefabricated panels for furnace construction are coming into greater use.

1.4 Possible Future Trends in the Industry. One of the most important developments would be brick of lower porosity to better withstand slag and abrasion. It is hoped that a fired structure could be reached which would give a performance comparable with that of fusion-cast structures. In talking with refractory users the author has found the desire for a low-porosity brick constantly expressed.

High-temperature firing will go hand in hand with high-purity materials, and shorter firing cycles will be used to increase kiln capacity.

Automation will find increased use, not only in batching but also in handling. There is no reason why brick cannot be set on the kiln car right from the press without labor.

Safeguards against air pollution, water pollution, etc., will be more rigidly enforced. While this may not directly affect the refractory manufacturer, it may alter the type and amount of refractories purchased by some users. For example, the Thomas converter used in many areas blows the fumes into the air; however, tighter controls are taking place so that it may not pay to use a number of small converters, but one large one with a connected electrostatic precipitator will be used instead.

CHAPTER TWO

Literature on Refractories

2.1 Introduction. The literature on refractories is voluminous, even when confined to the heavy classification. The more current material will be found, of course, in the periodicals and bulletins. One should not forget the excellent books with fine illustrations turned out by the principal manufacturers of refractories, such as "Modern Refractory Practice" by the Harbison-Walker Refractories Co.

2.2 Books. The following books deal with many phases of refractory production and use.

Salmang, H.: "Die physikalischen und chemischen Grundlagen der Keramik," Springer-Verlag OHG, Berlin, 1933.

Partridge, J. H.: "Refractory Blocks for Glass Tank Furnaces," Society of Glass Technology, Sheffield, England, 1935.

Buell, W. C.: "The Open Hearth Furnace," 3 vols., Penton Publishing Company, Cleveland, 1936–1937.

Letort, Y.: "Produits refractaires," Dunod, Paris, 1946.

Mantell, C. L.: "Industrial Carbon, Its Elemental, Adsorptive and Manufactured Forms," D. Van Nostrand Company, Inc., Princeton, N.J., 1946.

Rait, J. R.: "Basic Refractories," Iliffe and Sons, Ltd., London, 1950.

Konopicky, K.: "Feuerfeste Baustoffe," Verlag Stahleisen mbH, Düsseldorf, 1957.

Gunther, R.: "Glass Melting Tank Furnaces," Society of Glass Technology, Sheffield, 1958.

Harders, F., and S. Kienow: "Herstellung, Eigenschaften und Verwendung-Feuerfester Baustoffe," Springer-Verlag OHG, Berlin, 1960.

Kairnarskii, I. S.: "Silica Refractories," State Publications of Scientific and Technical Literature on Ferrous and Non-ferrous Metals, Moscow, 1961 (in Russian).

Litvakovskii, A. A.: "Fused Cast Refractories," National Science Foundation, Washington, D.C., 1961.

Budnikov, P. P.: "The Technology of Ceramics and Refractories," The M. I. T. Press, Cambridge, Mass., 1964.

Chesters, J. H.: "Steel Plant Refractories," United Steel Companies Limited, Broomhill, Sheffield, England, 1966.

There are a few books dealing with raw materials that are pertinent to the refractories industry.

Ries, H.: "Clays, Their Occurrence, Properties, and Uses," John Wiley & Sons, Inc., New York, 1908.

Mudd, S. W.: "Industrial Minerals and Rocks," 2d ed., American Institute of Mining and Metallurgical Engineers, New York, 1949.

Ladoo, R. B., and W. M. Myers: "Nonmetallic Minerals," 2d ed., McGraw-Hill Book Company, New York, 1951.

Many books are available on crystal structure and means of identification.

Sosman, R. B.: "The Properties of Silica," Reinhold Publishing Corporation, New York, 1927.

Winchell, N. H., and A. N. Winchell: "Elements of Optical Mineralogy," vols. I, II, III, John Wiley & Sons, Inc., New York, 1927.

Winchell, A. N.: "Optic and Microscopic Characters of Artificial Minerals," University of Wisconsin, Studies in Science, vol. 4, 1927.

Bragg, W. R.: "The Crystalline State," G. Bell & Sons, Ltd., London, 1933.

Pauling, L.: "The Nature of the Chemical Bond," Cornell University Press, Ithaca, N.Y., 1944.

Brindley, G. W.: "X-ray Identification and Crystal Structures of Clay Minerals," The Mineralogical Society, London, 1951.

Grim, R. E.: "Clay Mineralogy," McGraw-Hill Book Company, New York, 1953.

Smothers, W. J.: "Differential Thermal Analysis," Chemical Publishing Company, Inc., New York, 1958.

Grim, R. E.: "Applied Clay Mineralogy," McGraw-Hill Book Company, New York, 1962.

Swineford, A.: "Clays and Clay Mineralogy," Ninth National Conference, Pergamon Press, New York, 1962.

Voinovitch, I. A., J. Debras-Guedon, and J. Lourrier: "Analyse des silicates," Hermann & Cie, Paris, 1962.

A few of the many books in the high-temperature area are listed below.

Anon.: "Temperature: Its Measurement and Control in Science and Industry," Reinhold Publishing Corporation, New York, 1940.

Sosman, R. B.: "Pyrometry of Solids and Surfaces," American Society for Metals, Cleveland, 1940.

McAdams, W. H.: "Heat Transmission," 3d ed., McGraw-Hill Book Company, New York, 1954.

Kingery, W. D.: "Property Measurements at High Temperatures," John Wiley & Sons, Inc., New York, 1959.

Kingery, W. D.: "Kinetics of High Temperature Processes," John Wiley & Sons, Inc., New York, 1959.

Harrison, T. R.: "Radiation Pyrometry and Its Underlying Principles of Radiant Heat Transfer," John Wiley & Sons, Inc., New York, 1960.

Levin, E. M., C. R. Robbins, and H. F. McMurdie: "Phase Diagrams for Ceramists," American Ceramic Society, Inc., Columbus, Ohio, 1964.

A book everyone should have who is interested in refractories in the United States is:

Anon.: "Product Directory of the Refractories Industries in the United States," The Refractories Institute, Pittsburgh, 1961.

2.3 Bibliographies. By far the most useful and complete bibliography is that prepared jointly by the American Ceramic Society and the American Iron and Steel Institute, Columbus, Ohio. The first volume covers the years 1928–1947, inclusive, while the second volume covers the years 1947–1956. It is unfortunate that no plans seem to be under way to bring it up to date. This bibliography is particularly useful as each entry is abstracted.

2.4 Glossaries. The following glossaries will be found useful.

Anon.: "Glossary of Terms Relating to Refractory Materials," British Standards Institution, no. 3446, 1962 (700 terms).
Van Schoick, E. C. (ed.): "Ceramic Glossary," American Ceramic Society, Columbus, Ohio, 1963.
Anon.: "Terms Relating to Refractories," ASTM C 71–64.

2.5 Periodicals and Other Publications. A great many publications contain information on refractories, but only a few of them are devoted exclusively to refractories. The majority have only occasional articles dealing with this subject. From a scientific and technical point of view the *Journal of the American Ceramic Society* and the *Transactions of the British Ceramic Society* are of the greatest importance. The periodicals in the following list have important or frequent articles on refractories:

American Ceramic Society, *Journal, Bulletin, Abstracts*, Columbus, Ohio.
American Refractories Institute, *Technical Bulletin*, Pittsburgh, Pennsylvania.
American Society for Testing Materials, *Proceedings*, Philadelphia, Pennsylvania.
Berichte der deutschen keramischen Gesellschaft, Berlin-Halensee, Germany.
Brick & Clay Record, Chicago, Illinois.
British Ceramic Society, *Transactions*, Stoke-on-Trent, England.
British Clayworker, London, England.
Ceramic Age, The Ceramic Publishing Co., Newark, New Jersey.
Ceramic Industry, Chicago, Illinois.
Ceramique, Paris (10e), France.
Glastechnische Berichte, Deutschen glastechnischen Gesellschaft, Frankfurt am Main, Germany.
Iron and Steel Institute, *Publications*, Grosvenor Gardens, London.
National Bureau of Standards, *Circulars, Scientific Papers, Technical News Bulletins, Technical Papers, Journal of Research*, Washington, D.C.
Ogneupory, Russia (Translation Consultants Bureau—English).
Ohio Ceramic Industries Association, *Bulletins*, Ohio State University, Columbus, Ohio.

Ohio State University Engineering Experiment Station, *Bulletins*, Ohio State University, Columbus, Ohio.

Refractories Journal, Sheffield, England.

Sprechsaal, Müller und Schmidt, Coburg, Germany.

Steklo i Keramika (Glass and Ceramics), Translation Consultants Bureau, 227 West Seventeenth Street, New York, New York.

U.S. Bureau of Mines, *Bulletins, Circulars, Mineral Resources, Reports of Investigations, Technical Papers*, Washington, D.C.

University of Illinois Engineering Experiment Station, *Bulletins*, University of Illinois, Urbana, Illinois.

CHAPTER THREE

History of
Refractory Development

3.1 Introduction. The early history of refractories in the United States is shrouded in considerable obscurity. A number of references are given here that throw new light on the subject, but a tremendous field still remains to be explored. A few months spent in examining old publications and records of Boston, Baltimore, Philadelphia, and parts of New Jersey would certainly yield valuable information. A little archaeological study around the sites of old furnaces, kilns, and plants would surely produce specimens of the old refractories used.

3.2 Furnace Stone. Undoubtedly the earliest type of refractory used in this country was mica schist or siliceous rock. All the early iron furnaces or forges were constructed mainly of this material. Thus we must study the early history of the iron industry. Fortunately this is quite well known, for an iron furnace was much more likely to be a subject of comment than a brickyard. The first furnaces were built in Virginia, but it is quite certain that they were never completed. The first successful furnace was built in 1645 at Saugus, Massachusetts, and the next year another was operated at Braintree. They were probably made from local stone, although there are no very satisfactory deposits in this region. The building of furnaces and forges spread rapidly in the next hundred years, and considerable stone must have been quarried. Samuel Robinson[1,*] in 1825 states that furnace stone was quarried near Providence, Rhode Island, for forty or fifty years and carried long distances in

* Superscript numbers indicate references listed in the bibliography at the end of each chapter.

14

wagons for use in furnace hearths. There were probably other famous quarries at this time.

Stone was used for the building of furnaces and forges late into the nineteenth century because of the greater cost of firebrick. Even now, stone is used for lining Bessemer converters. A number of the old stone or stone and brick furnaces are still standing in various parts of the country. Two interesting ones stand on Furnace Brook in Bennington, Vermont.

Records[20] show that sandstone was used for the walls of kilns burning refractories. Probably many of the early glass furnaces were built of stone. In fact Robert Hewes of the Temple Glass Company, Temple, New Hampshire,[7] states in a letter dated 1781: "I shall have to send sixty miles for stones to build my melting furnace, which will take eight teams." It is believed that these stones came from Uxbridge, Massachusetts, which is about sixty miles from Temple. Samples from the old furnaces confirm this.

3.3 Glass Pots and Crucibles. Probably the first clay refractories used in this country were glass pots. J. B. Felt[5] in his "Annals of Salem" in 1827 mentions a glass factory in Salem, Massachusetts, in 1638 and again as producing in 1641. The only other information I have been able to obtain is the names of John and Ananias Conklin as glass blowers. All else seems to be lost in the obscurity of time. We may be sure, however, that glass pots were used. They were probably made from English or German clays as were all the early glass refractories.

A century later, the Wister glassworks were started in New Jersey. Gross Almerode or Klingenberg clay was used for the pots at first, but later New Jersey or Maryland clays were probably used to some extent. J. C. Booth[4] in 1841 states that a white, highly plastic clay was wrought only in one place, on the Delaware shore below New Castle, and exported for crucibles and glass pots. The clay had been wrought for manufacturing purposes for some forty years. Porcelain crucibles were made of kaolin from New Milford, Connecticut, by a goldsmith in about 1807.

A number of other early glass plants were operated in this country, mainly by Germans. We have the Germantown factory (Braintree, Massachusetts) and the factory of Baron Steigel in 1763, as well as a number of others. German pot clays were probably extensively used as they are even now. In fact, J. F. Amelung, who started a glassworks at Frederick, Maryland, in 1784, brought over pot makers and built a pot works in conjunction with his plant.

A note in the Gaffield Collection at the Massachusetts Institute of Technology library states that the Boston Crown Glass Company, which for a number of years had been using imported clays for pots, attempted between 1810 and 1812 to replace them with domestic (probably New Jersey) clays. The attempt was not successful.

A recipe book of the Boston Glass Mfg. Co. in the Boston Athenaeum Library has an item dated 1791 stating that 200 lumps of clay were received from Amsterdam, undoubtedly for making pots. Another item lists fuel for keeping the glass pots (green) from freezing.

The manufacture of glass pots as an industry separate from the glass plants was started in 1860 by Thomas Coffin in Pittsburgh.[12] In 1879, the Pittsburgh Clay Pot Company was organized, and after this many other concerns started in Pennsylvania and Ohio.

3.4 Lime. Lime, often made from shells gathered on the shore, was used in forge walls from very early times. There are references to a number of early lime kilns.

3.5 Fireclay Bricks. We are still in doubt as to the location of the first American firebrick manufacturer. It may have been New Jersey, Boston, or Baltimore.

New England. Owing to the early development of iron and glass manufacture here, it is to be expected that firebrick would be needed at an early date. Stourbridge firebricks (English) are known to have been imported to some extent, but bricks were also made from New Jersey clay. The recipe book of the Boston Glass Mfg. Co. states that in 1790, two helpers were paid to make furnace tile and, in 1791, two men were paid for burning brick for the glass furnace. In 1793, an item lists the payment of expenses for a man to go to New York to procure clay from South Amboy. This is undoubtedly the earliest direct reference to the manufacture of firebrick from native clay. According to Professor Rogers,[8] Mr. G. W. Price of New Brunswick, New Jersey, stated that his father carried a vessel load of fireclay from Woodbridge to Boston in 1816 for use in making firebricks. It was thought to be the first shipment for this use. An advertisement in the *Boston Commercial Gazette* of July 6, 1818, by the New England Glass Company, read, "Also—Fire Brick Clay, raw and prepared, constantly for sale." This would seem to indicate that firebricks were then being made in Boston. In 1820, Jacob Felt of Boston bought 50 tons of clay of Jeremiah Dully, and this was afterward sent regularly from Amboy. Firebricks were made in Boston at this time.

The following advertisement appeared in the *Boston Commerical Gazette* of Mar. 5, 1827. "_____fire bricks made to all dimensions. For sale by Deming Jarves, No. 88 Water St." (New England Glass Bottle Works.) This probably explains where at least part of the New Jersey clay was used.

In 1835, a patent was taken out by Joseph Putnam of Salem for firebricks and stove linings. We do not know whether or not he manufactured, but it is doubtful if he would have applied for a patent for this type of article without having previously made it. As early as 1829,

L. Hine used the kaolin of New Milford, Connecticut, for firebricks and furnace linings.[3] He evidently had a considerable business, as $6,000 per year is mentioned as the value of his product and six workmen were employed. The bricks sold for two-thirds the price of the Stourbridge firebricks and were considered nearly as good.

In 1839, Hiram Harwood of Bennington, Vermont, records in his diary,[18] "Visited works of Judge L. Norton—large low buildings in forwardness for drying newly invented fire bricks—erected N. old works." This business apparently prospered in the next few years and was carried on by Norton and Fenton. It is quite possible that the manufacture of these bricks was started from information obtained from the New Milford enterprise. We are fortunate in having an accurate account of the manufacture of these early bricks by Norton and Fenton. C. B. Adams, state geologist of Vermont, in his first *Annual Report* (1845), page 52, states:

> For firebrick, the kaolin is made into paste with water, from which bricks are formed and burnt. These bricks, retaining the whiteness of the kaolin, and becoming very hard, are called "clay bricks." They are next broken up by a mill and sifted, so as to be of the coarseness of fine gravel. This is mixed with unburnt kaolin and arenaceous quartz, pressed in moulds of the required form and size, and burnt in the same manner as before. These firebricks are very white and hard, and when fractured show their composition of broken claybrick and kaolin.

C. W. Fenton, in a letter to Mr. Adams at this time, gives more particulars:

> The character of our fire-bricks is also well known. They are a composition of materials which we find here, consisting of arenaceous quartz and kaolin. Being very pure, they make a good fire-brick, which will stand longer in a strong heat than any other brick known. They are used for blast furnace hearth, and in many places, where no other fire-bricks will endure.

The value of pottery and firebrick produced then was stated as $20,000 per year.

In 1841 (Feb. 27), the following advertisement appeared in the *State Banner* of Bennington, Vermont:

> Bennington
> Stoneware Factory
> Julius Norton
> Manufactures and keeps constantly for sale
> . . . .
> Also *Patent Fire Brick*
> (the best in the world) at $50 per thousand

These kaolin bricks were made in Bennington as late as 1855 or 1856 and had a very general sale as a good firebrick.   But owing to the lack of coal and the need of twice burning the highly shrinkable kaolin, they could not long compete in cost with bricks of New Jersey and Pennsylvania.

It is also stated that firebrick of kaolin, New Jersey fireclay, crushed brick, and sandstone from Willsboro, New York, was made at Monkton, Vermont, in 1846.   These bricks were undoubtedly used in the many charcoal iron furnaces operating in this region, although a large part of the lining was quartzite or mica schist.   A reddish firebrick taken from an old furnace standing on Furnace Brook in Bennington, Vermont, fulfills this description quite well.   I was also fortunate in finding at the site of the United States Pottery in Bennington a white brick made of coarse calcined grog bonded with kaolin.   This brick is marked:

> Fire Brick Co's.
> No. 1
> Bennington, Vt.

Considerable space is given to these early kaolin bricks, because we have a fairly complete account of their manufacture and because they are the first to utilize successfully the pure and refractory but high-shrinking kaolin by the method of double burning.   For their time, they were really remarkable bricks and have scarcely been mentioned in histories of refractories.

Firebricks and retorts were probably being made in Boston regularly before 1850.   In 1864, the Boston Fire Brick and Clay Retort Manufacturing Company was making bricks and retorts on the site of the present South Station.   New Jersey fireclay was brought to their own dock.   For some years at about this time, the Morton Fire Brick Plant on K Street at South Boston was in operation.   Apparently no firebrick has been made in Boston since 1895.

From the beginning of the nineteenth century, a number of stoneware potteries were operating in New England.   In every case, firebrick must have been used in their kilns.   It is reasonable to suppose that the bricks were made from stoneware clay, all of which at that time came from New Jersey.

The manufacture of stove linings and firebricks has been carried on at Taunton, Massachusetts, for many years.   Now Pennsylvania clays are mainly used.

New Jersey.   A stoneware pottery was operated by Coxe at Burlington, New Jersey, in 1684.   He probably used English firebrick for his kilns, but some local clay may have been used.

Professor Rogers[8] states that New Jersey clays were used for making firebricks soon after 1812.   We need not necessarily assume, as many have, that the bricks were made in New Jersey. The clay may have been taken to Boston or Baltimore. In 1825, a firebrick plant (Salamanda Works) was started at Woodbridge, New Jersey. In 1836, John R. Watson established a factory for making firebricks at Perth Amboy. From 1845 to 1865, a number of plants were started in the Woodbridge district, and most of them are still operating. Sayer and Fisher, Henry Maurer & Son, and Valentine & Bro. were well-known firebrick manufacturers.

Delaware and Maryland. In a report[2] on the state of Maryland, published in 1834, occurs a most significant statement:

> Fire brick so far exclusively made in the United States at Baltimore, has been pronounced by competent judges after repeated trials, to be fully equal, if not superior to the far-famed Stourbridge brick of the same nature.

Later it stated that these bricks were used in iron furnaces to replace silica stone.

We may judge from this statement that manufacture of a high-grade firebrick of the "same nature" as the well-known Stourbridge firebrick was an established industry in Baltimore at this time. The reference to repeated trials would indicate its use for a considerable number of years before. The statement that firebricks were made only in Baltimore at this time is perhaps founded on lack of knowledge of the industry in New Jersey and Boston but more probably refers to a time some years prior to this date when the Baltimore industry was started. Further investigation is greatly needed on this point.

As previously mentioned,[4] considerable white refractory clay was mined in Delaware mainly for crucible and glass pots, but some may well have served for bricks.

In 1837, the famous Mt. Savage fireclay was discovered, and the Union Mining Company soon began operations here.

Florida. Ries[12] states that records show a shipment of firebrick from Florida to New Orleans in 1827. No more is known of it.

Ohio. According to Stout,[20] firebricks were first made in Ohio during 1841 by Andrew Russell near East Liverpool. They were produced there continuously until about 1900 by Russell and later by N. U. Walker. The clay mined behind the plant was plastic and of medium refractoriness. In 1852, G. and M. Meyers started a firebrick plant near Toronto. A plastic, lower Kittanning clay was used.

The manufacture of firebricks from flint clay was begun during 1863 in Scioto County by Reese Thomas. This brick was a high-grade product, made in updraft kilns with sandstone walls.

During 1866, the Diamond Fire Brick Company was opened in Akron by J. P. Alexander. At about the same time, a three-kiln plant was built at Dover. The Federal Clay Products Company and a number of other plants started about 1872 in this district. J. R. Thomas founded the Niles Fire Brick Company at Niles in 1872.

The Oak Hill Fire Brick Company and Aetna Fire Brick Company started in the Oak Hill district about 1873. Other plants were opened here later.

Pennsylvania. It has generally been believed that the first firebricks were made in Pennsylvania about 1836 at Queens Run. However, the following advertising card[10] would indicate that they were made in Philadelphia before 1832.

American China Manufactury
S. W. Corner of Schuylkill Sixth & Chestnut Sts.,
or at the Depository
Where is constantly kept on hand,..
....
....
also offered for sale
Fire-brick & Tile
Of a superior quality, manufactured in part from the materials of which china is composed.—These have been proved, by competent judges, to be equal to the best Stourbridge brick.

Eight years later the following advertisement[10] appeared.

Abraham Miller
has removed his manufactory
From Zane Street to James, near Broad Street,
Spring Garden
where his works are now in full operation, conducted by his late Foreman, Mr. J. C. Boulter.
. . . .
A large Assortment of Portable Furnaces, Stove Cylinders, Fire Bricks and Slabs,
....
....
Philad'a December 22d 1840.

In 1842, James Glover started the manufacture of refractories at Bolivar. The clay is of good quality, and refractories are still made there. In 1845, Kier Brothers started to manufacture at Salina. In 1859, Soisson and Company began manufacturing at Connellsville; and in 1865, the Star Fire Brick Works were built at Pittsburgh.

After this, so many firebrick plants were opened in the state that it assumed the leading position in the manufacture of refractories. The factors contributing to this great development were the excellent deposits of clay, especially flint clay; the abundance of coal; and the ease of distribution.

open-hearth steel plant in the United States. However, it was not a commercial success. In 1888, magnesite was used successfully by Carnegie, Phipps & Co. at Homestead. From this time, the use of magnesite increased rapidly because the general advantages of the basic process over the acid process became evident to all.

In the year 1898, 16,000 tons of Austrian dead-burned grain were imported. Bricks were at first imported, but later they were made in this country by a few companies.

Production of magnesite from brines, bitters, and seawater developed rapidly during the 1940s, but actually there was a small seawater plant in southern France as early as the 1890s. The first commercial producer in this country was the Marine Magnesium Products Corp. in San Francisco. Now there are a dozen plants making magnesia for the refractories industry.

Chemically bonded magnesia, magnesia-chrome, and tar-bonded magnesia have been made for a long time, but the direct-bonded basic brick with a low amount of glass phase came into use in the late 1950s.

3.9 High-alumina Refractories. The first bauxite discovered in the United States was found near Rome, Georgia, during 1888.[14] The Arkansas bauxite deposits were discovered in 1891. The production from the latter source increased rapidly until now it supplies a considerable part of the domestic demand. Coastal bauxite of Georgia was discovered and described in 1909. A number of small deposits are now being worked.

Missouri diaspore was recognized in 1917 as a source of material for super refractories. This discovery made Missouri one of the most important producing states.

The importation of gibbsite from Dutch Guiana in the last few years has supplied a very high-grade material to the manufacturers of refractories.

The deposits of sedimentary kaolin on the southern coastal plain have been known from the time of the first settlers. The remarkable beds form probably the largest supply of uniform and pure high-grade clay in the world. The suitability of kaolin for refractories was known as early as 1837 in Bennington, but the southern kaolin is of a purer grade and possesses a high vitrifying point and a very large shrinkage. These properties for many years have prevented its use in a refractory. During the last forty years, however, it has been possible to manufacture a successful refractory out of this pure kaolin.

3.10 Insulating Refractories. One of the important milestones in refractories development was the hot-face insulation or insulating firebrick. This product was pioneered by the Babcock & Wilcox Co. in the mid-1920s. In the last 10 years, the use limit has been extended above 1650°C (about 3000°F).

Missouri. In the year 1846, the excellent clay deposits of t
district were opened up. In 1855, the Christy Fire Clay Co
started; and in 1855, both the Evans and Howard and the L
Brick Company followed. The St. Louis district was one of
producing centers at this time.

Kentucky. This district was not opened up until about 1{
fireclay from Lewis County was sent to Cincinnati. In 1884, th
Amanda Furnace and Bellport Furnace were worked. Two y
the Ashland Fire Brick Company was started at Ashland.

West Virginia. Firebricks were not produced in this state ui
when the clays of Marion County were worked.

Colorado. The first firebricks made in this state were manufac
Golden during 1866. Later clay deposits were opened up at Pue
Cañon City and now have a good reputation in the West.

The West Coast. Refractory manufacture was rather late in dev
here, but recently many excellent refractories are being made in bo
fornia and Washington.

3.6 Silica Refractories. The first silica brick are believed to hav
made by W. W. Young from Dinas rock in South Wales about 184
little later he made brick of ganister. Probably the first silica
made in the United States were manufactured by J. P. Alexand
Akron, Ohio, about 1866, but a patent was granted to Thomas Ja
in 1858 for a lime-bonded silica brick, and he may have manufact
before Alexander. Of course, in England lime-bonded silica brick
made even before this.[23] J. R. Thomas made silica brick for the
industry at Niles, Ohio, about 1872. He used quartz pebbles ar
Sharon conglomerate and called the bricks "Dinas Silica," as they v
similar to the European brick of that name. A. Hall of Perth Am
made some silica bricks in 1875, but apparently there was no gr
demand for them. The modern silica brick of lime-bonded ganister i
more recent development. In the year 1899, the first silica-brick pla
was started at Mt. Union, which later became a great center. Sili
bricks were made later in the Chicago district.[21]

3.7 Chrome Refractories. Chromite, mainly in the form of brick
began to be used by the steel manufacturers about 1896, chiefly as
neutral zone between the acid and basic courses. Only a few firms hav
been manufacturing chrome bricks.

3.8 Magnesite Refractories. Magnesite as a steel-furnace lining was
suggested in Europe as early as 1860 but did not come into regular use
until 1880, when it was found that the Austrian material could be fritted
down into a good bottom. Styrian magnesite was imported into this
country in 1885 and was used by the Otis Steel Company in the first

3.11 Plastics and Castables. These have been gradually developed to form monolithic structures. Carl Akeley, the great taxidermist, is credited with initiating the gunning technique in 1911 for plaster. Gunning of refractories was used from the early 1920s.

3.12 Fusion-cast Refractories. Alumina has been melted for abrasives since the beginning of the century. In the early 1920s Dr. Fulcher of Corning Glass Works started making glass refractories by fusion and casting. Alumina-silica refractories were made at first, but in the mid-1930s zirconia was added. This was disclosed in U.S. Patent 2,271,366 to T. E. Field in 1942. In the late 1950s basic fusion-cast blocks became available to the steel industry.

3.13 Refractory Fibers. Glass and mineral fibers have been in use for a long while as insulation, but the demand for a higher use limit encouraged experimentation with more refractory silica-alumina glasses. U.S. Patent 2,467,889 by Harter, Norton, and Christie seems to be the earliest discloser of a now highly successful method of forming wool from fused kaolin.

BIBLIOGRAPHY

1. Robinson, Samuel: Notice of Miscellaneous Localities of Minerals, *Am. J. Sci.*, **8**, 232, 1824.
2. Ducatil, J. T., and J. H. Alexander: "Report on Projected Survey of State of Maryland," Annapolis, 1834.
3. Shepard, C. U.: "A Report of the Geological Survey of Connecticut," B. L. Hamlen, New Haven, 1837.
4. Booth, J. C.: "Geology of Delaware," pp. 16, 40, 1841.
5. Felt, J. B.: "Annals of Salem," 1827.
6. James, Thomas: Lime Bonded Silica Brick, U.S. Patent 20,433, 1858.
7. Blood, H. A.: "History of Temple, New Hampshire," Boston, 1860.
8. Cook, G. H.: "Report on the Clay Deposits of Woodbridge, South Amboy, and Other Places in New Jersey," 1, 1878.
9. Hunt, A. E.: Some Recent Improvements in Open Hearth Steel Practice, *Trans. Am. Inst. Mining Met. Engrs.*, **16**, 718, 1887.
10. Barber, E. A.: "Pottery and Porcelain of the United States," G. P. Putnam's Sons, New York, 1893.
11. Ries, H., H. B. Kummel, and G. N. Knapp: The Clay and Clay Industry of New Jersey, *Geol. Survey of New Jersey*, vol. 6, 1904.
12. Ries, H.: "The History of the Clay Working Industry in the United States," John Wiley & Sons, Inc., New York, 1909.
13. Maynard, G. W.: Introduction of the Thomas Basic Steel Process in the United States, *Trans. Am. Inst. Mining Met. Engrs.*, **41**, 289, 1910.
14. Shearer, H. K.: A Report on the Bauxite and Fuller's Earth of the Coastal Plain of Georgia, *Georgia Geol. Survey Bull.*, 31, p. 22, 1917.
15. McDowell, J. S., and R. M. Howe: Magnesite Refractories, *J. Am. Ceram. Soc.*, **3**, 185, 1920.

16. McDowell, J. S., and H. S. Robertson: Chrome Refractories, *J. Am. Ceram. Soc.*, **5**, 865, 1922.
17. Anon.: Die Entwickelung der deutschen Industrie feurfester Erzeugnisse seit 1871 und ihre heutige Lage, *Tonind.-Ztg.*, **49**, 589, 1925.
18. Spargo, John: "Potters and Potteries of Bennington," Houghton Mifflin Company, Boston, 1926.
19. McDowell, J. S.: Progress in the Refractories Industry, *Blast Furnace Steel Plant*, **17**, 88, 1929.
20. Stout, W.: Refractory Clays of Ohio, *Bull. Am. Ceram. Soc.*, **9**, 29, 1930.
21. Kurtz, T. N.: History of Silica Brick, *Bull. Am. Ceram. Soc.*, **15**, 26, 1932.
22. Greaves-Walker, A. F.: History of Development of the Refractories Industry in the United States, *Bull. Am. Ceram. Soc.*, **20**, 213, 1941.
23. Searle, A. B.: A Brief History of Refractory Materials Prior to the 19th Century, *Refractories J.*, **21**, 145, 1945.
24. Dennis, W. H.: Refractories, *Refractories J.*, **39**, 420, 1963.
25. Roberts, J. E.: The History and Development of the Refractories Industry, *Refractories J.*, **41**, 90, 1965.

PART TWO

Product

CHAPTER FOUR

Heavy Refractory Brick

4.1 Introduction. Of the more than one hundred elements found in the earth's crust, only a few have both abundance and the ability to form stable refractory compounds. These are silicon (Si), aluminum (Al), magnesium (Mg), calcium (Ca), chromium (Cr), zirconium (Zr), and carbon (C). These form the useful oxides: SiO_2, Al_2O_3, MgO, and ZrO_2. The oxide of chromium is volatile and that of calcium is unstable in the atmosphere; however, they may be combined into useful materials such as Ca,MgO to form dolomite or $Cr_2O_3 \cdot MgO$, the basic spinel. Carbon may be used directly, after graphitization, or combined with Si to form silicon carbide.

4.2 Classification. The classification in Table 4.1 is generally accepted in the United States as it seems to give a logical arrangement of the commonly used refractories. Most of the classes are adopted by the American Society for Testing Materials in C 27–60, C 416–60, C 445–62, and C 545–64T.

4.3 Materials. The clays used for making firebrick consist of flint fireclays, semiflint fireclays, plastic fireclays, and kaolin to produce a composition of 18 to 44 percent Al_2O_3 and 50 to 80 percent SiO_2. The superduty and high-duty firebrick contain a considerable portion of the refractory flint or semiflint fireclays which have PCE's of 33 to 35 and a low drying and firing shrinkage. As the flint clays have little plasticity, they are bonded with plastic fireclays with a PCE of 29 to 33. In some cases the high-shrinking, but refractory, kaolin with a PCE of 34 to 35 is added to the bond. Also, calcined clay or grog is added to the brick

27

TABLE 4.1   Classification of Heavy Refractories

Fireclay brick:[1,2]
  Pouring pit (PCE* below 15)
  Low-duty (PCE 15)
  Medium-duty (PCE 29)
  Semisilica (min SiO_2, 72%)
  High-duty (PCE $31\frac{1}{2}$):
    Regular
    Spall-resistant
    Slag-resistant
    High-fired
  Superduty:
    Regular
    High-fired
Kaolin (high-grog, high-fired)
High alumina:
  50% Al_2O_3 (PCE 34)
  60% Al_2O_3 (PCE 35)   1. Sintered grain
  70% Al_2O_3 (PCE 36)   plus bond
  80% Al_2O_3 (PCE 37)   2. Fused grain
  85% Al_2O_3 plus bond
  90% Al_2O_3 3. Fusion cast
  99% Al_2O_3 (Al_2O_3, 97% min)
Silica:
  Conventional (0.5 to 1.0% Al_2O_3, TiO_2, and alkalies)
    Regular
    Hot-patch (more spall-resistant)
  Superduty (0.2 to 0.5% Al_2O_3, TiO_2, and alkalies)
    Regular
    Hot patch (more spall-resistant)
  Lightweight (lower thermal conductivity)
Basic:
  Magnesia   Fired, silicate-bonded
  Magnesia-chromite   Fired, direct-bonded
  Chromite-magnesia   Tar-bonded
  Chromite   Fired, tar-impregnated
  Forsterite (2 $MgO \cdot SiO_2$) Fusion cast
  Dolomite (CaO, MgO)   Steel cased
Carbon:
  Carbon
  Graphite
Special:
  Zirconia
  Zircon
  ZrO_2-SiO_2-Al_2O_3 (fused glass-tank blocks)
  Silicon carbide
    Clay-bonded
    Frit-bonded
    Nitride-bonded
    Oxynitride-bonded
    Recrystallized
  Acidproof brick (dense, resistant to acids)
* Pyrometric cone equivalent (ASTM Designation C 24).

mix to reduce the firing shrinkage and to give greater stability in use. In other words, the superduty and high-duty firebricks are carefully proportioned mixtures of a number of materials to give the best possible performance for a particular use, as no firebrick can be expected to be superior in every property. However, much effort has been made to produce bricks with lower porosity for slag resistance by careful grog sizing, high-pressure forming, and high firing.[3]

As the high-grade refractory clays are of rather limited occurrence, there has been interest in making more extensive use of the huge and uniform deposits of sedimentary kaolin in the south. Because of the high shrinkage of this clay, a considerable proportion must be fired to a stable grog and then bonded with raw kaolin. When fired to high temperatures of 3000°F or more, excellent bricks result.

The less refractory classes of firebrick may use a single fireclay or be a mixture of several clays. For example, ladle bricks are often made of a single plastic clay of low softening point, and semisilica brick are usually made from a single siliceous kaolin.

High-alumina brick are produced to give better service under severe conditions than fireclay brick. The alumina content may run from 50 to 99 percent, so that raw materials other than kaolins or fireclays are needed. Diaspore was formerly used, but now sources are practically exhausted, and bauxite or bauxite clays are commonly used in spite of their high firing shrinkage. Also, kyanite, sillimanite, and andalusite· are used, as they are quite volume-stable. In some cases synthetic mullite is used in these bricks. For higher alumina contents it is necessary to add chemically prepared alumina. Most of the materials used in high-alumina bricks are precalcined in order to give lower firing shrinkage.

Silica brick are made from ganister rock. Sometimes it is washed to reduce the alumina content for superduty brick.

Magnesite brick are made from dead-burned magnesia burned between 2800 and 3350°F, depending on the purity. In the United States magnesia is derived from seawater or brines. The bricks may be kiln-fired, chemically bonded, or enclosed in sheet-steel jackets. While a few basic bricks are made from chromite ore, the majority are a combination of chromite and magnesite in various proportions. These brick were formerly silicate-bonded by the fluxes in the chromite, but in the last few years this flux has been reduced and the brick high fired to give a direct-bonded structure of much greater hot strength.[4]

Tar-bonded basic brick, largely used in oxygen converters, are made from dead-burned dolomite and magnesite mixtures. Because of sensitivity to hydration, their shelf life is short. Also, fired basic brick are impregnated with tar to fill the pores.

A few bricks are made of zircon and stabilized zirconia. Silicon carbide is made with a glassy bond, nitride bond, oxynitride bond,[5] and a very few are recrystallized.   Carbon blocks are made in large quantities for blast-furnace linings.

4.4   Production Methods.   Today nearly all standard shapes are formed on the power press, as this method, using a low water content, simplifies the drying and handling and allows automation.   However, a few bricks are made by extrusion, often followed by re-pressing.   Almost no hand-molded standards are made in this country today.

In addition to molding of fired bricks, several types of unfired bricks are formed in the power press or hydraulic press.   Chemically bonded basic brick and tar-bonded basic brick are made in this way.

Impact and vibratory presses are often used for tiles and standard shapes, especially for mixes of low plasticity.

Slip casting is used for forming glass pots and feeder parts, using a heavy thixotropic slip.

Drying is only a small problem for power-pressed brick, but extruded brick are put through tunnel driers, and larger pieces are often dried on hot floors.

Firing of all standard fireclay brick is now carried out in direct-fired tunnel kilns.   Some silica brick are fired in long tunnel kilns, but many are still fired in periodic kilns.   Temperatures of firing for pouring pit and low-heat refractories are around 2000°F; other fireclay brick is now fired at about 2400°F, except the high-fired superduty brick, which goes up to 2700°F.   Kaolin brick are fired at 3050°F and basic brick go up to 3800°F.

The fusion-cast refractories made in the electric furnace have been used for many years in glass tanks, but recently the fused basic refractories are being used increasingly by the steel industry in places where the service is very severe.

4.5   Properties.   The properties of firebrick vary depending upon the clays and the nonplastics used, the density of forming, and the firing temperature.   The properties of most importance are:

Fusion point
Creep under compression
Spalling resistance
Slag resistance
Stability against gases and vapors
Abrasion resistance

Each application requires consideration of these properties to obtain the best service.   It is always best to consult experienced manufacturers in selecting a particular brand.

Somewhat the same list of properties is of interest when using other types of refractory. In Part 5, the proper use of refractories will be discussed in detail.

BIBLIOGRAPHY

1. Debenham, W. S., and G. R. Eusner: Some Considerations in the Classification of Fireclay Brick, *Bull. Am. Ceram. Soc.*, **32**, 272, 1953.
2. Eusner, G. R., and K. K. Kappmeyer: Ratings of Fireclay Brick, *Bull. Am. Ceram. Soc.*, **41**, 1, 1962.
3. Gugel, E., and F. H. Norton: High-density Firebrick, *Bull. Am. Ceram. Soc.*, **41**, 8, 1962.
4. Hubble, D. H., and W. H. Powers: High-fired Basic Brick for Open-hearth Roofs, *Bull. Am. Ceram. Soc.*, **42** (7), 409, 1963.
5. Washburn, M. E.: A Silicon-carbide Refractory Bonded with Silicon Oxynitride, *Refractories J.*, **39** (10), 412, 1963.

CHAPTER FIVE

Insulating Materials

5.1 Introduction. The early kilns and furnaces were built with very thick walls of heavy refractories for two reasons—first, because efficient insulation was not available and, second, because of the fear of over-heating the lining due to the backup insulation. However, as temperatures became higher and fuel more costly, insulation cautiously came into use as a backup material. This enabled the use of thinner walls with less heat loss. It also enabled a reduction in floor space and volume of refractory, as well as fuel saving due to less stored heat.

One of the great evolutions in refractories was the initiation of the insulating firebrick (IFB), perhaps not so much in the development of a high-temperature insulator as the revolutionary idea that insulation could be used on the inside walls of furnaces. This procedure completely changed the design of kilns and furnaces, allowing still greater reduction in wall thickness and weight. Now furnaces with 9- or even 4½-in.-thick walls are common. IFB have been produced for operation above 3000°F only recently, but excellent brick are now available in this range.

The IFB has another advantage in that the heat storage in the walls and crown is greatly reduced over that of heavy brick walls, thus permitting increased efficiency in intermittent furnaces.

The IFB is not useful when in contact with fluid slags or glasses, first because the pores are filled and the insulation value is lost, and second because the rather fragile structure is readily attacked.

5.2 Classification of Backup Insulation. Hundreds of backup insulators have been introduced in the last 25 years, but the general types in Table 5.1 are now well standardized in the industry.

TABLE 5.1 Classification of Backup Insulation

Type	Weight, psf	Use limit, °F
Diatomaceous, block with asbestos and lime.........	23	1800–1900
Fibers:		
Slag wool block.............................	15–20	1500–1700
Slag wool blanket...........................	10	800–1000
Glass wool blanket..........................	3	800–1000
Silica-alumina wool block....................	12–20	2000–2300
Silica-alumina wool blanket..................	3–8	2000–2300
Silica-alumina wool paper....................	8–10	2000–2300
Silica-alumina wool loose....................	3–10	2000–2300
Vermiculite:		
Block....................................	19	1500–1600
Loose....................................	10	1500–1600
Kaolin-gypsum...................................	30	1600
Foamed glass....................................	10	1000

5.3 Classification of Hot-face Insulation. The IFB (insulating firebrick) bricks for hot-face use have been classified by the ASTM in Table 5.2.

TABLE 5.2 Classification of Hot-face Insulation

Group	Use limit, °F	Maximum allowable bulk density
16	1550	34
20	1950	40
23	2250	48
26	2550	54
28	2750	60
30	2950	68

5.4 Production Methods. The IFB is made of clay porosified by a number of processes, but the most generally used consists of mixing wood flour with clay and molding into shape. The product is fired, whereby the wood is burned out and the clay framework sintered into a rigid mass. Under these conditions the shrinkage is large; so, in order to maintain accurate dimensions, the brick are put through an automatic sizing machine. After this they are packed into cartons for shipment.

The production of IFB of high use limit has been a problem for a number of years. The best solution seems to be the bonding together of tiny, hollow spheres of alumina, thus giving a cellular structure. The alumina spheres are produced by blowing high-pressure air or steam against a stream of molten alumina coming from an electric-arc furnace.

It might be asked why spheres are blown from molten alumina while fibers are blown from fused kaolin under much the same conditions. The answer is that molten alumina has a low viscosity, allowing surface-tension forces to form spheres immediately, whereas the silicate has a high viscosity and allows fibers as blown to keep their form. Actually, brick of the 33 class are on the market with a bulk density of 85 pcf.

Other methods have been tried, such as that in the patent of Holland,[3] whereby an organic cellular sponge with permeable cell wall is impregnated with a dispersion of refractory particles in water. After drying, the piece is fired to burn out the organic and sinter the refractory. This is much like the way old English potters produced ceramic lace in their figurines by saturating real lace with porcelain slip and firing.

Foaming has been used for making porous structures for many years, but there are problems in stabilizing the structure to give small and uniform pores. The patent of Griffith[2] describes a high-temperature insulator by foaming a dilute aqueous suspension of boehmite which is dried and fired at 3100°F. Again in the Powell[4] patent a stream of molten glass interacts with a foaming agent producing a pumicelike structure. Zirconia and zircon foamed structures are described by the Russian authors Gaodu and Kainarskii.[5] Macdonald[6] describes an IFB using an aggregate of porous zirconia. Another Russian reference applies to foamed silicon carbide bricks.

The insulating blocks are commonly made from calcined diatomaceous earth mixed with asbestos. The mixture is made into a suspension with water and filtered into a formed cake which can then be dried. The diatomaceous earth consists of minute, cellular skeletons of diatoms having very fine pores with thin separating walls, a structure that is resistant to heat flow. Some of these blocks have lime added and are then autoclaved to form a silica-lime bond.

The vermiculite is a type of mica that exfoliates on heating, giving particles divided into many cells by thin partitions. These exfoliated particles may be used as loose filling or may be bonded into brick or block.

The slag wool fibers are blown by steam from a stream of slag from a cupola, or may be spun from a revolving wheel. The glass wool is made in much the same way from a stream coming out of a glass furnace. The silica-alumina wool is made by steam blowing against a stream composed of 50 parts white alumina, 50 parts potter's flint, and 1½ parts borax glass coming from an electric-arc furnace at temperatures of about 3400°F.[1] Another type is made from kaolin fused in an electric furnace and blown into wool by compressed air. The wool is used largely in blanket form because of its flexibility and ease of application. It is also used for filling expansion joints in brickwork.

5.5 Properties. One of the more important properties of insulators, of course, is the thermal conductivity. As insulators go higher in use temperature, the bulk density increases and so does the thermal conductivity, as shown in Chap. 19.

Another important property is the heat-storage capacity, which is roughly proportional to the bulk density. Therefore, for intermittent furnace use light bricks are a real advantage.

Insulating materials have lower crushing strengths and abrasion resistance than heavy bricks, but this is a disadvantage only in a few cases. For example, insulation between the shell and blocks of a rotary cement kiln must not disintegrate under the vibration and repeated stresses. Manufacturers should be consulted when selecting insulation for this service.

It has been found that many types of IFB disintegrate when exposed to dry hydrogen in some types of metallurgical furnaces. The Babcock & Wilcox Company[7] have recently made a brick to resist this action by using a composition of 95 percent Al_2O_3 and 5 percent CaO, thus eliminating the silica, iron, and titania, which are reducible in this service.

BIBLIOGRAPHY

1. Straka, R. C., Jr.: Product Forms of Alumina-Silica Ceramic Fibers, *Bull. Am. Ceram. Soc.*, **40**, 493, 1961.
2. Griffith, J. S., R. S. Olsen, and H. L. Rechter: Compositions and Processes for Making Foamed Alumina Refractory Products, and Articles So Produced, U.S. Patent 3,041,190, 1962.
3. Holland, I. J.: Porous Refractory Production, British Patent 923,862, 1963.
4. Powell, E. R.: Process for Making Foamed Ceramic Products, U.S. Patent 3,133,-820, 1964.
5. Gaodu, A. N., and I. S. Kainarskii: Highly Refractory Lightweights from Zirconia and Zircon, *Refractories, Moscow*, (8), 403, 1964.
6. Macdonald, A. C.: Insulating Refractories, *Refractories J.*, **40** (2), 61, 1964.
7. Anon.: Insulating Firebrick Developed for Use in Dry Hydrogen at High Temperatures, *Ind. Heating*, **31** (7), 1328, 1964.

CHAPTER SIX

Standard and Special Shapes

6.1 Introduction. If bricklaying were always in the form of a uniform wall, the only refractory required would be a straight brick. However, refractories must often be set to form cylindrical linings, sprung arches, and other complicated constructions. It would, of course, be possible to cut a straight brick into keys, arches, or wedges, but this would be very costly and the firm skin of the brick would be lost. Therefore, it is desirable to supply shapes already formed for efficient construction. In this chapter shapes, both standard and special, will be discussed.

6.2 Standard Shapes. A standard shape is a refractory brick or tile having dimensions that are conformed to by all refractory manufacturers. The basic size in the United States has been the $9 \times 4\frac{1}{2} \times 2\frac{1}{2}$ in. straight, but this is being gradually displaced by the $9 \times 4\frac{1}{2} \times 3$ in. straight. The European straight is $250 \times 123 \times 65$ cm. The 9 in. series of standard shapes is shown in Fig. 6.1. Similar series are available in other basic sizes, as shown below:

$9 \times 4\frac{1}{2} \times 3$ in.	$12 \times 6\frac{3}{4} \times 2\frac{1}{2}$ in.	$13\frac{1}{2} \times 4\frac{1}{2} \times 2\frac{1}{2}$ in.
$9 \times 6 \times 2\frac{1}{2}$ in.	$12 \times 6\frac{3}{4} \times 3$ in.	$13\frac{1}{2} \times 4\frac{1}{2} \times 3$ in.
$9 \times 6 \times 3$ in.		$13\frac{1}{2} \times 6 \times 3$ in.

Circle brick (Fig. 6.2) come in $9 \times 4\frac{1}{2} \times 2\frac{1}{2}$-in. and $9 \times 4\frac{1}{2} \times 3$-in. series with outside-circle diameters of 33 to 129 in.

Cupola blocks (Fig. 6.2) are $9 \times 6 \times 4$ in. and $9 \times 9 \times 4$ in. with outside-circle diameters of 42 to 150 in. and 66 to 180 in., respectively.

Rotary-kiln blocks (Fig. 6.2), arch type, $9 \times 6 \times 3\frac{1}{2}$ in., have outside-circle diameters of 42 to 168 in. The wedge type are $9 \times 6 \times 3\frac{1}{2}$ in. with outside-circle diameters of 54 to 162 in.

Skewback shapes are standard in $9 \times 4\frac{1}{2} \times 2\frac{1}{2}$ in. and $9 \times 4\frac{1}{2} \times 3$ in. in 48° and 60° angles.

All manufacturers publish handbooks illustrating in detail the various shapes that they stock; so purchasers should refer to them for specific information.

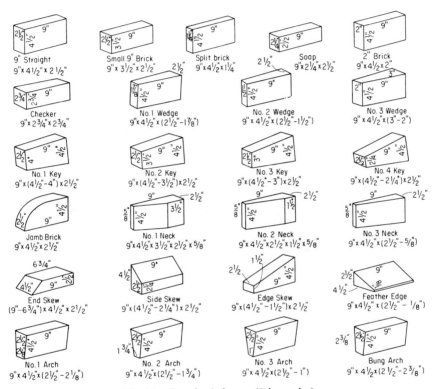

FIG. 6.1 Standard shapes (9-in. series).

Size tolerances on standard shapes are given by the ASTM as ± 2 percent on dimensions over 4 in. and ± 3 percent on smaller dimensions. ASTM method C 134–41 is used for measuring dimensions and method C 154–41 for warpage.

6.3 Special Shapes in Stock. Shapes may be divided into two groups; the first are shapes for which at least some of the manufacturers have molds for regular production, and the second are shapes on special order where molds must be made. The first group is often listed in manufacturers' handbooks, of which the following are examples.

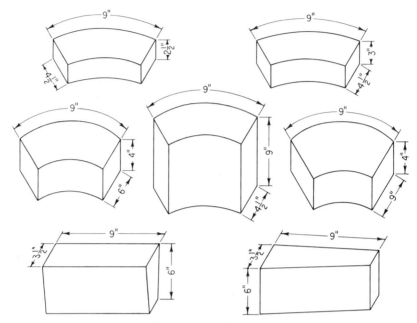

FIG. 6.2 Circle brick, cupola blocks, and rotary-kiln blocks.

Kiln-setting tile (silicon carbide) with lengths of 11 to 50 in., widths of 8 to 43 in., and thicknesses of $\frac{1}{2}$ to $2\frac{1}{4}$ in.

Kiln-setting tile (fireclay):
  Sanitary ware for straight TK:

     24 × 36 × 1¼ in.
     24 × 36 × 2¼ in.
     32 × 25 × 1¼ in.
     17 × 25 × 1¼ in.
     18 × 24 × ¾ in.

  General ware and electrical porcelain for straight TK:

     25½ × 21½ × 1 in.
     20½ × 22½ × ⅞ in.

  For circular TK:

     32 × 16½ × 18½ × 1 in.
     32 × 16½ × 18½ × ¾ in.
     28 × 16½ × 18½ × ¾ in.
     28 × 16½ × 18½ × ½ in.
     32 × 16½ × 18½ × ¾ in.
     32 × 16½ × 18½ × ½ in.

Square-edge fireclay tile are available in the following sizes:

12 × 6 × 2 in.	18 × 6 × 2 in.
12 × 6 × 2½ in.	18 × 6 × 2½ in.
12 × 6 × 3 in.	18 × 6 × 3 in.
12 × 9 × 2 in.	18 × 9 × 2 in.
12 × 9 × 2½ in.	18 × 9 × 2½ in.
12 × 9 × 3 in.	18 × 9 × 3 in.
12 × 12 × 2 in.	18 × 12 × 2 in.
12 × 12 × 2½ in.	18 × 12 × 2½ in.
12 × 12 × 3 in.	18 × 12 × 3 in.

Other special shapes often stocked are:

Malleable furnace shapes	Tap-out and slag-hole blocks
Blast-furnace linings	Burner blocks
Boiler-setting tiles	Suspended-arch tiles
Kiln-car shapes	Coke-oven shapes

Saggers for the pottery industry are gradually being replaced in this country by open settings using the tiles previously mentioned. On the Continent and in Japan, however, where hard porcelain is still made,

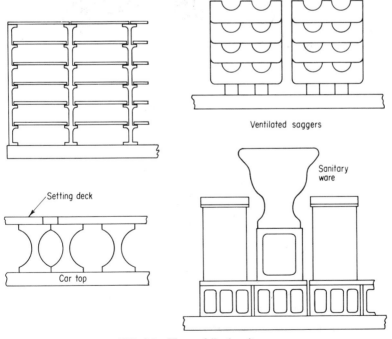

FIG. 6.3   Heavy kiln furniture.

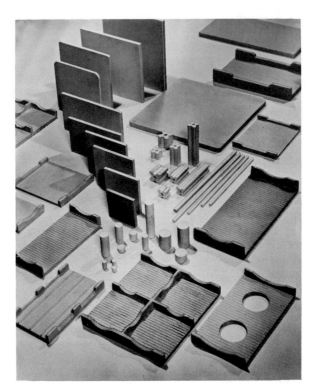

FIG. 6.4   Refractory shapes for setting whiteware.   (*Electro Refractories.*)

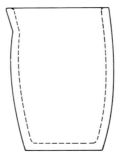

FIG. 6.5   A clay-graphite crucible.

saggers and cranks are used extensively.   Refractory bodies have been
developed to give good hot strength, good resistance to spalling, and
freedom from kiln dirt.   Fast-firing slab kilns are finding more use, so that
there is a demand for tiles to carry loads without deformation and yet
resist the fast heating and cooling.   Some sillimanite bodies have worked

quite well, even when the cycle time is only 15 hr. Sandwich-type setter plates having a core of silicon carbide and surface layers of fireclay, mullite, or alumina[1] have proved quite successful in some types of firing. For firing electronic ceramics, zircon and zirconia setters have been found quite useful. In Figs. 6.3 and 6.4 are shown several types of setter used for whiteware.

Retorts for gas production are built up in sections of fireclay tiles supplied by some manufacturers. The continuous retorts are made from interlocking silica shapes. Coke-oven shapes of silica are supplied to form the complete oven by many manufacturers.

Graphite crucibles[3,4] are extensively used for metal melting (Fig. 6.5). They are made from a mixture of plastic clays and flake graphite, the latter varying from 10 to 50 percent. The clay and graphite are mixed into a plastic mass and then jiggered inside a plaster mold. This operation aligns the flakes parallel to the surface and thus gives increased hoop strength. The graphite gives increased thermal conductivity and hot strength while the clay protects the graphite from rapid oxidation. Carbon-bonded silicon carbide crucibles are also made for metal melting.[2] Common sizes of graphite crucibles are shown in Table 6.1.

TABLE 6.1 Standard Bilge-shape Graphite Crucibles*

Size	Outside height, in.	Outside bilge diameter, in.	Working capacity, brass, lb
000	3	2$\frac{9}{16}$	1.2
2	4$\frac{9}{16}$	4$\frac{1}{16}$	4.7
6	6$\frac{9}{16}$	5$\frac{3}{8}$	15.4
10	8$\frac{1}{8}$	6$\frac{9}{16}$	25.7
14	9	7$\frac{1}{4}$	50
18	9$\frac{7}{8}$	7$\frac{15}{16}$	67
30	11$\frac{1}{2}$	9$\frac{1}{2}$	111
50	13$\frac{3}{4}$	11$\frac{3}{16}$	190
70	15$\frac{3}{8}$	12$\frac{1}{2}$	254
90	16$\frac{1}{8}$	13$\frac{3}{8}$	317
125	18	14$\frac{3}{8}$	396
175	19$\frac{1}{2}$	15$\frac{13}{16}$	553
225	20$\frac{15}{16}$	17	712
300	22$\frac{3}{4}$	18$\frac{7}{16}$	950
430	25$\frac{1}{2}$	20$\frac{1}{2}$	1,460
600	27$\frac{3}{8}$	24	1,990

* Joseph Dixon Crucible Company.

Fireclay crucibles are relatively inexpensive and find many uses. The common sizes are shown in Table 6.2.

TABLE 6.2   Clay Crucibles

Size*	Height, in.	Diameter, in.
C	5⅝	3⅝
H	5⅞	3¾
I	6	4
J	6⅝	4⅜
K	7¼	4¾
L	8	5¼
M	8¼	5¾
N	8⅜	6½
O	10	7
P	11	7¾
Q	12	8¼
R	13	9⅜

* Denver Fireclay Company.

Glass pots are made from grog and plastic clays. Some are hand-molded, but now most are slip-cast. A typical pot is shown in Fig. 6.6.

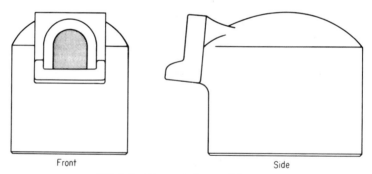

Front   Side

FIG. 6.6   Covered glass melting pot.

Glass-tank feeder parts are made in large quantities of high-alumina mixes, generally slip-cast in plaster molds from a highly thixotropic slip having water contents as low as 10 percent.[5]

Glass-tank blocks are made by slip casting and by fusion casting. Although sizes are not well standardized, manufacturers will give those sizes generally available. The slip-cast blocks are made from grog and clay fired to a dense structure. The fusion-cast blocks are mullite, alumina, or zirconia-alumina-silica compositions.

6.4 Special Shapes to Order. When designing new refractory structures, it often becomes advantageous to have special pieces made. It should be remembered that special shapes when ordered in small quantities are expensive because of the mold cost. Also special shapes may be hand-molded and therefore will not have as good a structure as machine-pressed ware. The shape should be kept as simple as possible, with the avoidance of undercuts and thin projections. In designing shapes to be used in conjunction with straight brick, it is advisable to have the vertical dimensions a multiple of $2\frac{9}{16}$ or $3\frac{1}{16}$ in. and the horizontal length a multiple of $4\frac{9}{16}$ in.

BIBLIOGRAPHY

1. Saunders, A. C.: Refractory Laminates—A Step Forward in Refractory Technology and Engineering, *Ceram. Age,* **78** (4), 51, 1962.
2. Anon.: Manufacture of Carbon-bonded Silicon Carbide Crucibles at Electrorefractories, *Ceram. Age,* **79** (6), 50, 1963.
3. Dixon, D.: Crucibles—White and Black, *Refractories J.,* **40** (11), 470, 1964.
4. Ito, U.: Studies of the Composition of Graphite Crucibles Consisting of Graphite, Silicon Carbide, Ferrosilicon, Borosilicate Glass and Fluoride, *J. Ceram. Assoc. Japan,* **73** (4), 67, 1965.
5. Babcock & Wilcox Co.: High Alumina Body for Glass Feeder Parts, British Patent 993,635, 1965.

CHAPTER SEVEN

Refractory Mortars, Plastics, Concretes, and Coatings

7.1 Introduction. Refractory mortars are used in laying refractory brick and shapes and serve the following purposes: to bond the brick-work into a solid unit so that it will be more resistant to shocks and stresses, to provide a cushion between the slightly irregular surfaces of the brick so that one course of brickwork will have a firm bearing on the course below it, and also to make a wall gastight or to prevent penetration of slag into the joints. The best mortars are combinations of a plastic clay and a volume-constant grog, as the use of raw fireclay alone as a mortar is confined to low-temperature applications.

Mortars are sometimes thinned with water and used as coatings for the face of the refractory walls in order to seal the joints further or to protect the wall from destructive elements in the furnace.

Plastic refractories are composed of a coarse aggregate of crushed firebrick or grog bonded with plastic firebrick and are used for three general purposes. The first is for making molded refractory shapes to be used in the furnace in the green state; the second is to form a molded monolithic wall or furnace structure; and the third, to repair and patch worn brickwork.

Refractory concretes contain an aggregate and a setting material, such as hydraulic cement, to give them cold-setting properties. The aggregate may be calcined fireclay, which will give a high density to the concrete, or it may be a porous grog, diatomaceous earth, vermiculite, or other porous material to give light weight and low heat conductivity to the resulting concrete.

Hot patching materials are not used very extensively as yet but are intended for forming a stable refractory while the furnace is at operating temperatures.

The general composition of these refractory materials varies a great deal depending upon the application for which they are intended. The majority are made from a fireclay base, but there are many special compositions in which chrome, silicon carbide, silica, and alumina are used. The manufacturer of the brick should be consulted when selecting a mortar or coating.

7.2 Mortars. *Heat-setting Types.* Mortars in this classification are generally composed of grog and a bond clay. No air-setting ingredient such as sodium silicate is added, and the mortar obtains its strength by the vitrification of the bond. For low-temperature service or where a strong mortar is needed, a clay high in fluxes should be used for the bond, whereas a mortar for high-temperature service would use a fireclay or kaolin. The nonplastic part is crushed to pass a 35-mesh screen and mixed with the bond. To reduce the shrinkage, the nonplastic ingredient may comprise as much as 60 percent of the total mortar. Sufficient raw clay must be used, however, to give proper plasticity and workability, and therefore the clays must be carefully selected for this purpose. The size distribution of the nonplastic grog also affects the working properties of the mortar. The ASTM Specification C 105–47 covers ground fireclay mortars specifying four classes, superduty, high-duty, medium-duty, and low-duty with minimum PCE of 31, 27, 26, and 16, respectively, depending on the refractoriness of the clay. Heat-setting mortars are usually shipped in 100-lb moistureproof bags.

Air-setting Mortars. Air-setting mortars are composed of a base of precalcined fireclay or raw flint clay crushed to pass a 35-mesh screen. To this is added a plastic fireclay and from 5 to 20 percent sodium silicate solution. The choice of plastic bond clay and the soda-silicate ratio of the sodium silicate solution will determine the properties of the mortar. To obtain a mortar for severe temperature conditions, a highly refractory clay and a sodium silicate solution that has a low ratio of soda to silica should be chosen. For more moderate temperature service, a more plastic but less refractory clay combined with a sodium silicate solution richer in soda will give a mortar of better working properties and lower vitrifying point, but it will, of course, have a lower temperature use limit. The grog base should be sized so as to produce in the finished mortar the least drying and firing shrinkage compatible with good workability.

The water content of an air-setting mortar is adjusted until it has the consistency of a thick batter. On air drying, these mortars set to a good strength and form an almost monolithic structure with the brickwork, in some cases the mortar joint being stronger than the surrounding brick.

A batch mixing process is generally used to ensure good control of the tempering water added. After mixing, the mortar is packed in steel drums and sealed with rubber-gasketed steel lids. These containers are sold in 50-, 100-, and 200-lb sizes. ASTM Specification C 178–47 specifies three classes of mortar, superduty, high-duty, and medium-duty with test temperature of 2910, 2730, and 2550°F, respectively. The mortar must be fine enough to pass 95 percent through a 35-mesh screen and when used in a joint should give a dry strength of over 200 psi modulus of rupture as measured by ASTM Specification C 198–47.

For some special applications, a dry, air-setting mortar is desired. In this case, dry, powdered sodium silicate is used instead of the wet solution, the rest of the mortar remaining essentially the same. The air-set strength of these dry mortars, however, is not so great as that obtained with the wet mortars. Dry, air-setting mortar is packaged in moisture-proof bags, usually in 100-lb sizes.

Some manufacturers have had trouble with air-setting mortars hardening in storage, and considerable research has been carried out to find the cause. It is generally believed now that the hardening is due to a base-exchange reaction between the sodium silicate and the natural clay with adsorbed calcium ions. This is brought out in Table 7.1, derived from some experiments made in the Ceramic Laboratory of the Massachusetts Institute of Technology. It has also been found that clays containing organic matter may develop an acid reaction in storage and thus cause setting.

TABLE 7.1   Hardening Properties of Air-setting Mortars

Adsorbed ions on clay	*Hardening*
Na^+	None
K^+	None
Ba^{++}	None
Ca^{++}	Severe
H^+	Severe

Special Mortars. For special applications, both heat-setting and air-setting mortars are made up using bases other than fireclay. Magnesite, chrome, silica, alumina, and silicon carbide are the more common ones and are used in laying up bricks of their respective materials. Alumina mortars are used where extreme temperature conditions are encountered and where freedom from iron is important. So-called "natural mortars" are much used in Europe. They consist of a naturally occurring clay and sand mixture with good working properties.

Application. Mortar is generally applied in one of three ways, but it should be emphasized that in most cases, the thinnest possible joint is desired. The first method results in the so-called "buttered joint," where the mortar is mixed to a batter consistency and the top of the exposed course of brickwork is spread, using a trowel, with a thin coating. The brick to be laid is given a trowel coating on the bottom of one end and then tapped or pushed into place.

The second is known as the "dip" method. The mortar is thinned out with more water than when using buttered joints. The brick to be laid is dipped into the thin batter on the bottom and one end and then pushed into place. The thickness of the joint can be regulated by the consistency of the batter. A combination method sometimes used consists of pouring a batter onto the exposed top course and then dipping the bricks to be laid on them. This results in more completely filled joints than the straight dipping method.

Approximately 400 lb of mortar is required for setting 1,000 bricks.

Tests for Refractory Mortars. A number of methods have been proposed for testing mortars,[1] but it is still difficult to get a quantitative measure of the illusive property called workability.

The fusion point can be determined

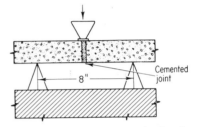

FIG. 7.1   Method of measuring bonding strength.

by the methods described in Chap. 14; but when silicate of soda is present, a preliminary calcining of the material to 1100°C is recommended by Phelps in order to obtain consistent results. However, the fusion point is not particularly valuable, and the block tests as used for glazes and enamels[1] or the amount of flow of the mortar when used as a bond is more important.

The bonding range is generally determined by cementing two half bricks together on the 2½- by 4½-in. faces with a definite thickness of joint. (ASTM Specification C 198–47.) After drying, they are broken as shown in Fig. 7.1, or they may be heated to various temperatures and broken while heated or when cooled to room temperature. Care must be exercised to prevent any stresses in the samples due to movement of the pieces while drying; otherwise low values will result from fine cracks.

The shrinkage can be measured on bars of the mortar in the usual way or by the characteristics of a layer dried and fired on a brick. The coefficient of thermal expansion can be measured by the methods described in Chap. 18.

The workability is an important property, but one difficult to measure. Troweling ability is measured by a machine developed by the National

Bureau of Standards,[1] but it is generally better to depend on the reports of experienced bricklayers. Rate of settling is important in mortars used in the pouring consistency and is simple to evaluate. Water-retention values, as measured by Phelps, are determined by cementing two half bricks together with no load and then, after a certain time interval, adding a load and noting the amount of mortar squeezed out. This test is particularly important for insulating refractories.

7.3 Plastic Refractories. Plastic refractories are now being used in somewhat larger quantities than in the past; in some instances, quite large furnaces have been constructed of monolithic plastic material with good results. It should be noted that the plastic material fires hard only on the hot surface and has a weaker and more flexible zone on the cooler side. Although this may sometimes be a disadvantage, in other cases it actually seems to give a greater life than the burned refractory because of the flexibility of the wall. However, success can be obtained with plastics in large monolithic structures only by skilled application.

Fireclay Plastics. Some of the plastic materials are cold-setting; i.e., they develop strength on drying. As in the case of cold-setting mortars, this strength is achieved by the addition of sodium silicate to the mass. Most of the plastics are made up with a base of fireclay grog, either specially prepared or made from waste brickbats and saggers. The maximum size of this grog runs from 4 to 2 mesh in order to reduce the drying and firing shrinkage. The bond clay used must be selected with great care to give good workability and low shrinkage.

It is particularly important that the plastics have a low drying and firing shrinkage; otherwise the structure will be badly cracked after the first firing. For this reason, the proportion of grog is comparatively large, running between 40 and 60 percent of the total. Not only the maximum size of the grog is important in achieving good results, but the grading of the grog to give the best characteristics is quite important. The plastic mass should be of such consistency that it is coherent and sticks to the brickwork. It should not be crumbly when worked up or get soft and wet when hard rammed. A good plastic can be rammed into a patch in the crown without slumping or pulling away. ASTM Specification C 176–47 calls for two types of plastic varying in refractoriness.

Ladle Mixes. A considerable amount of plastic refractory called "ladle" or "ramming" mixes is used for lining ladles in the iron and steel industry and for pouring spouts and parts of that nature. These refractories must have rather special properties, such as good workability and a softening point approaching the pouring temperature of the metal, so that the surface of the refractory becomes glazed and offers a good resistance to metal and slag attack. Whereas many of these mixes have bases of fireclay grog, others consist of a kaolin base; and in some cases, high-alumina clays have been used.

Special Plastics. Besides the fireclay-base plastics, considerable quantities of chrome-base plastics are used, particularly under conditions of severe slag erosion. The important field for this class of plastics is in the stud walls of boiler furnaces, where it can be rammed around the waterwall tubes to form a refractory capable of withstanding terrific punishment in the way of high temperatures and slag conditions. This type of plastic is usually made up of raw or calcined chrome ore with some clay bond and often silicate of soda. African or Cuban chrome ore is generally preferred for this use.

Silica ramming mixtures are made from crushed ganister or sandstone and fireclay as a bond. These mixes should be made up to have good workability and are used largely for patching cupolas and other metallurgical furnaces. Silicon carbide ramming mixtures are used where high heat transfer or good slag resistance is desired. Though comparatively expensive, they give excellent service under certain severe conditions. Magnesite ramming mixtures usually consist of dead-burned magnesite grain mixed with some organic bond like tar or dextrine. These are used for the basic electric furnaces, basic open-hearth furnaces, and some of the nonferrous constructions.

Application of Plastics. Satisfactory service cannot be expected from plastic materials unless they are properly applied. The directions given by the maker of the plastic used should be carefully followed in order to ensure a first-class job. Although the methods vary somewhat depending on the type of plastic, in general the following steps should be taken in the application of this material.

In laying a plastic furnace bottom, such as plastic chrome ore in a forging furnace, it is customary to use the following procedure: Take the material out of the cans and dump it on a clean spot on the floor near the furnace, cutting the lumps with a spade but adding no water. If the material has to be left for any length of time, it should be covered with damp burlap bags to prevent its drying out. Beginning on one side of the furnace floor, lumps of the material should be spread out in a layer approximately 50 percent thicker than the finished layer; i.e., if a 3-in. floor is going to be put in, the thickness of the lump material should be 4 or 5 in. It is well to work in a strip about 30 in. wide beginning at one side of the furnace. A heavy spade then thoroughly cuts through the lumps with deep strokes so that the blanket of material is thoroughly tied together. Then with a heavy tamper, the whole mass is consolidated with heavy strokes. The thickness of the layer should be tested with a rod. If it is too thin, the whole mass will have to be cut up again with a spade before adding more material; otherwise it will not consolidate. The final surface can be put on by laying boards on top and tamping on these. The next 30-in. strip can then be started, care being taken to ensure that it is well cut into the edge of the finished strip to make a bond. More boards can be laid on this, and the procedure repeated until the whole floor is done. The finished surface should not be walked on unless boards are put down.

It is good practice in all plastic construction to punch the layer with small vent holes about 2 in. deep to let out the steam on drying. This can be very conveniently done by making a rectangular frame into which spikes have been driven on about 3-in. centers. The frame can be readily pressed down onto the surface of the plastic.

In some cases, successful results have been obtained by tamping the hearth while it is hot with a heavy weight on the end of a pipe that can be introduced through the furnace door. If this tamping is carried out every weekend, the hearth soon becomes consolidated into an excellent structure.

When using plastic materials in side walls, wooden forms are required in order that the plastic can be rammed in to form a really monolithic structure. Sometimes plastic materials have been used for small furnace roofs where the ramming is done on top of a form. Generally, however, a refractory concrete would be used for this purpose.

Plastics are also used for patching walls; and under these conditions, great care must be taken to obtain a satisfactory piece of work. The wall itself must be carefully cleaned of any adhering slag, and the brickwork surface roughed up and, in some cases, undercut to hold the patch. This is usually hammered in with a heavy wooden mallet, precautions being used to build up the layer uniformly and without laminations.

Packing of Plastics. Plastics are usually shipped wet in tightly sealed cans, although manufacturers sometimes pack a cube of the mixture in waterproof paper and ship it in a carton. Occasionally, manufacturers have had trouble with air-setting plastics hardening in storage, as discussed for air-setting mortars.

Testing of Plastics. The important characteristics of plastic materials are the fusion point, the workability, the reheat shrinkage as determined by the panel test, and the strength when fired to various temperatures. The workability is measured as specified in ASTM C 181–47, which seems to give satisfactory results. The work of Heindl and Pendergast[3] may also be referred to.

7.4 Refractory Concretes. Probably there is no branch of the refractory industry that has developed so rapidly as refractory concretes since the Third Edition of this book. This is due to both improved quality of the product and labor saving in installation. The use has been greatly aided by improved gunning techniques.

Heavy Refractory Concretes. The early refractory concretes were made of crushed brickbats or flint clay as aggregate and 15 to 25 percent of commerical high-alumina cement as bond. This material was quite satisfactory for low- and medium-heat-duty use but failed at higher temperatures because of impurities in both the aggregate and cement. In 1946, Pole and Moore[2] described a pure calcium aluminate, $3CaO \cdot 5Al_2O_3$. This pure cement allowed the forming of concretes with excellent high-temperature properties, but Gitzen, Hart, and MacZura[6] believe a cement having a composition of $CaO \cdot 2.5Al_2O_3$ to be preferable. These pure cements better withstand CO disintegration than commercial cements.

The commercial high-alumina cement has typical compositions as shown in Table 7.2, while two high-purity cements are shown with it, as reported by Tseung and Carruthers.[9] The components in cement *d*, as determined by x-ray diffraction, are given in Table 7.3.

TABLE 7.2 Composition of High-alumina Cement

Constituent	Commercial high-alumina cement			High-purity cement *d*	High-purity cement *e*
	a	*b*	*c*		
Al_2O_3.......	39.2	41.8	50.8	72.75	80.0
CaO........	36.9	36.6	39.8	24.88	18.0
SiO_2........	9.3	8.3	5.8	0.45	0.05
Fe_2O_3.......	10.2	4.7	5.6	0.63	0.4
MgO........	1.0	1.1	....	0.69	0.4
TiO_2........	1.1	0.2	....	Trace	
Ignition loss.	....	1.1	....	0.46	

TABLE 7.3 Components in High-purity Cement *d*

Component	Amount
$CaO \cdot Al_2O_3$.....................	48.1
$CaO \cdot 2Al_2O_3$....................	36.1
Al_2O_3.........................	9.4
$2CaO \cdot Fe_2O_3$...................	1.1
$2CaO \cdot Al_2O_3 \cdot SiO_2$...............	1.8
$MgO \cdot Al_2O_3$....................	2.4

When these high-alumina cements are hydrated, the products in Table 7.4 are developed, according to Tseung and Carruthers.[9]

The concrete is formed by using an aggregate of high-burned fireclay or kaolin, sintered alumina, and possibly kyanite. The aggregate is

TABLE 7.4   Hydration Steps for Calcium Aluminate

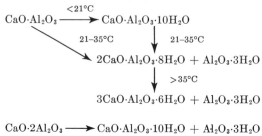

$$CaO \cdot Al_2O_3 \xrightarrow{<21°C} CaO \cdot Al_2O_3 \cdot 10H_2O$$

21–35°C $\quad\quad$ 21–35°C

$$2CaO \cdot Al_2O_3 \cdot 8H_2O + Al_2O_3 \cdot 3H_2O$$

$$> 35°C$$

$$3CaO \cdot Al_2O_3 \cdot 6H_2O + Al_2O_3 \cdot 3H_2O$$

$$CaO \cdot 2Al_2O_3 \longrightarrow CaO \cdot Al_2O_3 \cdot 10H_2O + Al_2O_3 \cdot 3H_2O$$

carefully sized to give a close-packed system with a 4-mesh particle the maximum size.   To this is added the finely ground cement in the amount of 15 to 20 percent.   Of course, the higher the cement content the lower will be the fusion point.   Using cement a of Table 7.2 and an aggregate of calcined flint clay, the relation of amount of cement to fusion point is shown in Table 7.5.   Of course, these concretes must be stored in air-tight containers to prevent hydration and special care must be taken to prevent segregation before use.   The water-cement ratio would be close to 0.4 for gunned mixes to 0.6 or 0.8 for cast concrete.

TABLE 7.5   Fusion Point of
Refractory Concrete

% Cement	PCE
5	33–34
10	31–32
15	29
20	27
25	17–18
30	16

The early refractory concretes were poured into forms in the same way as structural material, as shown in Fig. 7.2, but in the early twenties gun application began to be used—first for patching and later for complete linings.   With improved guns and better application techniques, it was found possible to achieve densities as high as or as much as 5 percent higher than cast material and strength from 50 to 100 percent higher.

The gun used may be the dry type,[13] as shown diagrammatically in Fig. 7.3.   However, it has been found desirable to moisten the mix previous to adding it to the gun chamber.   As much as one-fifth of the batch weight of water may be added and still keep the mix flowable. The remainder of the water is added at the nozzle at the control of the operator.   The dry gun gives a very dense concrete with low water-cement ratio.

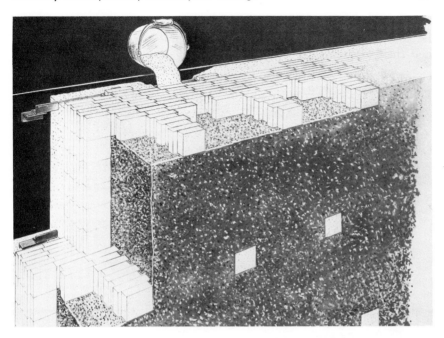

FIG. 7.2 Casting insulating refractory concrete back of a cast refractory wall. Fired anchor blocks are used on the hot face. (*M. H. Detrick Co.*)

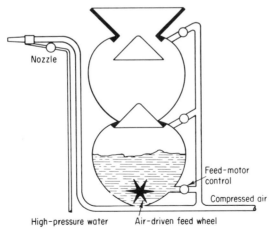

Nozzle

Feed-motor control

Compressed air

High-pressure water   Air-driven feed wheel

FIG. 7.3 Dry gun.

Less used is the wet-type gun,[13] shown in Fig. 7.4, where a slurry is formed by mixing a controlled amount of water with the dry material and then forcing it through the nozzle. The concrete placed with the

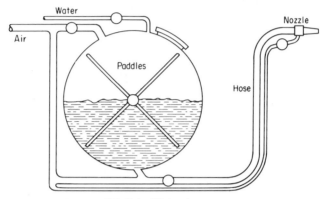

FIG. 7.4   Wet-mix gun.

FIG. 7.5   Gunning refractory concrete inside a sulfur plant converter. (*Babcock & Wilcox Company.*)

wet gun was less dense than that placed with the dry gun.[11]   Figure 7.5 shows a gun in use.

One of the problems is the considerable amount of rebound material, which normally runs as high as one-third that passing through the dry gun.   Livovich[14] has shown that small amounts of clay or bentonite added to the mix cuts down the rebound but often reduces the final properties. He finds that 6 percent of ball clay cuts down the rebound from 33 to 13 percent and has little effect on the final strength. As the rebound material is low in density, it must not be built into the structure. However, it can usually be salvaged and added to the gun supply if the water content is controlled. To form a uniformly dense structure requires great skill on the part of the nozzle man.

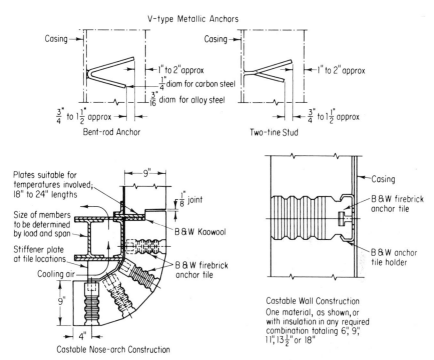

FIG. 7.6 Anchors for castable walls. (*Babcock & Wilcox Company.*)

Many types of anchor have been devised to hold monolithic gunned or rammed wall or roofs in place. Some are metal clips for lower-temperature use, but generally a refractory shape is fastened to the casing as shown in Fig. 7.6.

After the concrete is in place it is allowed to set with no loss of water. Formerly the concrete was frequently sprayed with water to prevent drying, but this was a time-consuming operation requiring the scaffolding to be kept in place until the set was complete. Now it is the general practice to spray the surface of the completed job immediately with an impermeable membrane of resin solution or special asphalt cutback to hold the moisture in place.

The final properties of some typical castables are given by Schneider and Mong[7] and Heindl and Pendergast.[3] The cured and fired strengths are excellent and there is no pronounced weak zone. It has been shown by Wygant[12] and others that these final properties are influenced by the curing temperature which, of course, rises in the large masses by the heat of hydration. It seems clear that this amount of heat up to 75°C (165°F) is not harmful, but if it goes higher, the final properties will suffer.

An excellent study of heat-setting and air-setting mortars by Eusner and Bachman[8] is summarized in Table 7.6.

TABLE 7.6   Range of Results for Properties of 32 Mortars for Fireclay Brick

Water required for optimum trowelability	21–38%
Workability, satisfactory	21 of 32
Linear shrinkage after 220°F	1.3–6.2%
Linear shrinkage after 2550°F	0–14.1%
Bulk density after 2550°F	1.22–2.05
Load test at 2500°F	0.3–33%

The gunned refractory concretes now find many applications, such as building up blast-furnace linings, placing linings in vessels for the petroleum industry, repairing cupolas, and many other purposes.

Insulating Refractory Concretes. These materials are now much used for insulating walls, since they consist of a lightweight aggregate such as vermiculite, diatomaceous earth, or porous grog particles, with high-alumina cement added. Most of the mixture may be applied with a gun to give a low-cost installation. Experience is necessary to produce a low-density structure in this way. Densities of 45 to 60 pcf are usual.

Phosphate Bonding. Phosphate bonding has been used for many years in dental porcelains using ZnO moistened with phosphoric acid. Since the Third Edition of this book, considerable study has been given to phosphate bonding of refractory mixes. A basic study of the subject was carried out in the author's laboratory at the Massachusetts Institute of Technology by Kingery.[4,5] Here the literature on the subject is summarized and experimental work carried out to find the best compositions for mortars. He found that the following composition was quite satisfactory with regard to workability and strength, in fact somewhat better than a comparable high-alumina-cement mortar:

Kaolin grog to pass 40-mesh screen	65%
Florida kaolin	27%
$Al(H_2PO_4)_3$ in solution	8%

One problem with phosphates is the migration of the soluble phosphates to the surface on drying. By the addition of ball-milled kaolin of high adsorption, the migration is prevented, but the drying shrinkage is increased and the low-temperature strength lowered.

Gilham-Dayton[10] described ramming mixes of silica, kaolin grog, mullite, and chrome magnesite bonded with 2 to 12 percent of $Al(H_2PO_4)_3$ solution. The properties of the set and fired specimens were excellent.

7.5 Coatings. Refractory coatings are used to some extent on the surface of certain types of brickwork to close the pores and present a more homogeneous surface, thus minimizing abrasion and slag erosion. However, unwarranted claims have been made for coatings, claims indicating that a good coating would make a lower-grade refractory give the service of a high-grade refractory. This is not generally true, and a coating is by no means a cure-all for refractory troubles.

Natural Coatings. Raw fireclay, which was used to some extent in the early days, has been more or less discarded because of its high shrinkage and tendency to peel off the brickwork. Natural coatings are used extensively in Europe containing clay, quartz, and feldspar in such proportions as to reduce the shrinkage and give a fluxing action. This material is very similar to the natural mortars previously discussed and generally has good working properties with a comparatively low cost.

Prepared Fireclay Coatings. For high-temperature conditions, the coating used most frequently is a mixture of calcined grog with a plastic bond clay. A satisfactory coating should have a low shrinkage and adhere well to the brick so that no peeling or checking is observable on heating to the working temperature. The brick manufacturers' recommendation should be taken, because a coating that may work very well with one type of refractory may not work at all with another type, even though of very similar properties.

Special Coatings. Many special coatings are available such as those with a chromite base or a chrome oxide base which are used to give some additional slag resistance to fireclay or insulating firebrick. There are also coatings having fused alumina or silicon carbide as a base for special purposes. They all must have the property, however, of low shrinkage and adherence to the brick with which they are used.

Application of the Coatings. Coatings can be applied to the brickwork with a brush or with a spray gun. The latter is generally more satisfactory in giving a really good job. The coatings must be applied in comparatively thin layers, $\frac{1}{16}$ to $\frac{1}{8}$ in. being the maximum generally used. Even then, a better job is obtained by applying it in several successive applications if this is possible. About $\frac{1}{4}$ to $\frac{1}{2}$ psf of material should be allowed for an average coating job. Particular care should be taken that the brickwork is clean and free from dust before the coating is applied.

BIBLIOGRAPHY

1. Heindl, R. A., and W. L. Pendergast: Properties of Air-setting Refractory-bonding Mortars of the Wet Type, *Natl. Bur. Std. (U.S.), Res. Paper* 1219, 1939.
2. Pole, G. R., and D. G. Moore: Electric Furnace Alumina Cement for High Temperature Concretes, *J. Am. Ceram. Soc.*, **29**, 20, 1946.
3. Heindl, R. A., and W. A. Pendergast: Reliability of "Workability Index" of Fire-clay Plastic Refractories, *J. Am. Ceram. Soc.*, **30**, 329, 1947.
4. Kingery, W. D.: Cold-setting Properties, *J. Am. Ceram. Soc.*, **33**, 242, 1950.
5. Kingery, W. D.: Fundamental Study of Phosphate Bonding in Refractories, *J. Am. Ceram. Soc.*, **35**, 61, 1952.
6. Gitzen, W. H., L. D. Hart, and G. MacZura: Properties of Some Calcium Aluminate Cement Compositions, *J. Am. Ceram. Soc.*, **40**, 158, 1957.
7. Schneider, S. J., and L. E. Mong: Elasticity, Strength, and Other Related Properties of Some Refractory Castables, *J. Am. Ceram. Soc.*, **41**, 27, 1958.
8. Eusner, G. R., and J. R. Bachman, Investigation and Testing of 32 High Grade Mortars for Fireclay Brick, *Bull. Am. Ceram. Ser.*, **37**, 12, 1958.
9. Tseung, A. C. A., and T. G. Carruthers: Refractory Concretes Based on Pure Calcium Aluminate Cement, *Trans. Brit. Ceram. Soc.*, **62**, 305, 1963.
10. Gilham-Dayton, P. A.: The Phosphate Bonding of Refractory Materials, *Trans. Brit. Ceram. Soc.*, **62** (11), 895, 1963.
11. Cook, M. D., C. P. Cook, and D. F. King: Pneumatic Placement of Refractory Castables, III, *Bull. Am. Ceram. Soc.*, **43**, 380, 1964.
12. Wygant, J. F., and M. S. Crowley: Curing Refractory Castables, *Bull. Am. Ceram. Soc.*, **43**, 1, 1964.
13. Young, W. J.: The Gunning of Monolithic Refractories, *J. Brit. Ceram. Soc.*, **2**, 262, 1965.
14. Livovich, A. F.: Properties of Pneumatically Placed Refractory Concretes, *Bull. Am. Ceram. Soc.*, **45**, 11, 1966.

PART THREE

Manufacture

CHAPTER EIGHT

Raw Materials
for Refractories

8.1 Introduction. Only a relatively few chemical elements having high fusing compounds are found in the earth's crust in sufficient abundance to be useful for heavy refractories. These include silicon, aluminum, magnesium, calcium, chromium, zirconium, and carbon. The compounds of these elements, particularly the oxides, are the refractory materials discussed in this chapter. Fortunately, many of them are found in deposits of sufficient purity to enable direct use. These include clays (Si, Al), ganister (Si), magnesia (Mg), dolomite (Mg, Ca), and chromite (Cr, Fe, Al). On the other hand, zirconium minerals need to be extracted from the naturally occurring ores. Carbon is obtained from coal, oil (petroleum coke), or as native graphite. In a few cases refractory materials are obtained by chemical treatment, such as magnesia from seawater and alumina from bauxite.

8.2 Clays. Clay deposits are widespread over the surface of the earth with great variations in refractoriness, extent, thickness of deposit, and amount of overburden. Here we are interested mainly in kaolins, fireclays, bauxites, and diaspores.

Definitions. It seems impossible to give a definition for a clay that is inclusive and yet free from exceptions. Many of the characteristics generally associated with clays, such as plasticity, water of hydration, and plate structure, are found in other materials. The best that can be done is to define a clay broadly as a hydrated earthy material, containing a considerable portion of alumina or silica and showing the property of plasticity.

More specifically the following definitions have been proposed by the Committee on Geological Surveys of the American Ceramic Society.

Bauxitic clay: A clay consisting of a mixture of bauxitic minerals, such as gibbsite and diaspore, with clay minerals, the former constituting not over 50 per cent of the total. (The opposite of this would be an argillaceous bauxite.)

Burley clay: A clay containing burls, oölites, or nodules, which may be high in alumina or iron oxide. As used in Missouri, it refers to a diaspore-bearing clay usually averaging 45 to 65 per cent Al_2O_3.

Diaspore clay: A clay usually containing over 60 per cent alumina with the mineral diaspore often in shotlike particles, or "oölites." There may be some gibbsite or cliachite present.

Fireclay: A clay either of sedimentary or residual character that has a PCE of not less than cone 19. It may vary in its plasticity or other physical properties, and while it often fires to a buff color, it does not necessarily do so. It is recommended that clays with a PCE from 19 to 26, inclusive, be called "low heat-duty fireclays" and that fireclays with a PCE of cone 27 or higher be designated "refractory."

The terms "Nos. 1, 2, and 3 fireclay" as sometimes used do not always refer to the same degree of fusibility.

While some fireclays are found underlying coal beds, many show no association with coal; in fact, some clays underlying coal beds do not conform to the description given.

Flint clay: A clay, usually refractory and hard, with a dense structure and conchoidal fracture. It is difficult to slake and has little plasticity under usual working conditions.

Kaolin: A white-firing clay which, in its beneficiated condition, is made up chiefly of minerals of the kaolinite type. Two types of kaolin may be recognized as follows:

1. Residual kaolin: A kaolin found in the place where it is formed by rock weathering.

2. Sedimentary kaolin: A kaolin that has been transported from its place of origin. Sedimentary kaolins show more pronounced colloidal properties than residual kaolins.

Clay Minerals. Contrary to the early conception that clay consisted largely of an amorphous colloid, we now know, by means of x-ray diffraction techniques, that substantially all the particles in clay are crystalline. A crystal is a piece of solid matter in which the atoms are packed together in an orderly array. Each atom species has a specific diameter, which in the case of clay minerals can be summarized as 2.64 Å* for oxygen, 1.14 Å for aluminum, and 0.78 Å for silicon. Because of its large size, the oxygen atom largely determines the type of packing in the crystal,[8] the smaller atoms fitting into the interstices. In ionic crystals, with which we are largely concerned, the atoms are charged, with the alumi-

* An angstrom unit, Å, is equal to 10^{-8} cm.

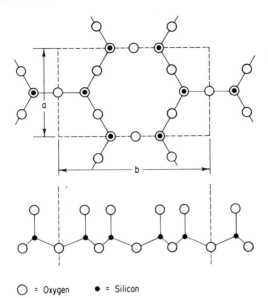

○ = Oxygen   ● = Silicon

FIG. 8.1   The tetrahedral sheet.

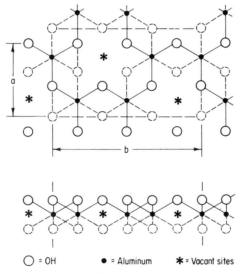

○ = OH   ● = Aluminum   ✱ = Vacant sites

FIG. 8.2   The octahedral sheet.

num and silicon atoms positive and the oxygen atom negative. The charged atoms, called ions, are held together firmly by the resultant attractive forces.[9]

The building block in clay mineral crystals is the Si-O sheet shown in Fig. 8.1 and the Al-OH sheet of Fig. 8.2. When one sheet is transposed on top of the other, they may be fitted together to form the unit cell of kaolinite as shown in Fig. 8.3 or, with a slightly different arrangement, the unit cell of dickite in Fig. 8.4.   These two minerals, which comprise the bulk of material found in kaolins, have the composition $Al_2(Si_2O_5)$

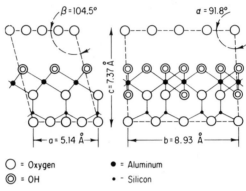

FIG. 8.3   Unit cell of kaolinite.

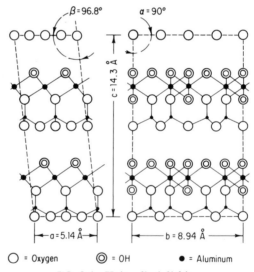

FIG. 8.4   Unit cell of dickite.

$(OH)_4$.   The electron photomicrograph shown in Fig. 8.5 shows the typical hexagonal plates of kaolinite.   For some reason that is not quite clear, these crystals seldom grow larger than 50 μ*[*] in diameter.[17]

Another type of rather rare kaolin mineral called halloysite is in the

* A micron, μ, is equal to 10^{-4} cm.

form of rolled-up tubes, as shown in Fig. 8.6. It is believed that a slight mismatch of Si-O and Al-OH sheets produces a curling of the very thin layers into these peculiar crystals.

For a long time fireclays were thought to be composed largely of kaolinite, but careful work with x-ray diffraction indicated a somewhat different structure called by Brindley the "fireclay" mineral.[7] It differs from kaolinite mainly in being less perfectly crystalline.

FIG. 8.5 Electron photomicrograph of kaolinite crystals.

A number of methods are available for the identification of the clay minerals both qualitatively and quantitatively. It is not intended to give a detailed description of these methods here as references are readily available.

The x-ray diffraction method, if used on properly prepared specimens, is able to distinguish most of the clay minerals with certainty. This is clearly brought out in the treatise, "X-ray Identification and Crystal Structures of Clay Minerals."* Another powerful tool for identi-

* C. W. Brindley, Mineralogical Society, London, 1951.

fication is the method of differential thermal analysis, where the sample is raised in temperature at a uniform rate, while heat evolution or absorption is observed.   References 5 and 16 give details of this method. A third method of use in some cases is the electron microscope[14,25] to show the morphology of the crystals.   Petrographic methods, while suitable for the larger and more perfectly crystalline particles, are of little use for the finer fractions or disordered structures.

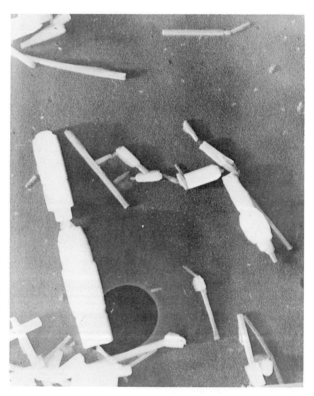

FIG. 8.6   Electron-microscope picture of halloysite tubes, 90,000×.   (*J. J. Comer.*)

The minerals of the montmorillonite group are of a three-layer structure, an Al-O sheet sandwiched between two Si-O sheets.   So loosely are the layers bonded together in some minerals that water molecules can enter into the lattice and cause swelling normal to the layers.   One member of this group is pyrophyllite, the unit cell of which is shown in Fig. 8.7.

The hydrated aluminas are interesting in that both hydrogen and hydroxyl bonding are found.   These bonds tie the oxygen ions together with H^+ as shown in Fig. 8.8.   The mineral gibbsite, $Al(OH)_3$, is a simple

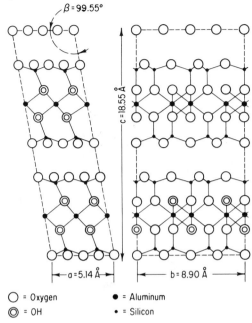

= Oxygen   ● = Aluminum
◎ = OH   • = Silicon

FIG. 8.7   Unit cell of pyrophyllite.

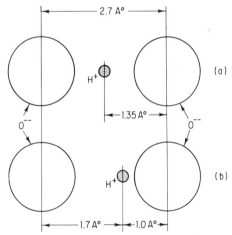

FIG. 8.8 (*a*) Hydrogen bond. (*b*) Hydroxyl bond. The location of the H^+ is the average position.

Al-OH sheet, as shown in Fig. 8.9.   On the other hand diaspore, $HAlO_2$, is more complicated.

In Table 8.1 are shown the clay minerals and their allies as known at present.   Most of these minerals may be produced in the laboratory by hydrothermal methods.[20]

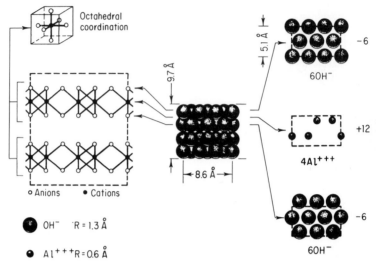

FIG. 8.9 The unit cell of gibbsite, 8 [Al(OH)₃]. (*"Elements of Ceramics,"* *Addison-Wesley Publishing Company, Inc., Reading, Mass.*)

Accessory Minerals. All clays contain mineral fragments of a nonclay type that have an important influence on the properties of the clay. The minerals commonly found are as follows:

Quartz
Feldspars (orthoclase, plagioclase)
Micas (muscovite and biotite)
Iron minerals (hematite, magnetite, limonite, pyrite, siderite)
Titanium minerals (rutile, anatase)
Limestone (calcite, dolomite)
Magnesite
Gypsum
Garnet
Tourmaline

A few of these minerals, such as quartz, act as a refractory nonplastic, whereas the others act more or less as fluxes and reduce the refractoriness of the clay.

Organic Matter. Nearly all clays contain a trace of organic matter, but many of the fireclays, ball clays, and glass-pot clays contain a great deal. In the last few years our knowledge of these organics has been strengthened by several investigations, particularly those of Worrall.[15] He showed that most of the organic matter was a lignite very similar to those found in nearby beds. This lignite may be separated into waxes,

TABLE 8.1 The Clay Minerals

Kaolinite group	Montmorillonite group	Micaceous group	Aluminous group
Kaolinite, $Al_2(Si_2O_5)(OH)_4$	Pyrophyllite,* $Al_2Si_4O_{10}(OH)_2$	Muscovite,* $Al_4K_2(Si_6Al_2)O_{20}(OH)_4$	Gibbsite,* $Al(OH)_3$
Dickite, $Al_2(Si_2O_5)(OH)_4$	Talc,* $Mg_3Si_4O_{10}(OH)_2$		Diaspore,* $HAlO_2$
		Bravaisite, $Al_4K_x(Si_{8-x}Al_x)O_{20}(OH)_4$	Boehmite, $HAlO_2$
	Montmorillonite, $(Al_{1.67}Mg_{0.33})Si_4O_{10}(OH)_2$ $\overset{\uparrow}{Na_{0.33}}$		
Nacrite, $Al_2(Si_2O_5)(OH)_4$		Brommallite, $Al_4Na_x(Si_{8-x}Al_x)O_{20}(OH)_4$	
	Beidellite, $Al_{2.17}(Al_{0.83}Si_{3.17})O_{10}(OH)_2$ $\overset{\uparrow}{Na_{0.33}}$		
Anauxite, $Al_{2-n}(Si_{2+n}O_5)(OH)_4$		Attapulgite, $(Mg_5Si_8)O_{20}(OH)_2 2H_2O$	
Endellite, $Al_2(Si_2O_5)(OH)_4 2H_2O$	Nontronite, $(Fe_{2.00})Al_{0.33}Si_{3.62})O_{10}(OH)_2$ $\overset{\uparrow}{Na_{0.33}}$		
	Saponite, $Mg_3(Al_{0.33}Si_{3.62})O_{10}(OH)_2$ $\overset{\uparrow}{Na_{0.33}}$	Ordovician bentonites	
Halloysite, $Al_2(Si_2O_5)(OH)_4$	Hectorite, $(Mg_{2.67}Li_{0.33})Si_4O_{10}(F, OH)_2$ $\overset{\uparrow}{Na_{0.33}}$	(Most of the minerals in this group are not very specific)	
Allophane, amorphous	Sauconite, $Zn_3(Al_{0.33}Si_{3.67})O_{10}(OH)_2$		

* These minerals are not usually considered among the clay minerals but when finely ground behave like clays in ceramic processes.

resins, and lignin-humus. Most of the organics occur in the coarser clay fractions; so they may be readily separated. However, the humus is adsorbed on the particle surfaces and therefore influences the colloidal properties, as will be discussed later.

Adsorbed Ions. All the clay minerals have the ability to adsorb ions on their surfaces. The amount that can be adsorbed is called the base-exchange capacity. The finer clays, because of their greater surface area, have larger values, as shown in Table 8.2.

TABLE 8.2 Base-exchange Capacity

Clay	Mean spherical diameter, μ	Surface area sq m per g	Base-exchange capacity, milliequivalents per 100 g
Kaolinite fraction.............	10	1.1	0.4
Kaolinite fraction.............	4	2.5	0.6
Kaolinite fraction.............	2	4.5	1.0
Kaolinite fraction.............	1	11.7	2.3
Kaolinite fraction.............	0.5	21.4	4.4
Kaolinite fraction.............	0.2	39.8	8.1
Raw Georgia kaolin...........	....	....	1
Pennsylvania flint fireclay.....	....	....	5
Kentucky flint fireclay.........	....	....	7
Plastic fireclay...............	....	....	8
Kentucky ball clay...........	....	....	12
Bentonite...................	....	....	100

Normally the clay particles have negative charges and so adsorb cations. These adsorbed ions influence the properties of clays when in the form of slips or plastic masses.

Kaolins. This white-burning clay with the formula $Al_2(Si_2O_5)(OH)_4$ is largely used for paper fillers and coatings. A smaller amount is used in the whiteware industry for tableware, tiles, and electrical insulators. Its use in refractories has been rather late in developing because of the large firing shrinkage, the high-temperature kilns required for firing, and the use of large proportions of grog, in the mix. At present about 10 percent of the kaolin mined in the United States is used for refractories.

There are two types of kaolin deposit; one is residual, where the kaolin is in the place where it was altered from the parent rock, while the other is a sedimentary deposit. The latter is the type used for refractories. On the map of Fig. 8.10 are shown deposits in the United States, the most useful of which are in the broad band below the fall line in North Carolina, South Carolina, and Georgia.

In general the kaolins are derived from feldsparthic rocks by leaching, weathering, or hydrothermal processes. In the residual deposits where the kaolin is formed in place, the steps of metamorphosis are believed to be as follows:

$$KAlSi_3O_8 + H_2O \rightarrow HAlSi_3O_8 + KOH \qquad \text{(Hydrolysis)}$$
Orthoclase

$$HAlSi_3O_8 \rightarrow HAlSiO_4 + 2SiO_2 \qquad \text{(Desilication)}$$

$$2HAlSiO_4 + H_2O \rightarrow (OH)_4Al_2Si_2O_5 \qquad \text{(Hydration)}$$
Kaolin

However, there is no evidence that these reactions occur in the step-by-step manner shown. More probably they occur simultaneously and the chemical means may well be augmented by colloidal transfer.

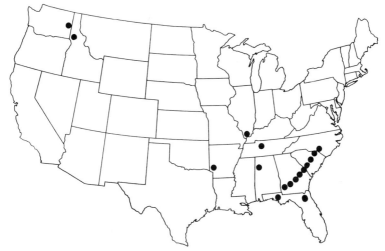

FIG. 8.10 Deposits of sedimentary kaolin in the United States.

In the case of sedimentary deposits, the products of decay are washed out by moving streams with some alteration during transport. Natural classification allows the finer particles to settle in the still waters of lakes and lagoons where the final kaolination takes place, often by the aid of organic acids. Often brackish water acts as a flocculant and causes the fine particles to settle, which accounts for the band of sedimentary kaolins in the Southern states below the fall line.

Kaolins for use in whitewares are treated to remove most of the accessory minerals, but this is seldom done for refractories as the additional cost is not justified. Chemical analyses of a few typical kaolins are given in Table 8.3.

As kaolins are never used in refractories completely in the raw form, the physical properties of the kaolin itself are of no practical importance.

However, it may be noted that, compared with fireclays, the kaolins have low green strength and high firing shrinkage.

The mineral pyrophyllite[1] has been used in refractories as it is found in a few deposits with good uniformity. However, the formula $(OH)_2Al_2(Si_2O_5)_2$ shows that the alumina content is not so high as in kaolin.

Ball and Glass-pot Clays. These clays contain organic matter, are fine-grained, highly plastic, and have high green strength. They have been used for glass pots, retorts, and similar types of refractories.[18] They also may be a useful ingredient in casting slips.

In the United States these clays are found in Kentucky and Tennessee,[19] while in Europe sources are in southern England and Germany. A chemical analysis of two typical ball clays is shown in Table 8.3.

TABLE 8.3   Analyses of Kaolin, Ball Clays, Halloysite, and Pyrophyllite

Constituent	Raw Georgia kaolin, %	Washed English China clay, %	Washed North Carolina kaolin, %	Raw Tennessee ball clay, %	Raw Kentucky ball clay, %	Raw halloysite, %	Raw pyrophyllite, %
Silica..........	45.8	48.3	45.8	53.9	56.4	44.3	63.5
Alumina........	38.5	37.6	36.5	29.3	35.0	37.4	28.7
Titanium oxide..	1.4	....	0.0	1.6			
Ferric oxide.....	0.7	0.5	1.4	1.0	1.0	0.4	0.8
Calcium oxide...	Trace	0.1	....	0.7	0.4	0.2	Trace
Magnesium oxide.........	Trace	....	0.5	0.4	Trace	0.1	Trace
Alkalies.........	....	1.6	0.3	....	5.3	0.2	0.4
Combined water.	13.6	12.0	13.4	12.8	7.9	15.1	5.9

Fireclays. These clays[10,22] are the backbone of the refractories industry. They include all clays having a fusion point above 2570°F (cone 15) that are not white-burning. There are many types, some being highly plastic while others are hard and difficult to make plastic with water. Some are high in silica, while others contain up to 38 percent of alumina. Fireclays have not had nearly so much study as kaolins, perhaps because there is such a large variation in the clay mineral. In some, we find perfectly crystalline kaolinite, while in others the structure is almost completely disordered, and many fall in between these two end points. Since the fireclays are not washed, they carry with them accessory minerals and often organic matter.[23]

The fireclays are derived from pegmatite rocks in the same way as kaolins and are always of sedimentary character. In the United States the fireclays are found from the Carboniferous to the Tertiary horizons. The fireclays of the Ohio-Pennsylvania region are of the Carboniferous Age, while most of the fireclays from New Jersey are from the Lower Cretaceous Age without association with coal. Fireclays from the Kentucky-Tennessee region are of the Tertiary Age. Many of the fireclays of the Middle West are associated with coal seams, but we are not sure why this is so. An excellent chart by Chelikowsky[2] shows the geological age of our fireclay deposits.

Chemical analyses of representative fireclays are shown in Table 8.4. It will be seen at once that there is a considerable variation in composition, not only in the major constituents but also in the impurities.

TABLE 8.4 Analyses of Fireclays

Analyses by Downs Shaaf

	Plastic, Lawrence, Ohio	Flint, Cambria, Pa.	Flint, Carter, Ky.	Semi-flint, Jackson, Ohio	Semi-flint, Clearfield, Pa.	Plastic, Vinton, Ohio	Flint, Montgomery, Mo.	Siliceous, New Jersey*
SiO_2...	58.10	44.43	44.78	50.32	43.04	46.72	44.04	59.93
Al_2O_3..	23.11	37.10	35.11	31.53	36.49	33.06	38.03	26.95
Fe_2O_3..	1.73	0.46	1.18	1.02	1.37	0.68	0.63	1.24
FeO...	0.68	0.55	0.74	0.35	0.83	0.55	0.22	1.24
FeS_2...	0.55	0.22	0.14	0.12	0.24	0.34	0.01	
MgO..	1.01	0.19	0.55	0.18	0.54	0.19	0.12	0.07
CaO...	0.79	0.60	0.77	0.80	0.74	0.61	0.40	
Na_2O..	0.34	0.10	0.29	0.07	0.46	0.42	0.10	Trace
K_2O...	1.90	0.55	0.44	0.05	1.10	1.53	0.22	Trace
H_2O-.	2.27	0.80	0.84	2.47	0.82	2.21	0.78	
H_2O+.	7.95	12.95	13.07	11.25	12.44	11.50	13.55	9.63
CO_2...	0.05	0.11	0.07	0.14	0.05	0.02	0.04	
TiO_2...	1.40	1.84	2.22	1.45	1.79	2.20	1.82	1.90
P_2O_5...	0.17	0.21	0.02	0.48	0.10	0.12	0.28	
SO_3....	0.03	0.01	0.01	0.01	0.01	0.01	0.01	
MnO..	0.01	0.01	0.02	0.02	0.01	0.01	0.01	
ZrO_2...	0.01	0.01	0.01	0.01	0.01	0.01	0.01	
Org. C.	0.22	0.10	0.11	0.07	0.22	0.04	0.01	
Org. H.	0.03	.....	.....	.....	0.03			

* Ries.

The softening point, of course, will be higher for the compositions high in alumina and low in fluxes. The maps of Figs. 8.11 and 8.12 show the important deposits of plastic and flint fireclays in the United States.

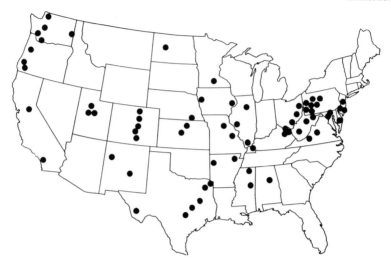

FIG. 8.11   Deposits of plastic fireclays in the United States.

FIG. 8.12   Deposits of flint fireclays in the United States.

The physical properties of the fireclays vary as widely as the composition.   The flint clays[11] are hard with a conchoidal fracture, will not slake in water, and when fired show little or no shrinkage.   On the other hand, the plastic clays will readily work up with water into a plastic mass or slip and show high drying and firing shrinkage. As today's refractory is almost never made from a single fireclay, it seems unnecessary to give any detailed physical properties here.

8.3 High-alumina Materials. As the service conditions for fireclay refractories have become more and more severe, interest has turned toward raw materials of higher alumina content.

Diaspore ($HAlO_2$). During the First World War this mineral was recognized in Missouri and from then on it has been mined extensively for high-alumina refractories. Unfortunately, the best material is located in isolated pockets, and unless some new deposits are found, this excellent material will no longer be available. Surrounding the diaspore are found burly clays intermediate in composition between diaspore and flint clay. In Table 8.5 are shown some typical analyses and in the map of Fig. 8.13 the location of deposits.

Table 8.5 High-alumina Clays

Clay	Dutch Guiana gibbsite	First-grade diaspore, Mo.	Bauxite, Ga.	Burly, Flint, Mo.	Second-grade diaspore, Mo.
Chemical Analyses					
Silica..................	4.5	10.9	26.0	33.8	29.2
Alumina..............	58.4	72.4	54.0	49.4	53.3
Titanium oxide.........	2.9	3.2	2.1	2.6	2.7
Iron oxide............	3.2	1.1	1.0	1.8	1.9
Lime.................	0.4				
Combined water........	30.6	13.5	16.1	12.0	12.0
Total...............	100.0	101.1	99.2	99.6	99.1
Thermal Analyses					
Kaolinite..............	6	None	67	71	18
Diaspore..............	6	90	None	26	53
Gibbsite..............	88	None	38	None	None

From a manufacturing point of view, diaspore is excellent as it has enough plasticity to be workable and yet has excellent firing properties. Geologists believe that diaspore is derived from kaolin by leaching out the silica under warm, humid conditions that were known to exist in Missouri at one time.

Gibbsite [$Al(OH)_3$]. This mineral is found only in tropical or subtropical regions, for under these conditions the silica in kaolin is removed by leaching. In fact, most tropical soils contain this mineral. Material with the highest alumina content occurs in nodules. Bauxites[3] are considered to be mixtures of gibbsite and kaolinite. They are not so satisfactory as diaspore for manufacturing refractories because of the high firing shrinkage. Locations of the important deposits in the United States are shown in Fig. 8.14 and for other countries in Fig. 8.15.

FIG. 8.13   Deposits of diaspore in the United States.

FIG. 8.14   Deposits of bauxite in the United States.

A large part of the gibbsite used in the United States is imported from Dutch Guiana.

Sillimanite Types of Minerals.   There are a number of minerals having a composition Al_2SiO_5, which gives a silica content of 37 percent and an alumina content of 63 percent. Sillimanite occurs in a number of small deposits in this country, such as California, North and South Carolina, and Georgia, but the largest source of supply is in India. Andalusite has a different crystal form from sillimanite and is found in small deposits in California and in considerable quantities in Spain.

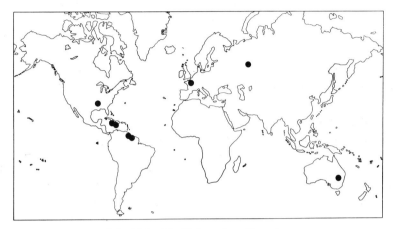

FIG. 8.15   World deposits of bauxite.

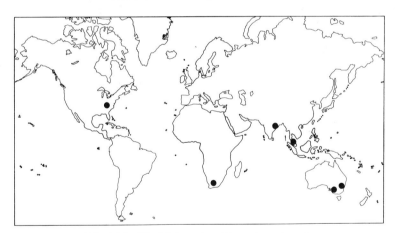

FIG. 8.16   World deposits of kyanite minerals.

Kyanite occurs in India, in some of the Southern states, and on the West Coast of the United States, as shown on the map of Fig. 8.16. All these minerals are decomposed to mullite and silica on heating above 1555°C (about 2830°F) when pure and at a lower temperature when fluxes are present. These materials are being used to a considerable extent as refractories at the present time but commercially are often confused one with another.   The great problem is to concentrate the ore efficiently and at the same time provide fairly coarse grains.

Topaz.   This mineral has been found in India in some quantity, and recently an important deposit has been discovered in South Carolina. Topaz has the composition $Al_2SiO_4(F,OH)_2$ and thus will readily transform to mullite on firing, hence offering attractive possibilities as a raw material.

Corundum (Al_2O_3).   Natural corundum is found in many small deposits in Asia Minor, South Africa, Greece, Ontario, and North Carolina.   At present it finds no use as a refractory.   However, alumina made in the electric furnace is used as fusion-cast blocks and as grain in high-alumina refractories.

8.4   Silica Minerals.   *Ganister.*   The most commonly used silica mineral for refractories is ganister, a common name for true quartzite. Not all quartzites are suitable for making refractories, as they must be of high purity and have good mechanical strength.   The principal deposits of ganister in this country are the Medina sandstone of Pennsylvania, the Baraboo quartzite of the Devil's Lake region, Wisconsin, and the quartzites of Alabama and Colorado.   A quartzite is usually mined from the bodies of talus, or slide rock, occurring on the mountain slopes.   A few typical analyses of quartzites are shown in Table 8.6 and location of deposits in the United States in Fig. 8.17.

TABLE 8.6   Analyses of Quartzites

Constituent	Medina quartzite, %	Baraboo quartzite, %	Alabama quartzite, %	Sharon con- glomerate	Washed ganister
Silica, SiO_2	97.8	98.2	97.7	98.0	98.6
Alumina, Al_2O_3	0.9	1.1	1.0	0.9	0.3
Ferric oxide, Fe_2O_3	0.9	0.2	0.8	0.9	0.8
Lime, CaO	0.1	0.0	0.1	0.3	0.1
Magnesia, MgO	0.2	Trace	0.3	Trace	0.1
Alkalies, K_2O, Na_2O	0.4	0.1	0.3	0.2	0.1
Total	100.3	99.6	100.2	100.3	100.0

The fusion point of quartzite is very close to 1700°C (about 3090°F); and like mullite, it shows little softening below its melting point.   This is its chief value when used as a refractory, as it retains its resistance to load at higher temperatures than fireclay materials.   Owing, however, to their high thermal expansion, silica materials are very sensitive to temperature changes in the low-temperature ranges.   Most silica refractories are made in the form of bricks or special shapes, but some ganister is

used in rammed furnace bottoms.   Recently quartzites have been washed
to bring the total impurities below 0.5 percent for superduty silica brick.

Firestone.   Natural stone refractories are still used to some extent.
They consist of an easily cut sandstone or a mica schist.   In either
case, the stone must have a certain flexibility of structure to reduce
the tendency to spall.   Firestone is rather wide in occurrence, working
deposits being found in Pennsylvania, Tennessee, Ohio, and several other
states.

Sand.   A small amount of beach sand is used for furnace bottoms, as,
for example, in the hearth of malleable-iron melting furnaces, but the

FIG. 8.17   Ganister deposits in the United States.

volume used is very small compared with the refractories made from
quartzite.   In Europe, glass sand is ground and added to quartzite in
the making of silica brick, mainly to increase the purity.

Diatomaceous Earth.   Another type of siliceous mineral, used mainly
as a heat insulator, is diatomaceous earth (infusorial earth or kieselguhr),
a widely distributed material.   It is composed of the skeletons of diatoms,
which are microscopic organisms.   The desirable characteristics of this
material are its closed cells and high porosity, which give a low density
and a low thermal conductivity.

8.5   Materials for Basic Refractories.   Basic refractories are those
used for resistance to basic slags, and more recently for strength at very
high temperatures.   The common basic materials are magnesia, lime,
and chromite, which may be used separately or as mixtures.

Magnesite (MgCO₃).   This mineral occurs in two types of deposit. One is known as crystalline, from its content of fine to coarse crystals and conchoidal fracture.   The other is called cryptocrystalline with a fine grain structure and no cleavage.   The crystalline type is found in larger deposits, but the cryptocrystalline magnesite is more widespread.

It is believed that the crystalline type has been derived from dolomite by alteration caused by percolation of magnesium solutions which leached out the calcium.   On the other hand, the cryptocrystalline type is formed by alteration of serpentine or similar rocks.

In the United States deposits of the crystalline variety have been worked in the state of Washington and in Nevada, but at present little,

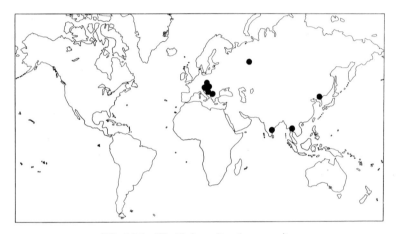

FIG. 8.18   World deposits of magnesite.

if any, goes to the refractories industry.   In Europe, there is a band of crystalline magnesite running through Austria, with huge lenses of ore in Styria and Radenthein.   These deposits have supplied the European steel industry for many years.   Because of the high iron oxide content, the ore can be dead-burned at comparatively low temperatures.   There are huge deposits of crystalline material in Manchuria which are believed to serve the Russian steel industry.

The only large working deposits of cryptocrystalline magnesite are in Greece and India—both of high purity.   There are undoubtedly many other deposits, but their distance from large steel plants has not favored their exploitation.   Figure 8.18 shows the important world deposits of magnesite, and the analyses are shown in Table 8.7.

Magnesium oxide is now being extracted from seawater, bitterns, and inland brines.   In fact, at the present time practically all magnesia used by the refractories industry in this country comes from the half

TABLE 8.7

	Dead-burned Magnesites (Seawater or Brines)					
Constituent	Michigan	Cape May	Cape May	Lavino	Kaiser	Kaiser
SiO_2........	0.6	4.6	1.7	1.1	2.1	0.4
Al_2O_3.......	0.2	0.3	0.3	0.3	0.7	<0.05
TiO_2........	....	....	....	Trace	Trace	0.01
Fe_2O_3.......	0.4	0.6	0.6	0.2	0.5	0.11
Cr_2O_3.......	....	....	....	.....	0.5	0.15
CaO........	0.7	1.3	1.1	1.0	1.1	1.1
MgO........	98.1	93.1	96.3	97.2	95.0	98.1

	Dead-burned Natural Magnesites				
Constituent	Austrian	Austrian	Grecian	State of Washington	Manchurian
SiO_2........	5.8	1.0	6.6	4.9	3.7
Al_2O_3........	1.7	1.0	4.4	1.5	1.0
Fe_2O_3.......	4.0	6.9	4.4	3.4	1.5
CaO........	5.0	2.1	2.4	2.8	1.6
MgO........	85.0	88.6	86.4	87.1	92.0
Ignition loss.	0.2	0.3	0.2	0.1	0.1

dozen seawater plants. The process of extraction, described in a number of references[4,24] consists essentially of precipitating as the hydroxide the 0.013 percent of magnesium in seawater by means of calcium hydroxide. The calcium is obtained from oystershells or dolomite. The precipitate is filtered out, dried, and burned in a rotary kiln. The process is highly mechanized, as great quantities of seawater must be handled to obtain a ton of magnesite. In Fig. 8.19 is shown a plant at Moss Landing, California.

Magnesite for refractories is dead-burned at temperatures of 1550 to 2000°C (about 2820 to 3600°F) in shaft or rotary kilns. The final product is quite dense and consists of periclase (MgO) crystals. Magnesite may also be fused in the electric-arc furnace to form large periclase crystals. The dead-burned material is then crushed to the desired size and rebonded into bricks.

Lime ($CaCO_3$). When lime is dead-burned at 1800°C (3270°F), it becomes dense and fairly stable in air for short periods. However, this material does not seem to have any advantages over dead-burned magnesia as a refractory and has the great disadvantage of instability in long-period storage.

Dolomite. This plentiful mineral consists of mixed crystals of $MgCO_2$ and $CaCO_2$. When pure, there are an equal number of molecules of both, although all ratios occur in nature.   Dolomites are very widely distributed in the earth's crust, and there is little difficulty in finding deposits of high purity.

Great quantities of dolomite have been used in the bottoms of open-hearth furnaces, and now tar-bonded dolomite bricks are extensively used in basic oxygen converters.

Olivine (Mg_2SiO_3).   This magnesium silicate usually has 5 to 8 percent of the MgO replaced by FeO in the theoretical formula.   By mixing

FIG. 8.19   Seawater magnesite plant at Moss Landing, California.   (*Kaiser Refractories Co.*)

with the correct amount of magnesite, this material has been used in the last few years to make forsterite refractories.   Olivine deposits are widespread, but excellent material occurs in the dunite rock of North Carolina.

8.6   Chromite.   The commercial chrome ore contains a solid solution of spinels closely approximating

$$\begin{matrix} Cr_2O_3 & \diagdown \diagup & FeO \\ Al_2O_3 & \diagup \diagdown & MgO \end{matrix}$$

In most cases, the spinel is balanced, but some cases have been reported where either the R_2O_3 or the RO is somewhat in excess; however, this may be due to difficulty in obtaining uncontaminated chromite grains for analysis.   The chromite is always associated with a gangue chiefly of serpentine but also containing small amounts of feldspars, silica, carbonates, etc. Chrome ores are believed to originate both from magmatic crystallization and by formation from hydrothermal solutions. Concentration methods are being used much in separating the gangue from the ore, but the addition of MgO will cause the formation of forsterite on

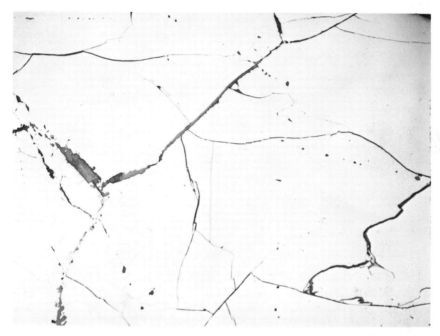

FIG. 8.20 Philippine chrome ore, massive type, with silicate gangue veins filling fractures; 125× under reflected light. (*Harbison-Walker Refractories Co.*)

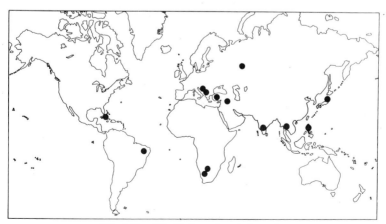

FIG. 8.21 World deposits of chromite.

firing and thus remove the low-softening siliceous compounds.   A thin
section of a typical chrome ore is shown in Fig. 8.20.

The location of the chief ore bodies is shown in the map of Fig. 8.21.
At present the largest producers (except the U.S.S.R.) are Albania,
Philippines, Turkey, Republic of South Africa, and Southern Rhodesia.
The metallurgical and chemical grades of chromite require a high Cr_2O_3
content, while the refractory grades should have a high percentage of
$Cr_2O_3 + Al_2O_3$, but a low SiO_2.   Typical analyses of chromites are given
in Table 8.8.

TABLE 8.8   Chromite Ores, Refractory Grade

Constituent	Philippine		Transvaal	
	Lump	Concentrate	Lump	Concentrate
SiO_2..........	5.1	2.8	2.8	0.8
Al_2O_3.........	27.9	29.5	15.0	17.4
FeO..........	13.0	13.9	24.1	24.6
CaO..........	0.5	0.4	0.3	0.2
Cr_2O_3.........	33.2	34.4	46.3	47.3
MgO..........	18.7	17.3	10.6	9.7
Ignition loss....	1.1	1.0	0.3	0.1

8.7   Other Refractory Materials.   There are many other refractory
materials besides those previously mentioned in this chapter, but only
a few are plentiful enough to be considered for heavy refractories.

Zirconia Minerals.   The common zirconia mineral is baddeleyite
(ZrO_2), large deposits of which are found in Brazil.   Zircon ($ZrSiO_4$)
is widely distributed, as shown in the map of Fig. 8.22.   Some zircon is
recovered by concentration of beach sands.

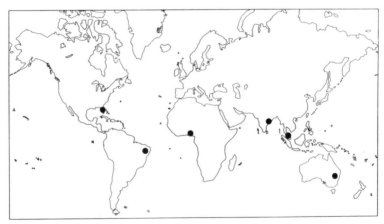

FIG. 8.22   World deposits of zirconium minerals.

8.8 Coal, Oil, and Graphite. Carbon and graphite are important refractory materials, for they are extremely refractory, very strong when hot, resistant to slags at all temperatures. Natural graphite[12,13,22] is found in many places, as shown in Fig. 8.23. Both amorphous and crystalline forms are mined. Ash contents[13] vary from 2 to 34 percent. Flake graphite is often mixed with clay to form melting crucibles for metals. Carbon is obtained by coking coal, although the coal ash acts as an impurity, so that high-purity carbon is made from petroleum coke.

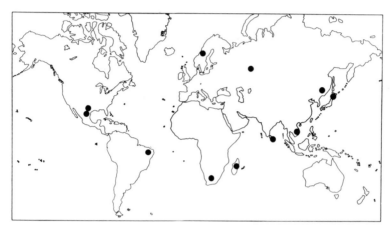

FIG. 8.23 World deposits of graphite.

Carbon may be converted to graphite by long heating in a reducing atmosphere.

Organics. It has often been found desirable in the forming operation of some types of refractory to add a small amount of organic material to act as a lubricant, binder, or plasticizer. In Table 8.9 are given some

TABLE 8.9 Organic Additives

Type	*Name*
Flours and starches	Corn flour
	Cornstarch
	Corn dextrine
	Potato starch
	Casein
Alcohols and cellulose derivatives	Methyl cellulose
	Polyvinyl alcohol
	Sodium carboxymethyl cellulose
	Cellulose glycolate
Wood extracts	Lignin extract
	Goulac
	Sodium alginate

TABLE 8.10 Production Statistics for Refractory Materials*

Year	Bauxite (crude ore)				Chromite				Dolomite (dead-burned)				Fireclay				Graphite (natural)			
	U.S. production, M long tons	Value of U.S. production, dollars per long ton	Importations, M long tons	World production, M long tons	U.S. production, M short tons	Value of U.S. production, dollars per short ton	Importations, M short tons	World production, M short tons	U.S. production, M short tons	Value of U.S. production, dollars per short ton	Importations, M short tons	World production, M short tons	U.S. production, M short tons	Value of U.S. production, dollars per short ton	Importations, M short tons	World production, M short tons	U.S. production, M short tons	Value of U.S. production, dollars per short ton	Importations, M short tons	World production, M short tons
1946	1,104	6.24	852	4,518	4	25.58	757	1,257	1,078	9.37	...	...	7,908	2.63	...	...	6	45.30	33	86
1947	1,202	5.73	1,822	6,367	1	...	1,106	1,828	1,395	10.25	0.05	...	9,039	2.90	...	...	4	50.44	44	105
1948	1,457	5.97	2,489	8,227	4	...	1,542	2,300	1,545	11.55	2	...	9,850	2.99	...	...	10	45.30	52	147
1949	1,149	5.90	2,688	8,344	...	26.93	1,204	2,300	1,319	12.08	2	...	8,572	2.96	...	...	6	77.88	32	180
1950	1,335	5.76	2,516	8,238	...	...	1,304	2,500	1,759	12.35	2	...	9,536	3.04	...	...	5	83.87	44	165
1951	1,849	6.73	2,820	10,811	7	72.38	1,428	3,100	1,966	14.31	3	...	11,853	4.11	...	...	7	108.12	54	190
1952	1,667	6.46	3,498	12,634	21	83.41	1,709	3,700	1,928	13.54	2	...	11,285	4.29	...	...	6	106.07	43	190
1953	1,580	8.51	4,391	13,600	59	58.37	2,227	4,300	2,294	13.71	4	...	10,267	3.75	...	...	6	...	51	200
1954	1,995	8.22	5,259	15,600	163	43.91	1,471	3,600	1,521	14.44	4	...	8,814	3.82	...	...	...	...	41	185
1955	1,788	8.13	5,225	16,400	153	43.35	1,834	3,800	2,129	14.76	8	...	10,840	3.89	...	...	...	...	49	290
1956	1,744	8.66	5,670	18,540	208	41.96	2,175	4,565	2,424	15.57	9	...	11,803	4.55	...	...	...	...	48	290
1957	1,416	9.09	7,089	20,150	166	47.02	2,283	5,110	2,251	15.93	10	...	10,805	4.75	...	...	...	...	42	410
1958	1,311	9.77	7,915	21,020	144	43.03	1,263	4,165	1,659	16.50	7	...	8,808	4.59	...	...	...	...	27	350
1959	1,700	10.43	8,149	22,600	105	35.86	1,554	4,350	1,988	16.64	8	...	9,862	4.58	...	...	...	...	37	410
1960	1,998	10.56	8,739	27,205	107	35.64	1,387	4,885	1,949	16.66	13	...	9,915	4.56	...	...	...	...	48	480
1961	1,228	11.35	9,206	28,895	...	...	1,329	4,650	1,983	16.40	4	...	8,690	4.46	...	...	...	...	30	455
1962	1,369	11.40	10,575	30,895	...	...	1,446	4,845	1,857	16.73	9	...	8,065	4.44	...	...	...	...	40	590
1963	1,525	11.30	9,212	30,220	...	...	1,391	4,355	1,949	16.96	9	...	8,390	4.71	...	...	...	...	52	780
1964	1,601	11.16	10,180	33,145	...	...	1,428	4,720	2,168	17.51	29	...	8,549	4.80	...	...	...	...	47	600

*"Minerals Yearbook," U.S. Department of Commerce.

TABLE 8.11 Production Statistics for Refractory Materials*

Year	Zirconium concentrates U.S. production, M short tons	Value of U.S. production, dollars per ton	Importations, M short tons	World production, M short tons	Kaolin (processed) U.S. production, M short tons	Value of U.S. production, dollars per ton	Importations, M short tons	World production, M short tons	Kyanite, etc. U.S. production, M short tons	Value of U.S. production, dollars per ton	Importations, M short tons	World production, M short tons	Magnesite (crude) U.S. production, M short tons	Value of U.S. production, dollars per ton	Importations, M short tons	World production, M short tons	Pyrophyllite, talc, and ground soapstone U.S. production, M short tons	Value of U.S. production, dollars per ton	Importations, M short tons	World production, M short tons
1946	...	...	17	...	1,322	10.25	89	...	...	...	11	...	325	6.85	0.06	1,599	461	...	18	1,004
1947	...	...	31	...	1,425	12.00	83	...	...	...	13	...	376	6.91	...	2,041	516	...	18	1,081
1948	...	...	18	...	1,569	12.59	100	...	...	...	17	...	...	...	0.10	2,206	528	...	18	1,434
1949	...	...	21	...	1,416	13.43	77	...	...	...	12	...	287	6.79	...	2,978	459	...	19	1,434
1950	...	...	17	...	1,751	13.68	123	...	...	...	17	...	429	7.20	...	3,309	617	...	23	1,572
1951	...	...	27	...	1,866	13.57	110	...	...	...	20	...	670	6.72	...	4,191	640	...	21	1,820
1952	...	...	23	...	1,829	13.78	104	...	...	...	9	...	511	5.62	0.02	4,100	601	...	20	1,550
1953	...	...	25	...	1,884	14.77	119	...	...	...	7	...	553	5.83	...	4,400	632	5.58	23	1,630
1954	...	...	19	...	1,873	14.96	134	...	...	...	5	...	284	4.90	...	4,700	619	5.64	22	1,590
1955	...	...	29	...	2,166	14.72	152	...	...	...	8	...	486	5.58	0.01	4,700	726	6.24	29	1,770
1956	44	...	31	...	2,250	15.34	145	...	...	...	7	...	687	3.64	0.10	5,400	739	6.58	23	1,930
1957	57	55.00	42	...	2,184	16.30	...	...	...	...	6	...	678	4.81	0.01	5,650	684	7.01	20	2,080
1958	30	41.00	19	...	2,222	16.39	...	...	...	...	2	...	493	4.89	0.01	6,100	718	6.57	23	2,030
1959	...	47.25	55	...	2,535	16.44	...	...	...	...	6	...	594	4.04	0.03	6,660	792	7.09	25	2,260
1960	...	...	34	...	2,730	16.73	...	...	...	...	6	...	499	4.11	0.02	7,850	734	7.33	24	2,780
1961	...	...	34	...	2,740	17.13	...	...	...	...	5	...	604	5.18	0.06	8,300	762	6.86	27	2,990
1962	...	...	31	...	2,998	17.84	...	...	...	...	5	...	492	4.64	1.70	8,750	772	6.84	26	2,960
1963	...	...	53	...	3,164	18.89	...	...	...	...	3	...	528	3.37	...	9,200	804	6.85	26	3,330
1964	...	...	44	...	3,331	19.39	...	...	...	...	2	...	...	...	...	10,000	890	6.99	23	3,760

*"Minerals Yearbook," U.S. Department of Commerce.

of the common types used in ceramics.[6] The lower-cost products, such as dextrine and the wood extracts, are usually used in molding of special shapes from mixtures lacking in plasticity.

8.9 Production and Price of Minerals in the Refractories Industry. In Tables 8.10 and 8.11 are given the production figures and prices of many materials used for refractories as taken from the "Minerals Year-book." Most of these figures are rounded off, and it should be realized that in some cases, restrictions prevent publication of complete values. In other cases, it is impossible to separate the figures into refractory and nonrefractory uses. Nevertheless some interesting deductions may be drawn from these data.

BIBLIOGRAPHY

1. Bell, W. C., I. W. Gower, and J. R. Hart: Properties of Pyrophyllite as a Refractory Raw Material, *Brick Clay Record,* **123** (6), 62–63, 65–66, 1935.
2. Chelikowsky, J. R.: Geologic Distribution of Fire Clays in the United States, *J. Am. Ceram. Soc.,* **18,** 367, 1935.
3. Thoenen, J. R., and E. F. Burchard: Bauxite Resources of the United States, *U.S. Bur. Mines Rept. Invest.* 3598, 1941.
4. Desch, C. H.: Magnesia from Sea Water, *J. Roy. Soc. Arts,* **91,** 273, 1943.
5. Speil, S., et al.: Differential Thermal Analysis—Its Application to Clays and Other Aluminous Materials, *U.S. Bur. Mines Tech. Paper* 664, 1945.
6. Treischel, C. C., and E. W. Emrich: Study of Several Groups of Organic Binders under Low Pressure Extrusion, *J. Am. Ceram. Soc.,* **29,** 129, 1946.
7. Brindley, G. W., and K. Robinson: An X-ray Study of Some Kaolinitic Fireclays, *Trans. Brit. Ceram. Soc.,* **46,** 49, 1947.
8. Rigby, G. R.: The Application of Crystal Chemistry to Ceramic Materials, *Trans. Brit. Ceram. Soc.,* **48,** 1, 1949.
9. Hauth, W. E., Jr.: Crystal Chemistry in Ceramics, *Bull. Am. Ceram. Soc.,* **30,** 5, 47, 76, 137, 140, 165, 203, 1951.
10. Bole, G. A., C. H. Bowen, and J. O. Everhart: Refractory Material of Southeastern Ohio and Northeastern Kentucky, *Bull. Am. Ceram. Soc.,* **30,** 323, 1951.
11. Halm, L.: Comparative Study of American and French Flint Clays, *Bull. Am. Ceram. Soc.,* **31,** 79, 1952.
12. Howe, J. P.: Properties of Graphite, *J. Am. Ceram. Soc.,* **35,** 275, 1952.
13. Mackles, L., R. A. Heindl, and L. E. Mong: Chemical Analyses, Surface Area, Thermal Reactions of Natural Graphite, *J. Am. Ceram. Soc.,* **36,** 266, 1953.
14. Taggart, M. S., et al.: Electron Micrographic Studies of Clays, *Natl. Acad. Sci. Publ.* 395, 1955.
15. Worrall, W. E.: The Organic Matter in Clays, *Trans. Brit. Ceram. Soc.,* **55,** 689, 1956.
16. deJosselinde, Jon G.: Verification of Use of Peak Area for Quantitative Differential Thermal Analysis, *J. Am. Ceram. Soc.,* **40,** 12, 1957.
17. Keeling, P. S.: A New Concept of Clay Minerals, *Trans. Brit. Ceram. Soc.,* **60,** 449, 1961.
18. Worrall, W. E.: The Mineralogy of Some German Clays, *Trans. Brit. Ceram. Soc.,* **60,** 291, 1961.

19. Holdridge, D. A.: Mineralogy of Some American (U.S.) Ball Clays, *Trans. Brit. Ceram. Soc.*, **62**, 857, 1963.
20. DeKimpe, C., M. C. Gastuche, and G. W. Brindley: Low Temperature Formation of Kaolin Minerals, *Am. Mineralogist*, **49**, 1, 1964.
21. Tron, A. R.: "The Production and Uses of Natural Graphite," H. M. Stationery Office, London, 1964.
22. Smoot, T. W., and D. O. McCreight: Refractory Clays: Their Mineralogy and Chemistry, *Brick Clay Record*, **146** (2), 36, 1965.
23. Worrall, W. E., and D. A. Holdridge: Carbonaceous Matter in Fireclays, *Trans. Brit. Ceram. Soc.*, **64**, 159, 1965.
24. Anon.: Sea-water Magnesia, *Ceramics*, **16** (194), 34, 1965.
25. Bates, T. F., and J. J. Comer: Microscopy of Clay Surfaces, *Proc. 3d Natl. Clay Conf.*, 1963.

CHAPTER NINE

The Mining
and Preliminary Treatment
of Refractory Raw Materials

9.1 Introduction. Since the Third Edition was written, there have been many changes in the procurement of raw materials. Diaspore in the United States is no longer available, and many of the high-grade fireclays are being depleted. At the same time more and more demands are being made on the quality of the materials, such as magnesite, ganister, and chromite, which lead to the necessity for beneficiation. In the years to come we shall see more treatments used to upgrade the raw materials.

On the whole there has been little change in the mining methods, except the more general use of heavier equipment. Grinding equipment has not changed, except for production of very fine particle sizes where more efficient machines have been developed. Storage methods have been made more fully automatic to save labor costs.

9.2 Clay Mining. *Open-pit Methods.* Many refractory clays are mined by open-pit methods. Usually the bed of clay is covered by an overburden of gravel or earth, which may run as thick as 10 to 20 ft. If the thickness of overburden is much greater than 20 ft, it is seldom economical to mine a low-cost clay by this method.

The first process consists in stripping off the overburden in the most economical manner. In larger operations, the overburden is stripped by power shovels, tractor-pulled wheel scrapers, bulldozers, or shovel dozers. In other workings, it is found economical to strip the overburden with dragline scrapers with a capacity of ⅓ to 4 cu yd. This type of scraper is very flexible and can be made to cover a large area by using a movable cable anchor. Lines as long as 500 ft are used; and with a 1 cu yd bucket, 20 to 75 cu yd per hr can be moved, depending upon the length of the haul. A modern bulldozer for stripping the overburden from a kaolin deposit is shown in Fig. 9.1, and in Fig. 9.2 is shown a self-powered scraper of large capacity. In Europe stripping

FIG. 9.1 A rubber-tired bulldozer for clay stripping. (*The Le Tourneau Company.*)

FIG. 9.2 A self-powered scraper for stripping clay. (*The Le Tourneau Company.*)

is often done by a chain-bucket excavator working from a track on top of the deposit.

The method of removing the clay itself varies with the conditions. If the clay is fairly soft, the face can be broken with picks or bars and the broken material loaded by hand into carts or cars. In the larger plants, however, this method is not economical, and power shovels are generally used. The shovel capacity varies from ½ to 2 cu yd. The larger shovels are chosen not only for their greater capacity but also for their greater strength and longer reach. Caterpillar treads are almost universally used.   A 1 cu yd shovel will load, under good conditions, 500 tons of clay in an 8-hr day.   Recently, lighter and more agile shovels are being favored.

The softer clays can be taken out with dragline scrapers or bucket excavators.   Often bulldozers are used, as shown in Fig. 9.3.

Provided the clay is not too hard and yet is firm enough to hold a permanent face, as in the case of many shales, it has been found economical to remove the materials by a planer, which is an endless chain with cutting knives working vertically on the whole face as shown in Fig. 9.4.   This machine has the advantage of removing the material in small pieces and of giving a good average composition by taking the material from the whole face.   The cost of removing the material with a

FIG. 9.3   Mining clay with a Caterpillar tractor.

machine of this type is comparatively low, as one man can usually operate it and obtain a production of 200 to 400 tons a day.

In Europe, the endless-chain excavator working from the top of the bank is often used.   Although the initial investment is large, the low labor and maintenance cost justifies it on large operations.   One man can remove 100 tons per hr with an average-sized machine.

In the case of hard clay, it is generally necessary to drill and blast the face in order to obtain small enough pieces to be readily handled.   The drilling is done in the smaller pits with hand drills of 1 to $1\frac{1}{2}$ in. in diameter.   In the larger mines, however, electric and air drills are used. Recently, portable well drills have been used to a considerable extent to drill deep holes behind the working face so that larger and deeper charges of explosives can be set off.   These drills will produce holes from 3 to $5\frac{1}{2}$ in. in diameter, which will allow the use of a considerable quantity of explosive. The use of ammonium nitrate as an explosive has been employed in some mines.[13]

When the clay has been loosened from the face, it is loaded into cars or trucks by power shovels. Figure 9.5 shows an excavator much used in Europe for removing clay.

Underground Methods. When the clay deposit is too deep for stripping, underground mining is used. Most of the productive clay veins are from 5 to 15 ft thick. The room-and-pillar method is generally used, with advance from the entrance, which consists of cutting a mine entry and advancing galleries from that out to the edge of the property. This

FIG. 9.4　Removing kaolin with a shale planer made by the Eagle Iron Works at the Georgia Kaolin Company. (*Georgia Kaolin Company.*)

method has the advantage of not requiring extensive entry development when the mine is started, but it is not so economical as the retreating method, which consists of cutting an entry to the boundary line and then starting the galleries at the farther end. Work is then carried toward the entrance. The pillars can be drawn only where the subsidence of the ground is unimportant. Whenever underground mining is used, it is important to plan out the working carefully and have a systematic method before starting to mine.

Practically all underground clay is taken out by drilling and blasting. Hand, electric, or pneumatic drills are used, and the lumps of clay are loaded on the car by hand or, in more modern mines, by car loaders.

The same factors of economical mining apply to clay mines as to other

FIG. 9.5 Mining with a chain excavator. (*Rieterwerke.*)

FIG. 9.6 A tractor loader filling a dump car. (*Caterpillar Tractor Co.*)

mines. In the first place, drainage must be well cared for, as muddy floors and tracks increase the haulage costs as well as make the timbering difficult. Next, the timbering should be well done with generous posts. The tracks should also be graded and kept in good condition, as the haulage cost mounts up when dirty, poorly laid tracks are used.

Clay is usually transferred to the plant on rails with dump cars ranging in capacity from $\frac{1}{2}$ to 6 cu yd. The cars are propelled by steam or diesel locomotives or sometimes by cable haulage. In a few cases where the distance from the mines to the plant is considerable and the country is rough, it has been found economical to transfer the clay by aerial cableways. At present trucks are coming into more general use than formerly.

It is hard to give any definite figures on the cost of mining clay; but in general, clay can be loaded on cars or trucks from the open pit under the best conditions for $1 a ton. Where the overburden is heavy or where hand selection is used, the cost may run as high as $2 per ton at the plant. In the case of special high-grade clays, the cost may run even higher. It may be said, however, that a great number of plants are getting out their refractory clays for around $1 per ton. In underground mining, the cost varies according to the thickness of the vein, the hardness of the clay, the amount of timbering required, and the distance to the plant—the variation in most cases being from $3 to $5 per ton on cars. It may be mentioned that in many of the underground clay mines, coal is obtained at the same time as clay, which, of course, reduces the total cost considerably.

Attempts have been made to beneficiate fireclays to raise their grade. Powell[6,9] reports on some work by the U.S. Bureau of Mines. Wet-table treatment proved most successful. It is certain that more work of this type will be carried out as high-grade clays become less plentiful.

9.3 The Mining of Ganister. Since the mining of ganister is rather different from clay mining, it will be described separately. The ganister is usually found in flows of loose rock on the mountainsides. Where the rocks are large, they are broken up by mud capping and blasting; in only a few cases is drilling necessary. The smaller rocks are broken up by sledges into sizes that can be lifted by hand. The broken rock is loaded into dump cars by hand, as some sorting is necessary. One man will load from 15 to 20 tons of rock a day.

The dump cars are trammed either by hand or by locomotive, as the case may be, to a gravity plane and dumped into cars to be lowered down the plane. In a few cases, ganister is carried long distances by railroad to reach the plant.[10]

In some localities, the ganister is obtained from bedrock, because the flows have been used up or because the rock is more uniform. The ledge rock, however, is usually not thought to be so satisfactory as the flow rock, because the soluble constituents have not been leached out by weathering.   Others claim that bedrock gives superior bricks.   Regular quarrying methods are used in removing this rock from the hillside.

Ganister for low-alumina superduty silica brick must be treated to lower the impurities[4] by scrubbing the grains to remove clay.   Many ganisters can be successfully treated in this way.

9.4   Crushing and Grinding.   Practically all clay must be crushed before it is taken to the tempering and brickmaking machinery.   The softer

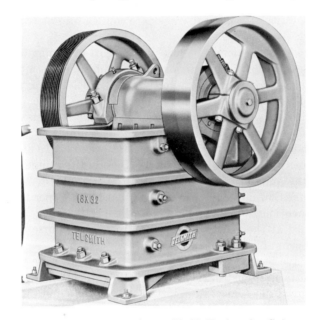

FIG. 9.7   Jaw crusher.   (*Smith Engineering Co.*)

lump clays are usually passed through single or double roll crushers, which rapidly break down the lumps to sizes small enough for wet pans or pug mills.   The wear and tear on rolls of this type is very small; so the cost of breaking up the softer types of clays is not great.

Jaw (Fig. 9.7) or gyratory crushers are usually used for the harder clays, rocks, and grog.   They will crush down to 1 in. with capacities of $\frac{1}{2}$ ton to 500 tons an hour.   These crushers cannot be used for the softer clays containing moisture because of packing in the jaws.

Roll crushers (Fig. 9.8) are used for crushing hard materials from 1 in. down to $\frac{1}{4}$ or $\frac{1}{8}$ in.   For materials like grog, the rolls should be supplied with tires of manganese steel, as the wear is considerable.   In order to obtain fine grinding, it is necessary to keep the rolls fairly true cylinders, as flanges tend to form on the ends and the material will be passed through the center uncrushed. Plain rolls should never be used for crushing clays with any moisture, because the material builds up on the rolls and there will be no crushing action.

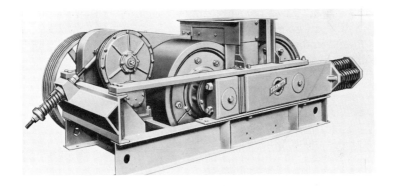

FIG. 9.8 A roll crusher. (*Smith Engineering Company.*)

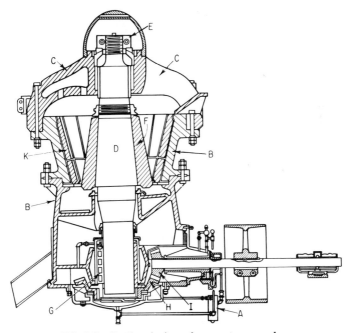

FIG. 9.9 Sectional view of a gyratory crusher.

Gyratory crushers (Fig. 9.9) are often used for crushing magnesite and ganister. These crushers have a large capacity but will not take such large rocks as jaw crushers.

Dry pans (Fig. 9.10) consist of a heavy revolving bed in a horizontal position on which the material is fed; two heavy rolls (mullers) rest on this bed with their own weight, crushing the material underneath. The portion of the bed not directly under the mullers is usually supplied with screens or slits so that crushed material can pass through, permitting continuous operation. The dry pan is used for grinding flint clay, ganister, and burned grog. Wet pans are similar machines with a solid bottom so that the mix can be ground wet.

For finer grinding, it is often desirable to use crushers of the B & W, ring-roll, or hammer-mill type, as shown in Figs. 9.11, 9.12, and 9.13.

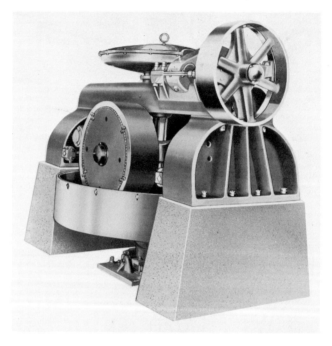

FIG. 9.10   A heavy dry pan for crushing clay and grog.   (*Clearfield Machine Company.*)

Machines of these types will pulverize down to 100 or 150 mesh and can be used for clays with some degree of moisture.

Ball mills are much used for fine grinding. For large production, the continuous type of mill should be selected. The fine material is removed with an air separator, and the oversize is automatically returned to the mill. A production of several tons per hour of 200-mesh material can be obtained with this type of installation. Even when coarser material is required, the continuous ball mill is often found suitable. In Fig. 9.14 is shown a typical ball mill for continuous grinding. The capacities of some ball mills are shown in Table 9.1.

Vibratory ball mills are used for the smaller sizes with high efficiency.[14] An interesting vacuum impact mill is described by Planiol[15] for fine sizing, and Rockwell and Gitter[16] describe fluid energy mills.

9.5 Screening. It is usually necessary to screen the materials being crushed so that the required product can be passed on to the brick-

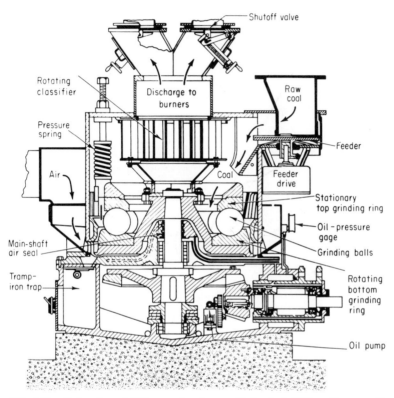

FIG. 9.11   Babcock and Wilcox pulverizer.   (*Babcock & Wilcox Company.*)

making machinery and the coarser material returned to the crushers. Various types of screens are used.

A common type of screen is the shaking screen operated by a weight (Fig. 9.15).   Often several layers of screens of various sizes can be superimposed in order to classify a product into a number of different sizes. Closed-in screens (Fig. 9.16) are used with dusty materials.   For screening materials below 60 mesh, it has been found desirable to use air separators rather than screens, and very satisfactory classification devices can now be obtained to produce material down to 200 mesh or even finer.[12,17]

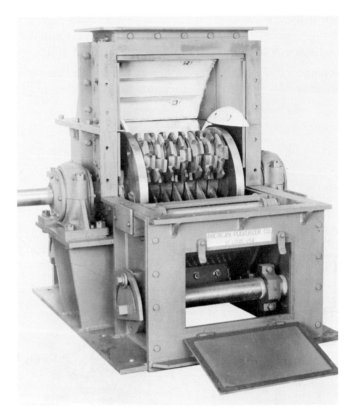

FIG. 9.12   A ring-roll crusher.   (*American Pulverizer Company.*)

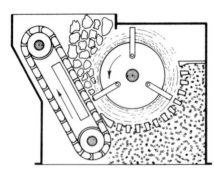

FIG. 9.13   Schematic section through a hammer mill.

FIG. 9.14   Conical ball mill.   (*Hardinge Co.*)

TABLE 9.1   Capacity of Ball Mills*

Material	Size and type	Tons per hour	Circuit	Horse-power	Size of feed	Fineness of product	Pebble load, lb
Silica (dry)......	7 × 22 ft tube	3	Closed	210	−20 mesh	98 % minus 200 mesh	42,000
Sillimanite.......	3 × 2 ft conical	¼	Closed	15	−¼-in. mesh	80 % minus 100 mesh	14,000
Grog...........	5 × 22 ft conical	4½	Closed	40	−1½-in. mesh	100 % minus 4 mesh	6,000
Feldspar.........	7½ × 10 ft cylindrical	1.05	Closed	85	−¾-in. mesh	90 % minus 325 mesh	

* From G. F. Metz.

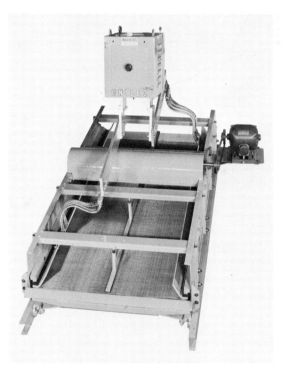

FIG. 9.15   An open-type shaking screen with electric heating. (*Universal Vibrating Screen Company.*)

9.6   Mineral-dressing Methods.

Many refractory raw materials can be purified by simple mineral-dressing methods.   As the demand for refractories becomes more exacting and our supplies of the better raw materials less plentiful, concentration methods will be increasingly used.

Tabling.   Minerals of a different specific gravity from the unwanted

FIG. 9.16   A closed-type screen.   (*Deister Concentrator Company.*)

material with which they are mixed can often be efficiently separated by an air or water table.   The raw material is ground fine enough to unlock the individual grains and then passed over an inclined vibrating table, which will give a more or less clean separation.   Although this method has been used in the past mainly for the treatment of ore minerals, it has been quite successful in the concentration of kyanite from our North Carolina deposits.

Settling.   Dorr-type thickeners and centrifuges are used to remove coarse particles from clay slips as well as to dewater these slips.

Magnetic Separation.   Powerful magnetic separators are now available that will remove all minerals containing even small amounts of iron. These include the iron oxides, sulfides, and more complex minerals such as garnet, hornblende, and micas.

The magnetic method of purification is used quite generally with feldspars and other materials for whitewares, but it has not as yet been used extensively for refractory materials.   However, there is opportunity of employing it with some refractories, such as those used in melting glass or enamel frits.

Flotation.   Froth-flotation methods have been developing with such rapidity since 1920 that many separations can now be made that would have been thought impossible in the past.   The method, in general, consists in stirring the finely crushed mineral in water with a frothing agent, which provides a multitude of fine bubbles.   By selecting the proper agents, it is possible to get a selective action of the bubbles so that they adhere to certain types of particles and float them to the top of the water where they are carried off. Although considerable progress has been made with flotation methods in purification of minerals like feldspar, little has been done yet on refractory materials,[11] but it would seem quite possible that this method would be well adapted to the concentration of chrome ores.

A little laboratory work has been done by flotation methods in removing iron minerals from fireclays, but as yet the results have not been shown to be economically feasible. Nevertheless, there is certainly promise that with further development along this line, second-grade clays can be so beneficiated that first-grade refractories can be made from them.

Electrostatic Separation. The electrostatic separator has been used to purify both feldspar and kyanite. The method consists in separating one type of mineral from another by virtue of the different attractive forces in a high-voltage field. This is a dry method and should not be particularly expensive except that the particles must be fairly fine before treatment. A patent by Weis[1] describes a method for removing quartz from kyanite.

Chemical Methods. Chemical methods can be used for the higher-priced refractories. For example, pure alumina refractories are produced from $Al(OH)$ obtained from bauxite by the Bayer or other of the many processes[5] proposed for this purpose. Magnesium ore is treated by flotation and magnetic separation to reduce the silica and iron.[18] Chrome ore is treated to remove silica by various processes.[7,8] Magnesia is recovered from seawater by precipitation by lime.[2]

9.7 Clay Storage. It is usually necessary to store a certain amount of clay to provide for interruptions in mining due to cold or rainy weather or holidays. The amount of needed storage depends on the length of shutdown in the mining operation. To provide sufficient storage for a large clay plant is rather costly, and it is unusual to store more than a few days' supply.

In many of the ceramic industries, the practice is to weather, or age, the clay supply for a considerable length of time to develop plasticity and disintegrate the lumps; but in the case of refractories, the amount of clay used is so large that this cannot be done economically, and only in the case of special high-grade clays can aging be resorted to.

Some of the high-grade refractory clays must be shipped a considerable distance by railroad. In these cases, it is desirable to dry the clay partially in order to reduce the shipping weight. This can be done by storing in open sheds or racks, where the moisture in good weather can be reduced to 5 or 10 percent, or by passing it through a rotary drier.

Storage bins are made in various styles.[3] Some are cylindrical bins having conical bottoms; others are of a rectangular shape with a V bottom with provision for removing the clay from any portion desired (Fig. 9.17); or silos as shown in Fig. 9.18 may be used. The clay is usually put in the bin with a bucket elevator, but other handling methods such as grab buckets are occasionally used. In Europe material is often stored in bins with a bucket excavator for removal as in Fig. 9.19.

When crushed materials such as clay or grog are to be stored, precautions must be taken to prevent segregation of the material in the bin. In

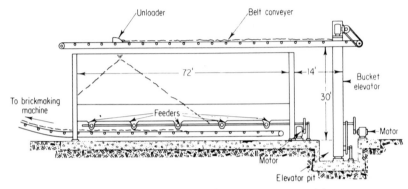

FIG. 9.17   A method of handling clay in storage bins.

FIG. 9.18   Modern materials storage in a refractories plant. (*Harbison-Walker Refractories Co.*)

nearly all manufacturing processes, it is highly desirable to have the crushed materials of a proper and uniform screen analysis.[19]   When the material from the crusher falls into the bin, the larger particles roll to the outside and the finer ones build up into a central cone.   The material drawn from the bottom of the bin will contain at first mostly fine and later

FIG. 9.19   Clay storage in aging bin and excavator for removal, used in European plants.   (*Rieterwerke*.)

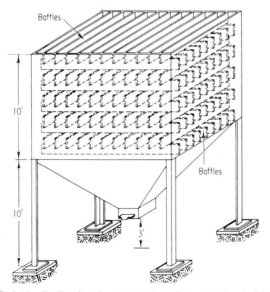

FIG. 9.20   Baffles for the prevention of segregation in bins.

mostly coarse material.   This segregation can be minimized by a number
of methods.   One consists in moving the point of supply over the surface
of the bin so that no cone can form.   Another consists in a number of
vertical vanes in the bin to prevent the heavy particles from rolling
to the outside of the bin (Fig. 9.20).

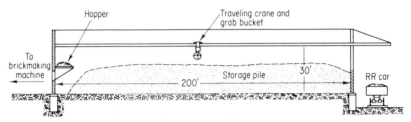

FIG. 9.21   An overhead crane and grab bucket for handling clay in storage.

FIG. 9.22   Aerial view of a plant for calcining and dead-burning dolomite.   (*Kaiser
Refractories Co.*)

In some cases, clay is stored in open pits or on floors, from which it is
handled with a locomotive crane and grab bucket, a dragline scraper, or in
more recent installations, by a tractor loader.   In other cases, the clay is
stored in sheds, and the handling is done by an overhead crane and grab
bucket (Fig. 9.21).   This gives a very flexible arrangement, as the clay can
be dropped onto a conveyor to be taken directly to the crushing or mixing
machinery.[20]   The clay-storage system of a modern refractory plant is
shown in Fig. 9.22.

BIBLIOGRAPHY

1. Weis, J. H.: Producing Kyanite, U.S. Patent 2,094,440, 1937.
2. Stedman, G. E.: Magnesium from Sea Water, *Metals & Alloys*, **20**, 941, 1944.
3. Korn, A. H.: Practical Pointers in the Design of Storage Bins, *Chem. Eng.*, **53**, 100, 1946.
4. Rao, P. D.: Attritional Scrubbing of Ganister, Thesis, Pennsylvania State University, 1961.
5. Peters, F. A., P. W. Johnson, and R. C. Kirby: Methods for Producing Alumina from Clay: An Evaluation of Five Hydrochloric Acid Processes, *U.S. Bur. Mines Rept. Invest.* 6133, 1962.
6. Powell, H. E., and W. A. Calhoun: The Hydrocyclone in Clay Beneficiation, *U.S. Bur. Mines Rept. Invest.* 6275, 1962.
7. Anon.: Removal of Silica from Chrome Ore, *Can. Dept. Mines Memo* 1962, p. 11, 1963.
8. Didier-Werke, A. G.: Chromite Purification Process, British Patent 914,461, 1963.
9. Powell, H. E., W. A. Calhoun, and C. K. Miller: Beneficiation of Refractory Clay, *U.S. Bur. Mines Rept. Invest.* 6142, 1963.
10. Polson, B.: Ganister Quarrying, Engineered for Quality Plus Economy, *Brick Clay Record*, **143**, 36, 1963.
11. McVay, T. L., and J. S. Browning: Flotation of Kyanite-Quartzite Rock, Graves Mountain, Lincoln County, Ga., *U.S. Bur. Mines Rept. Invest.* 6268, 1963.
12. Anon.: Classifying Refractory Powders by Air Separation, *Ceramics*, **15**, 14, 1964.
13. Dunlop, M. S.: Blasting Plastic and Flint Clays with Ammonium Nitrate, *Brick Clay Record*, **144**, 60, 1964.
14. Lissenden, A.: Vibratory Grinding-mill, *Chem. Process Eng.*, **46**, 203, 1965.
15. Planiol, R.: Centrifugal Vacuum Mills, *Silicates Ind.*, **30**, 129, 1965.
16. Rockwell, P. M., and A. J. Gitter: Fluid Energy Grinding, *Bull. Am. Ceram. Soc.*, **44**, 497, 1965.
17. Treasure, C. R. G.: Fine Particle-size Classification, *Trans. Inst. Chem. Engrs.*, **43**, T199, 1965.
18. Kent, G. A.: Production of High-purity Magnesia, *Can. Dept. Mines Res. Rept.* R163, 1965.
19. Williams, J. C.: Design of Storage Hoppers for Bulk Solids, *Chem. Process Eng.*, **46**, 173, 1965.
20. Bohme, J.: Bunker Control in the Preparation and Working of Ceramic Mixes, *Keram. Z.*, **17**, 452, 1965.

CHAPTER TEN

Molding Methods

10.1 Introduction. Heavy refractories are now seldom formed from a plastic clay alone, but rather from a mixture of plastic clay and a non-plastic such as grog or flint clay. This is true in the case of casting slips as well as for soft or stiff bodies. The reason for this is the need for a mixture that has good firing properties and good use properties. While a great amount of study has been given to the clay-water system, very little has been published on the fundamentals of the clay-grog-water system. Nevertheless, in this chapter the latter system will be treated to some extent, especially in considering the dry-press process.

10.2 Workability of Clay. *Types of Flow.* Before entering into the discussion of workability of clays, it will be well to review a few definitions as adopted by the Society of Rheology.[3]

> *Fluid:* A substance that undergoes continuous deformation when subjected to any system of finite shearing stress.
>
> *Solid:* A substance that undergoes permanent deformation or rupture only when subjected to a system of shearing stresses that exceed a certain minimum value.
>
> *Simple (Newtonian) liquid:* A liquid in which the rate of shear is proportional to the shearing stress (for laminar flow).
>
> *Elastico-plastic solid (the clay-water system):* A solid that obeys the law of an elastic solid for values of the shearing stress below the critical stress corresponding to the elastic limit in shear and that deforms plastically when the shearing stress exceeds that value.
>
> *Viscosity:* In a simple liquid, the constant ratio of shearing stress to rate of shear.

Plasticity: That property of a body by virtue of which it retains a fraction of its deformation after the deforming stress is zero.

Elasticity: That property of a body by virtue of which it recovers its original size and shape after deformation.

Thixotropy: That property of a body which causes a decrease in shearing stress with time at a constant rate of shear.

Rheopecsy: That property of a body which permits agitation to accelerate the setting up of a gel.

Dilatency: That property of a body which causes an increase in shearing stress with time at a constant rate of shear.

The following symbols are generally used in flow problems:

F_0 = the yield stress in shear, dynes/sq cm
F = the shearing stress, dynes/sq cm
dv/dr = the rate of shear strain, cm/sec/cm
η_s = the coefficient of apparent viscosity of a suspension (dynes, sec, cm^{-2})($ML^{-1}T^{-1}$)(poise)
η_1 = the coefficient of viscosity of suspending medium
η = the basic viscosity of suspension (at infinite rate of shear strain)
C = the volume concentration of solid in relation to total volume
V = the volume flow, cc/sec
v = the velocity of flow, cm/sec
T = the time
θ = the coefficient of thixotropy ($ML^{-1}T^{-1}$)

In considering the clay-water system it is usual to separate it into two parts: the slips of low clay content and the plastic masses of high clay content. This is done because the sticky region between is incapable of accurate measurement at present and also because this region is not used industrially, but a more fundamental reason is that different laws seem to apply to the two cases. The slips will be considered first.

Clay Slips. A simple liquid like water obeys the following relation in the viscous-flow region:[8]

$$\eta = k\frac{F}{v}$$

The quantity η can be measured in many ways, but the revolving-cup viscosimeter, as shown in Fig. 10.8, is very satisfactory for ceramic work. Here the viscosity is determined by the twist of calibrated wires, and thus a wide range of values can be covered. The rate of shear can also be varied over a wide range for the measurement of thixotropy in suspensions.

The viscosity of a suspension of spherical particles seems to show no change in apparent viscosity with varying rate of shear strain, but more work needs to be done to confirm this similarity to a liquid.   Suspensions of anisotropic particles, such as clay particles, show a pronounced change in apparent viscosity with change in shear strain rate as shown in Fig. 10.1, and the coefficient of thixotropy θ is determined from the slope A/B of the curve at infinite shear rate.

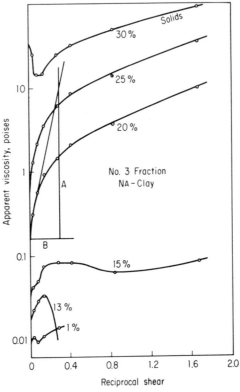

FIG. 10.1   Apparent viscosity of slips at different rates of shear.   (*J. Am. Ceram. Soc.*)

The apparent viscosity of a suspension is due to the sum of

1. The viscosity of the suspending medium
2. The energy absorbed by the individual particles revolving in the velocity gradient of the sheared fluid
3. The energy absorbed by the interference of the particles with each other

It is found that suspensions of monodisperse kaolinite[6] in either water or a dispersing electrolyte solution obey the following law:

$$\eta_s = (1 - C)\eta_1 + k_1C + k_2C^3$$

where k_1 and k_2 are constants varying with the particle size and electrolyte in the suspending medium. The first and second terms predominate at low values of C, while the last term increases rapidly at higher concentrations where the particles interfere. In Fig. 10.2 is shown the apparent viscosity of several deflocculated (Na-kaolinite) monodisperse suspensions on a basis of center-to-center distance of the particles, and it is clearly evident that as each fraction reaches a concentration such that the center-

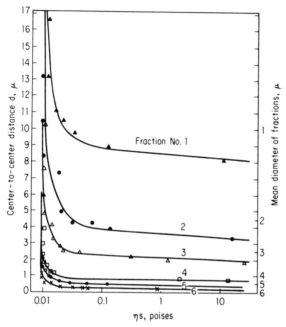

FIG. 10.2 Relation between viscosity and center-to-center distance of particles of six fractions containing Na-kaolinite. (*J. Am. Ceram. Soc.*)

to-center distance is equal to the mean diameter of the particle, the viscosity increases rapidly. As a visual guide to the suspension, there are shown to scale in Figs. 10.3 and 10.4 two suspensions represented in Fig. 10.2.

The thixotropy of a suspension is given by

$$\theta = k_3C + k_4C^3$$

The yield point of suspensions in water (H-kaolinite) may be shown to follow

$$F_0 = k_5C^3$$

over a wide range of concentration, k_5 varying with particle size. Well-deflocculated suspensions (Na-kaolinite), even in high concentration, show no measurable yield point.[6]

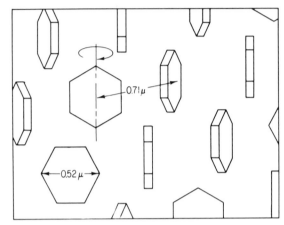

FIG. 10.3   Cross section of particles of No. 5 fraction containing Na-kaolinite at a viscosity of 0.017 poise and a volume concentration of 0.05 solid.   (*J. Am. Ceram. Soc.*)

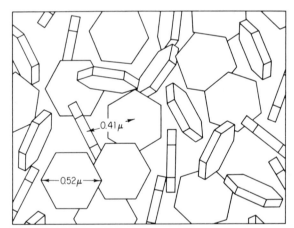

FIG. 10.4   Cross section of particles of No. 5 fraction containing Na-kaolinite at a viscosity of 3 poises and a volume concentration of 0.28 solid.   (*J. Am. Ceram. Soc.*)

Figure 10.5 shows the influence of particle size and amount of ions added to a slip of constant solid content. The various-size fractions were separated from a kaolin with a centrifuge and cleaned. Then definite amounts of sodium hydroxide were added in milliequivalents per 100 g of dry clay. The enormous and sudden reduction in apparent coefficient of viscosity on adding sodium hydroxide is evident in all cases. This is the principle of deflocculation, which permits fluid slips with low water content. The exact mechanism of deflocculation is not clearly understood at present, but it is believed that free OH ions over and above those absorbed on the clay are required to lower the particle charge and thus minimize the attraction forces between the particles, as discussed by Johnson and Norton.[6] These OH ions can be produced by the ionization of bases like NaOH or by the hydrolysis of salts like Na_2SiO_3. The positive ion accompanying the OH ion must be monovalent; therefore, calcium or magnesium salts will not deflocculate.

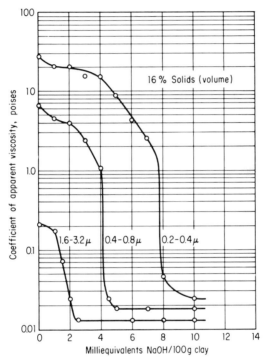

FIG. 10.5 Viscosity of kaolin slips.

Until recently it was difficult to explain flocculation, as charged particles would be expected to repel rather than attract each other. Now it is believed that the charges are not evenly distributed over the particle surface, so that the edge of one plate may be attracted to the face of another plate.[15]

The work of Coughanour and Norton,[11] who studied the effect of particle shape on the suspension properties, showed that when dispersed, the more elongated particles gave higher viscosities. As would be expected, the difference was small in dilute suspensions, but at higher concentrations, where particle interference occurred, the elongated particles gave ten to twenty times the viscosity of the isometric particles.   On the other hand, when the suspensions were flocced there was no measurable difference due to particle shape.   In other words, the floc shape is independent of the particle shape.   It was also noticed that the dispersed suspension of isometric particles showed no thixotropy effect.

An excellent discussion of rheology of clay slips is presented by Moore,[15] who gives a fine picture of thixotropy, a property of great interest to refractory castings.

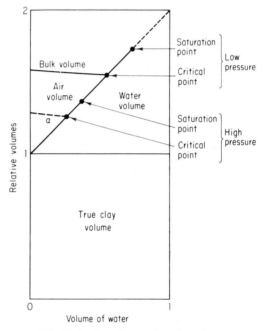

FIG. 10.6   The clay-water-air system.

Plastic Masses.   Turning now to the plastic masses with a high enough yield point to resist gravity forces, it is well to consider some fundamental principles.   In general, the three-component system, clay-water-air, may be expressed in the somewhat conventionalized volume diagram for a low molding pressure shown in Fig. 10.6.   Such a diagram is based on thoroughly mixed and kneaded clay and water batches of successively increasing water content with each molded in a cylindrical steel die at the given pressure.   Provision is made for the escape of excess water.   The wet and dry volume and the wet and dry weight will readily allow the volumes of air, water, and clay to be computed, provided the specific gravity of the clay is known.

As the amount of water is increased, the total volume of the mix, which originally consisted of clay and air, shows little change as the pores become filled. At the critical point, all the voids in the clay are filled with water and the air phase vanishes. As the water is increased still further, the particles are separated by a film of increasing thickness until the maximum for stability is reached at that pressure. This may be termed the "saturation point." If more water than this was originally in the clay, it will be squeezed out in permeable pressing.

The critical point is a very important one, as above this point the mass shrinks on drying owing to the removal of the water films, as discussed in Chap. 11. The thicker and more numerous the films per unit length of clay specimen the greater the shrinkage. Below the critical point, there is little or no drying shrinkage. The critical point also is the dividing

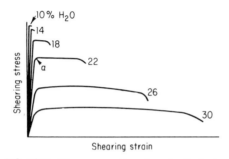

FIG. 10.7 Flow properties of a plastic body.

point between the consistency of a damp powder and a homogeneous plastic mass.

At a higher molding pressure such as would be obtained in a dry press, for example, the diagram would appear as shown in *a*, Fig. 10.6. Both the critical and saturation points are moved to low values of water content.

There are often slight exceptions to the simplified diagrams shown here. For example, the total volume sometimes decreases slightly as water is added up to the critical point, for the water acts as a lubricant and allows closer packing. On the other hand, if the clay contains bentonite, the particles of which swell in water, drying shrinkage may occur below the critical point.

The flow characteristics of plastic masses are shown in Fig. 10.7 for various water contents.[4] The flow is elastico-plastic; and as the water increases, the yield point becomes lower while the maximum strain becomes greater.

Measurement of Workability.   The consistency of slips is best measured in a concentric-cylinder viscosimeter as shown in Fig. 10.8.

In the case of plastic masses, not only the deforming force is required but also the maximum deformation at fracture.   The most useful method for evaluating the workability factor is to twist a hollow cylinder and obtain the stress-strain diagram in pure shear.   A satisfactory machine

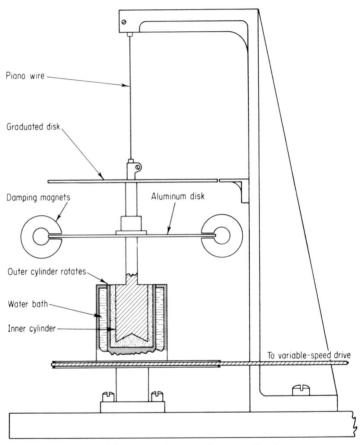

FIG. 10.8   Revolving-cylinder viscosimeter.

for accomplishing this, which was developed some time ago in the Ceramic Laboratory at the Massachusetts Institute of Technology,[4] is shown in Fig. 10.9 and since then a great deal of work has been done by various experimenters in torsion.

The specimen B, with its brass end pieces, is clamped in the torsion head C by means of a wedge, as shown more clearly in the detailed sketch. This offers a simple and secure locking device, which prevents any back-

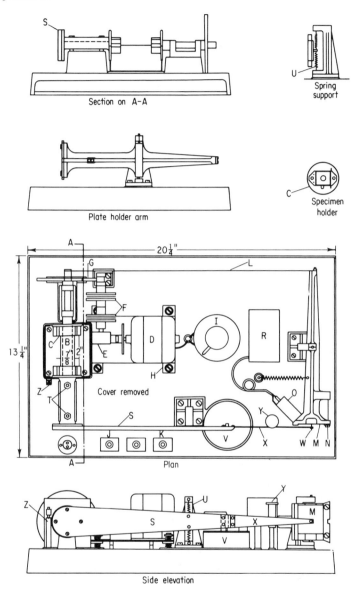

FIG. 10.9 Apparatus for measuring workability. (*American Ceramic Society.*)

lash during the test.   The torsion head is rotated by the motor D through
the reduction gear E, the flexible couplings F, and the gears G.   The
motor and reduction unit are mounted on spring suspensions H to prevent
any vibration from being transmitted to the remainder of the apparatus.
This drive gives a constant speed with sufficient power for any of the

tests with plastic materials.   The speed of the motor can be altered by the voltage regulator I, and its direction is reversed by switches J and K, which independently control the separate field coils.   In addition, the speed can be changed by various sets of gears at G.

The angle of rotation of the specimen is recorded by the steel tape L, one end of which is wound around a drum on the torque shaft and the other attached to the plate-holder arm so that the angular rotation is proportional to the movement of the smoked plate M.   This plate is held in a spring clip N and is illuminated from the back by the light O supplied by the 6-volt transformer R.   This illumination is necessary for close observation of the stylus in relation to the zero line, as will be described later.

The torque on the specimen is measured by the deflection of the light aluminum arm S, which is pivoted on the centerline of the specimen with the flexible steel emery knife edges T.   The deflection of the end of the arm is therefore free from any friction effects and is proportional to the torque.   In the case of the stiffer materials, the springs U are necessary to supplement the knife edges.   Under these conditions, the vertical movement of the end of the arm S is proportional to the torque for the small angles employed.   The movement of S is controlled by the dashpot V, filled with a light oil of a viscosity to give critical damping.   Such damping is necessary when rapid movements of the arm occur in the initial portion of the stress-strain diagram.

The recording of the diagram takes place on the smoked glass plate M by means of a pointed steel stylus W held against the plate lightly by the thin steel spring X so that very little friction results.   The stop Y allows the stylus to be removed from the smoked-glass plate after the completion of a record.

Referring back to Fig. 10.7, typical stress-strain diagrams of plastic clay at various water contents will be seen.   The points to be noted are the yield point a, the twist at the breaking point, and the maximum torque. No one figure will represent the workability, but a high yield point and high deformation will give the best value.   Roughly, the product of the yield point and the maximum deformation may be taken as the workability.   If this product is plotted against water content, it will show a maximum value at the consistency of best workability, as shown in Fig. 10.10.

The stress-strain diagram is altered by changing the rate of shear;[4] so it is well to perform the tests at a shear rate comparable with that encountered in a particular forming process.   Clay masses are elastic up to the yield point only under rapid deformations.

Numerous other methods have been mentioned for measuring workability, such as the orifice flow of Stull and Johnson, the tension test of

Hyslop, and the cylinder-compression method of Roller, as well as the many impression methods similar to the Vicart needle. Moore[15] recently published some interesting results using the torsion test. Baudran and Deplus[16] also give values from the torsion test.

Mechanism of Workability. A new theory explaining the workability of clay masses has been proposed by Norton,[9] which may be called the "stretched-membrane" theory. The yield point exhibited by the plastic mass is explained by the compressive force exerted on it by the surface layer of water on the outside of the mass. This layer, as represented in Fig. 10.11, may be thick in soft masses (*a*), in which case the surface curvature between the particles is of long radius and the surface tension forces due to capillarity are low.

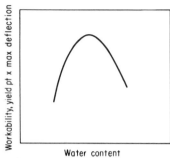

FIG. 10.10 Workability of a plastic body vs. water content.

At the same time the particles are widely separated by water films as evidenced by the great drying shrinkage from this condition. On the other hand, if some of the water is removed from the mass (*b*), it becomes stiffer because the surface film

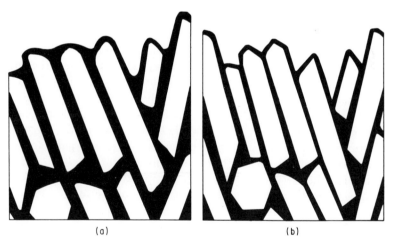

(a)   (b)

FIG. 10.11 Enlarged cross section of a plastic mass near the surface with varying amounts of water. (*J. Am. Ceram. Soc.*)

becomes thinner and draws down into the capillaries with a smaller radius of curvature. This puts the mass of particles under higher compression and thereby forces them closer together. Now unless this compressive force is balanced by an interparticle repulsion force, the particles will come together until they touch. But we know they do not touch, because drying shrinkage occurs, indicating that the separation is something of the order of 100 molecular layers.

In Fig. 10.12 is shown the force field about a hydrogen clay particle as postulated by Hauser and Hirshon.[5] A flocced suspension settled out of water would be represented by point B, where the attractive and repulsion forces just balance. Under the forces of shearing the particles will be separated slightly, bringing attractive forces into play, which accounts for the yield point found in such a system. On the other hand, if a little water is removed from the system by compressing between porous pistons (filtering) or by drying, the particles will come closer together, such as at point C, so that interparticle repulsion forces are built up just to compensate for the external compression produced either by the pistons or by the surface capillarity.

It is rather difficult to visualize the forces acting in the clay mass as produced by the stretched membrane; however, the simple mechanical diagrams shown in Fig. 10.13 should make the picture clearer. In diagram a there is shown a series of spheres held apart by compression springs loosely fitting inside a tube. The spheres represent the clay particles, the springs the repulsion forces, and the water surrounding them, the water in the clay mass. If two tightly fitting pistons close the ends of the tube and P_1 forces them

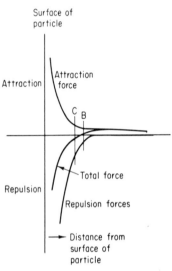

FIG. 10.12   The field of force about a clay particle when surrounded by water.

together, it will readily be seen that the forces between the particles are not affected, and therefore, because of the incompressibility of the liquid, force P_1 is resisted entirely by hydrostatic compression forces in the water itself, thus allowing no appreciable contraction of the total volume.

Suppose, however, that we substitute porous pistons as shown in diagram b, which represents the case of permeable pressing. In this case, it will easily be seen that now the force P_2 is not resisted by the water, as this can readily flow through the pistons. The force is consequently taken by the springs between the spheres; and if the pressure is large enough, these springs will be compressed until the spheres are actually touching.

Diagram c represents the condition occurring in the free plastic mass. Here tight pistons are used; and instead of the force P_1, a capillary force

P_3 acts to put the water in hydrostatic tension. This action on the pistons will draw the particles together, compressing the springs, and gives closer packing *only insofar as water is removed by the capillary.*

Therefore, it can be seen that in the case of the plastic clay mass the capillary forces of the stretched membrane act to draw the particles closer together and that this force is balanced by the repulsion forces between the particles.

The high capillary forces in wet clay are clearly brought out in the article by Westman,[1] where layers of clay were exposed to water on one side and nitrogen pressure on the other to balance the capillary forces.

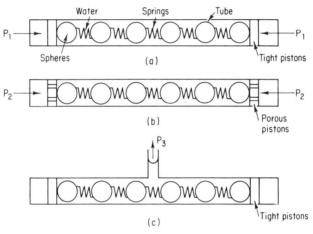

FIG. 10.13   Mechanical model of the clay-water system.

In this way, it was unnecessary to have a large column of water in tension, which was a limiting factor in some of the older tests of this type. Westman found, as we should expect, rather low capillary forces in the coarser-grained flint or feldspar but higher ones in kaolin and very high forces in fine-grained ball clay. His values in pounds per square inch are shown in Table 10.1.

This theory will account for the increasing yield point with decreasing

TABLE 10.1   Capillary Forces

Material	psi
Flint...............	5
Feldspar............	10
Kaolin..............	263
Ball clay...........	880

water content, as well as for the greater workability in finer-grained masses where the capillary forces are higher.

A very simple direct proof of this theory is to fill a rubber balloon with dry, powdered clay. The material then will act like a dry powder; but if the balloon is slowly evacuated, so that the atmosphere, pressing on the rubber, forces the clay particles together like the outer water film, then the mass assumes the identical plastic properties of wet clay.

Of course, the particles making up the mass have an important influence on workability. They should have a small size to produce small capillaries at the surface for a high yield point and numerous water films for

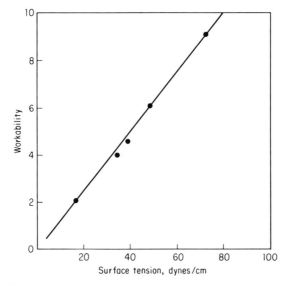

FIG. 10.14   Effect of surface tension of the liquid on workability.

good extensibility. Of all the natural materials, clay minerals have the smallest dimensions, a plate thickness of less than 0.1 μ. Then again an active surface to stabilize the water film is necessary, and here again the clay minerals seem to be unique in possessing large negative charges.

If the clay should be mixed with a liquid of lower surface tension than water, we should expect a lowering of workability, for the stretched-membrane forces would be lowered. This is just what Kingery and Francl[11] found when they made up specimens with various liquids. In fact, the workability was exactly proportional to the surface tension, as shown in Fig. 10.14. It was also shown that as the surface tension increased, the drying shrinkage and the firing shrinkage increased.

An excellent discussion of workability is given by Bloor[13,17] and Allison, Brock, and White.[18]

10.3 The Casting Process. *Casting Slips and Deflocculants.* The important properties of slips include specific gravity and viscosity. In the case of the high grog slips, which concerns us here, the specific gravity is high compared with many whiteware slips, generally running above 2.0. The viscosity, too, is generally high as thick sections are usually cast. The viscosity is difficult to measure in high grog slips, not only because of the coarse particles but also because the viscosity is generally variable with length of time and velocity of agitation.

One property of interest is the reduction of viscosity as the slip is stirred more rapidly due to thixotropy. Use is made of it by agitating or vibrating the slip as it is poured into the mold in a fluid condition. On resting quietly in the mold, the slip sets up so that very little water removal by the mold is needed to form an article firm enough to stand by itself.

Heavy refractories are made only in solid castings, for drain castings with slips containing coarse grog are so rough on the inner side as to be quite unsatisfactory.

Slips. A good solid casting slip should have the following properties:

1. A low enough viscosity to flow into and completely fill the mold
2. A high enough specific gravity so that the grog will not settle out
3. Capability of giving sound casts without shrinkage pockets
4. A reasonable stability in storage
5. Clean release from the mold
6. Rapid setting up in the mold
7. Low drying shrinkage
8. High cast strength
9. Freedom from entrapped air
10. Large thixotropy

Considerable experience is needed to make up a good casting slip. The type and amount of the clays are important, as well as the sizing of the grog. The amount of deflocculent must be carefully adjusted. While in the past the normal deflocculent was a mixture of sodium silicate and sodium carbonate, today Calgon (sodium polyphosphate) is generally used.

It is desirable to develop a high degree of thixotropy in slip so that it may be kept fluid by vibration in the feeder and also flowable in the mold by vibrating this also. In Table 10.2 there is given the composition of a high-grog thixotropic slip.[8]

Molds. The molds for slip casting are generally made from gypsum plaster in the same way as in the pottery. However, the low-water-content slips require very little water removal and so the molds may be quite thin. A typical solid-cast mold is shown in Fig. 10.15. There have been numerous experiments made to produce a more durable mold than

TABLE 10.2   Grog Casting Slip for Glass Pots

Material	*%*
North Carolina kaolin	13
Georgia kaolin	9
Tennessee No. 5 ball clay	10
Kentucky No. 4 ball clay	10
Maine feldspar	5
Grog:	
10–20 mesh	18
20–40 mesh	16
40–100 mesh	10
Through 40 mesh	9
Silicate of soda (S brand)	0.025
Sodium carbonate	0.012

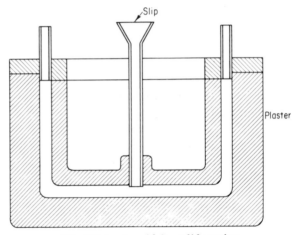

FIG. 10.15   A mold for solid casting.

plaster.   One approach is to form a mold with epoxy-resin-coated grains which are pressed into shape and then heat-treated.   In this way light and strong molds are produced, but there are problems of maintaining sufficient permeability.

Casting rate can be speeded up by supplying the slip to the mold under pressure.   Also, high slip temperatures are helpful as the viscosity of the water passing through the mold is decreased.

Casting Large Pieces.   An excellent example of slip casting is the manufacture of refractories for the glass industry, such as flux blocks, melting pots, and feeder parts.   The batch is made up from various kaolins, fireclays, pot clays, and ball clays.   A portion of the batch is calcined and carefully sized.   Mixing of the slip is carried out in a vacuum to remove all entrapped air, an unusual precaution for casting slips. The slip is heavy and thixotropic, so that vibration is used to cause flowing into the plaster mold, as shown in Fig. 10.16. After the slip has stayed in the mold a short time, it stiffens up and only a little water need be withdrawn by the plaster to enable it to be placed in a humidity drier. After drying, the blocks are set in periodic kilns and fired to the desired porosity of 16 to 18 percent, as shown in Fig. 10.17. After firing, the blocks are accurately sized on four or five sides by grinding with large cup

FIG. 10.16   Casting slip into a plaster mold.   (*Findlay Refractories Co.*)

wheels. They are then packed on skids for shipment, as shown in Fig. 10.18, where a vacuum pad is used for easy handling.

10.4 The Soft-mud Process. This process is now used for a very small percentage of refractories because of greater labor costs and the difficulty of holding close tolerances. However, some shapes are still made in this way. For example, the large floater in Fig. 10.19 is laid up by working in rolls of very carefully prepared mix until the whole structure is built. Then the sides and top are trimmed to size. Drying such a large piece takes several months under carefully controlled conditions. Pieces as long as 16 ft have been shipped.

The plastic mix of clay and grog is batched in a weighing hopper as in Fig. 10.20 and mixed in a wet-pan mixer (Fig. 10.21) or a pug mill (Figs. 10.22 and 10.23). Aging of the mix is common in European plants, but in the United States the mix is used at once.

Silica brick used to be made by the soft-mud process, but now are usually dry-pressed, with only 4 to 5 percent of water. The bond in the United States is 2 percent of milk of lime, but in Europe some sulfite liquor is also used.

10.5 The Stiff-mud Process. The stiff-mud brick are made with a consistency just above the critical point; therefore, considerable force is

FIG. 10.17 Setting large blocks in the kiln. (*Findlay Refractories Co.*)

required to form the clay. Most of the refractories made by this process are formed by forcing the plastic material through a die from which it emerges as a more or less homogeneous column that can be cut off into definite lengths. This column is generally produced by an auger, consisting of a propeller-shaped screw running in a trough, which forces the clay with high pressure through a die. Steam-pressed brick are forced through a die by a piston operated by steam pressure, an intermittent process.

A flow sheet of the stiff-mud process is shown in Fig. 10.24, in which the clays are prepared in much the same way as for the handmade bricks. Water is added to the mixture with either a wet pan or a pug mill. The latter is generally preferred because it is a continuous process and better

FIG. 10.18   Palletizing tank blocks for shipment.   (*Findlay Refractories Co.*)

FIG. 10.19   Hand molding a debiteuse block.   (*Findlay Refractories Co.*)

adapted to feeding the auger.   The pug mill is a long, trough-shaped container with one or two horizontal shafts running down the center, having attached to them suitable blades for kneading and mixing the clay and propelling it gradually toward the exit end. Water can be added to the material in the pug mill to bring the mixture to the proper consistency. A complete unit for making stiff-mud bricks is shown in Fig. 10.25.

FIG. 10.20   Batching with a weigh car in a refractories plant.   (*Findlay Refractories Co.*)

The auger, as shown in Fig. 10.26, must be a well-designed machine in order to produce a uniform column   It is important in designing the die to assure an even velocity of the clay in all parts of the stream to prevent strains.   Also the auger must be designed so as to prevent laminations, which often occur in the center of the column as an S-shaped crack.   The die itself is generally lubricated with oil to reduce the friction, and often it is steam-heated for the same purpose.   It has been found that if the clay is mixed in the auger in a vacuum chamber, the air is readily removed and a more dense and homogeneous column is produced in passing through the die.   One method of deairing in the auger is shown in Fig. 10.27.

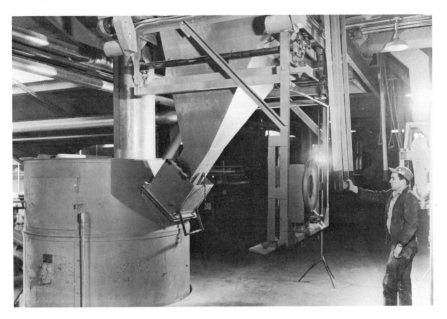

FIG. 10.21 Weight-controlled batching into a covered mixer. (*Harbison-Walker Refractories Co.*)

FIG. 10.22 Pug mill with shredder. (*The Bonnot Company.*)

Deairing has now become a very common practice in stiff-mud operation and enables the production of sound bricks from clays that previously could not be readily handled. It also permits a denser structure to be obtained. However, deairing is not necessary for all bodies, as many of them can be handled quite satisfactorily without it. In fact, deairing seems to make some plastic mixes harder to dry and burn.

The column of clay from the auger is cut into uniform sections with a wire cutter, as shown in Fig. 10.28. Many designs of cutter are available, but they all operate on the principle of passing one or more steel wires

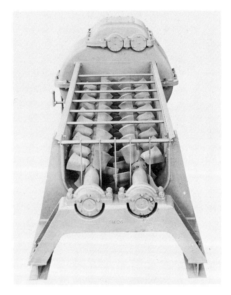

FIG. 10.23 The inside construction of a double-shaft pugging chamber to feed the deairing machine. (*Fate-Root-Heath Company.*)

through the column. It is difficult to operate a wire cutter satisfactorily if the grog content of the body is very high or if the size of the grog is very large.

There are few cases where the wire-cut brick produced are sufficiently uniform in size for use as refractories. Therefore, the general practice is to pass them through a machine re-press, where they are actually formed to size and the brand put on. After being re-pressed, the brick are quite firm and can be readily handled and stacked on the drier cars.

Since stiff-mud brick have a strong, tough structure, are good bricks for cutting, and are quite resistant to abrasion and slag, they can be used wherever a dense, strong brick is needed.

It is estimated that approximately 10 percent of the fireclay refrac-

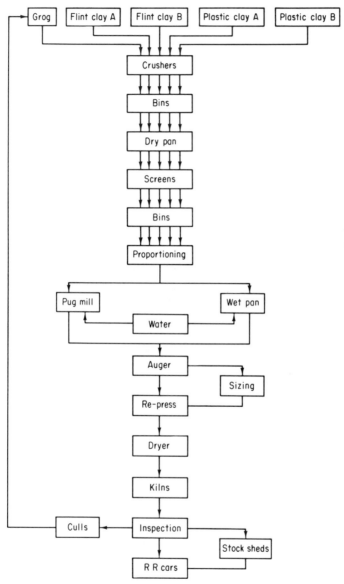

FIG. 10.24 A typical flow diagram for the manufacture of stiff-mud brick. (*E. H. Van Schoick, American Refractories Institute.*)

tories made in this country are produced by the auger and re-press. Inasmuch as an auger will produce as many as 6,000 bricks an hour and one machine re-press will turn out from 1,500 to 2,200 bricks per hour, four re-presses are generally set up with one auger.

A good many shapes, especially the medium and large ones, are made with a stiff-mud mix containing a high grog content because of the small total shrinkage and low warpage, even though the labor cost is higher than for the soft-mud method. Hard maple molds lined with $\frac{1}{16}$-in. steel are generally used, often with loose pieces. The mix is fed in slowly and continuously by one man while the other consolidates it with an air hammer having a corrugated foot.

FIG. 10.25 A unit for making extruded re-pressed firebrick. The raw materials come up to the pug mill by bucket conveyor, and then the batch drops down to the vacuum auger on the main floor. At the right are four machine re-presses.

Levitskii[25] worked out a mathematical analysis of the auger, while Pels Leusden[28] made a study of flow patterns. Capriz[24] produced a theory of the flow pattern, and Goodson[19] reported numerous experiments on the auger operation, while Parks and Hill[20] present a rigorous theory of extrusion forces.

Recently simple shapes like tile are being made on the impact press—an offshoot of the foundry molder. Here heavy steel molds are used which are filled with the mix and then repeatedly rammed by impact blows from an air-operated platen. A press of this type is shown in Fig 10.29. The structure of the molded piece is of more uniform density than that obtained from straight pressing, for the wall friction is broken down. Stoops[29] gives excellent data on this process.

Automatic Ramming. Ramming refractory mixes by hand requires great skill to produce a uniform piece. Therefore, considerable thought has been given to automating this process. One of the most successful

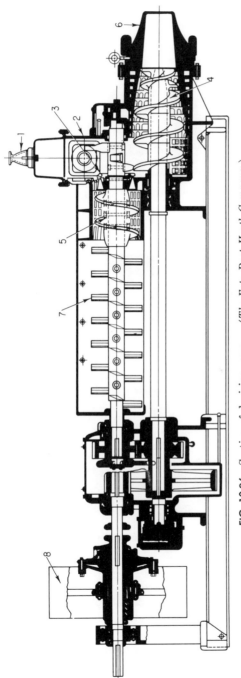

FIG. 10.26 Section of deairing auger. (*The Fate-Root-Heath Company.*)

133

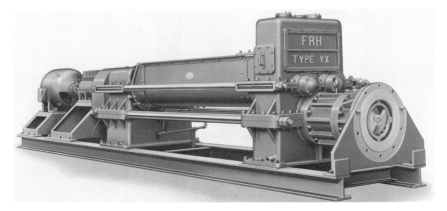

FIG. 10.27   Heavy deairing auger for making extruded brick.   (*Fate-Root-Heath Company.*)

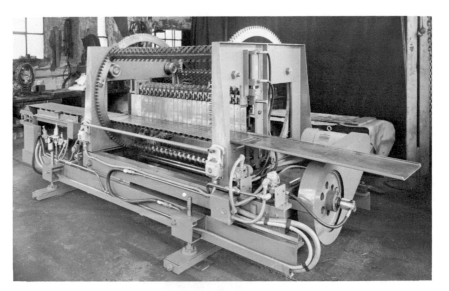

FIG. 10.28   Wire cutter for forming extruded brick.   (*The Bonnot Company.*)

methods for producing dense, pure oxide crucibles was developed in the author's laboratory at the Massachusetts Institute of Technology.   The equipment shown in Fig. 10.30 consists of a slowly revolving mold with an air-driven ramming tool projecting down into the annular space.   As the mold revolves, a uniform stream of feed falls to the bottom and is continuously compacted into thin spiral layers, the tool automatically rising on a carriage until the mold is filled.   Teeth on the ramming tool break up the layers and prevent lamination.   This method has been used

FIG. 10.29 Forming shapes on impact press. (*Harbison-Walker Refractories Co.*)

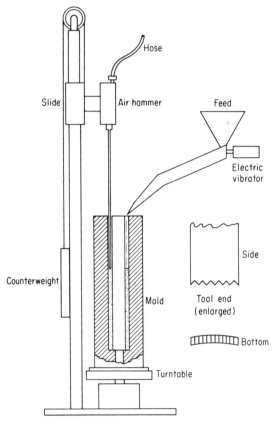

FIG. 10.30 Automatic tube molding.

for ramming long tubes and should be useful in ramming the linings of large crucibles.

There has been some effort made to automate the ramming of larger shapes by controlling the path of the ramming tool over a predetermined path with a continuous stream of feed preceding it. The problem is to build a machine with enough flexibility to handle a large range of shapes. It certainly can be done with our present knowledge.

10.6 The Dry-press Process. In the dry-press process of making refractories, the clay has the consistency of a damp powder containing perhaps

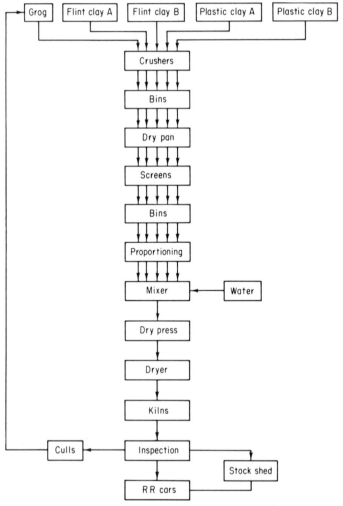

FIG. 10.31 Typical flow diagram for the manufacture of dry-press refractories. (*A. C. Hughes, American Refractories Institute.*)

FIG. 10.32 A toggle press operating at North American Refractories Co. (*Chisholm, Boyd, and White Co.*)

7 to 10 percent water. Only by high pressure can it be consolidated into a homogeneous mass. The flow sheet for the dry-press method is shown in Fig. 10.31, in which the preparation of the clays is very much the same as for the preceding processes. In the Missouri district, the clays and grogs are often coarsely crushed, stored in bins, and recombined by feeders into dry pans, where they are further ground, screened, and tempered. The mix is aged for 24 hr and then fed to the presses. Modern practice seems to be favoring the more careful control of the sizing of the grog and flint clays by screening and recombining in definite proportions. The mixture of dry materials is then moistened in a pan or a pug mill and brought up to the proper consistency. Certain types of special mixers are used to a considerable extent for this purpose. The mixed material is delivered to a hopper over the dry press, where the mixing action is continued, and permitted to flow into the dry press as needed.

The Dry Press. The dry press itself (Fig. 10.32) is usually the toggle type, pressing four standard bricks at one time. However, hydraulic presses are used for certain types of refractory (Fig. 10.33). The operation of the toggle press is shown clearly by the excellent diagrams prepared by Dr. G. E. Seil for the Third Edition of "Refractories," one of which is shown in Fig. 10.34. In order to make uniform brick, it is essential that the mix be uniform, that the same weight of the material be always charged into the dies each stroke, and that it be evenly distributed in the die box. For nonplastic mixes like magnesite and

FIG. 10.33 Hydraulic toggle-press forming refractories. (*Chas. Taylor's Son's Co., and Chisholm, Boyd, and White Co.*)

chromite, a small amount, about $\frac{1}{4}$ of 1 percent, of organic binder such as dextrine is used.

Until recently, the exact pressures in the toggle press were not known precisely. Now they can be determined by the stretch of the side arms, which indicate pressures as high as 14,000 psi. With many materials, the maximum pressure is limited by the expansion, which causes pressure cracking, of the entrapped air on releasing the pressure. Recently, many presses have been fitted with vacuum attachments that pull the air out of the die box and mix as soon as the top pad enters the box. Slots about 0.007 in. wide in the pads connect to the vacuum system.

The brick discharged from the dry press can usually be set on the tunnel-kiln cars with little or no drying. The rate of forming dry-press brick on the toggle press runs between 1,000 and 2,000 brick per hr. The dry-press process is used very extensively in the industry, and it is

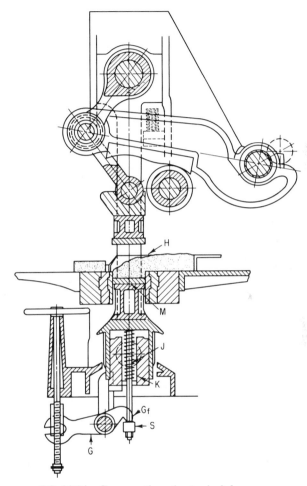

FIG. 10.34　Cross section of a typical dry press.

estimated that 80 percent of the fireclay refractories and practically all the magnesite and chrome refractories are made by the dry-press process.

The bricks produced by dry pressing are very uniform in size, with strong corners and edges and little tendency to warp. They have good spalling resistance and generally good resistance to load.

The thickness of the dry-pressed piece is limited by the side-wall friction to about 4 in. in the conventional press, but numerous attempts have been made to improve this condition. In one method developed in Germany, the bottom platen is vibrated while pressure is applied to the top one. In another, the sides of the mold box are oscillated up and down so that pressure is maintained between the platens in an effort to minimize the wall friction. Much work has been done by the powder metallurgist in studying the pressure distribution in pressed powders by incorporating a lead grid into the mass and later studying its deformation by x-ray pictures. Here it was found that the proper wall lubrication did much to equalize the pressure.

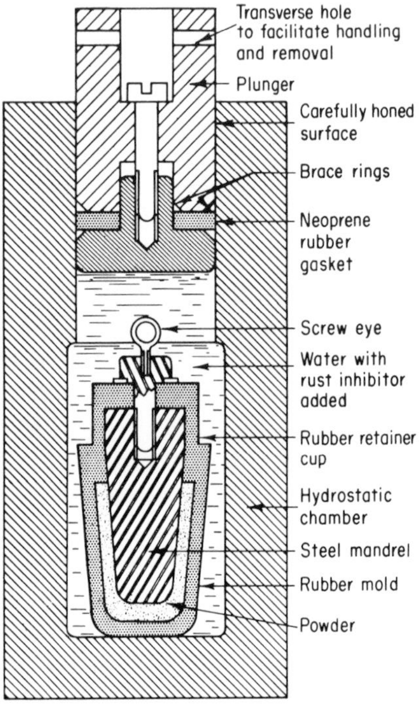
Transverse hole to facilitate handling and removal
Plunger
Carefully honed surface
Brace rings
Neoprene rubber gasket
Screw eye
Water with rust inhibitor added
Rubber retainer cup
Hydrostatic chamber
Steel mandrel
Rubber mold
Powder

FIG. 10.35 Hydrostatic chamber for molding.

Leuking and Bradley[26] investigated the effect of pressure on the density of green bricks. Freebury[27] discusses various types of presses, giving the comparative merits of the power press and the hydraulic press. Franke[30] gives data on the effect of using lubricants with nonplastic mixes in the dry press, while McRichie[31] gives data on the pressure distribution in dry pressing. Noble, Williams, and Clews[14] show a very interesting picture of the role of water in pressing.

Mixes of the dry-press consistency can be molded by hand ramming under special conditions. A patented process known as the Scheidhauer and Giessing process uses a deflocculated slip of clays mulled into carefully sized grog. The resultant mix contains about 90 percent grog and 10 percent clay, with a total water content of 5 to 6 percent. By heavy ramming in rigid steel molds, large shapes can be made with a total shrinkage of not more than $\frac{1}{4}$ of 1 percent.

Isostatic Pressing. As is well known, pressing from one direction in a die does not give a uniform density over the bulk of the piece because of the wall friction. This results in uneven density, uneven shrinkage, and a tendency to warp or crack. If pressure can be applied to the piece from all sides, the resulting density is uniform and the firing properties are improved. The simplest method of doing this is to fill a rubber mold with the granulated mix having 0 to 15 percent moisture. This mold can then be immersed in a fluid at high pressure, as shown in Fig. 10.35. Fine-grained materials such as oxides can be formed in this way with a very low water content, but clays press better with some moisture. Not only is the density uniform, but in many cases the product can be sintered at lower temperature levels than by other means of forming. Pressures of 2,000 to 20,000 psi are used. The greatest commercial use of this method is in the making of spark-plug cores.

FIG. 10.36 A large isostatic pressing unit used for forming zircon blocks, for glass-tank paving. (*Midvale-Heppenstall, Pressure Equipment Division.*)

There are now available pressure chambers for large pieces, perhaps 30 in. across. However, bodies containing relatively large grog particles tend to wear the rubber mold rather rapidly, so that this process is still confined to special pieces.[32]

Several refractory producers are now forming finely pulverized zircon into tank blocks by isostatic pressing in rubber molds (Fig. 10.36). The structure of these blocks is dense and uniform, but the present pressing cycle seems slow for efficient production.

Tar-bonded Dolomite Blocks. Steelmaking in the oxygen converter has greatly increased the use of tar-bonded dolomite brick. As these units have a short shelf life of 2 or 3 weeks, the manufacture in most foreign countries has taken place in the steel plant[33] so that the freshly pressed brick can go right into the converter lining. In the United States[34] the manufacture has been carried out in the refractories plant, but a number of steel companies have given careful consideration to an in-house product.

The product[21] is made according to the flow sheet of Fig. 10.37. The dolomite is calcined to a dense grog and in some cases MgO fines are added to increase the shelf life. Six to ten percent of tar is added, and the

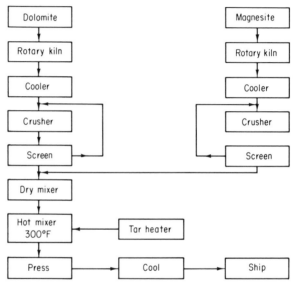

FIG. 10.37   Flow sheet for tar-bonded dolomite blocks.

batch is mixed in a heated mixer. The hot mix is pressed in a power press, allowed to cool, and then packed for shipment. Large blocks up to $18 \times 6 \times 3$ in. are commonly produced, rather than 9-in. brick.[22] Because of the low-cost raw material and the elimination of firing, this is an inexpensive refractory. A typical analysis is given in Table 10.3.

TABLE 10.3   Tar-bonded Dolomite Brick

*Constituents**	%
Silica.................	1.5
Alumina..............	0.7
Iron oxide............	3.1
Lime.................	32.8
Magnesia.............	61.6

* Inorganics.   Bulk density = 185 pcf.

Cold-pressed brick, using an asphalt emulsion, have been made but are not in general use. Tar-bonded magnesia brick are also made. These are air-stable but more costly than those of dolomite.

The properties of the tar-bonded refractory depend on the heating schedule, as the tar must be carbonized rapidly and yet not destroyed by oxidation.

10.7 Molding Insulating Firebrick.

The production of highly porous, lightweight refractories requires somewhat different methods from those employed for the heavier materials. Therefore, it was thought advisable to devote a special section to the molding of this product.

Method of Obtaining Pores. Perhaps the most common method of producing pores in a fireclay product is to introduce in the plastic mix an organic material that later on can be burned out to leave voids. Sometimes this organic material may be peat, which is naturally mixed with the fireclay, but usually it is necessary to add granular organic material such as ground wood, cork, or other materials of this type. Of course, the properties of the finished product will depend not only on the clay but on the size, shape, and amount of the organic particles.

A somewhat different method of producing pores consists in adding to the clay mix granular sublimable material such as flake naphthalene. When the ware is dried at high temperature, the naphthalene can be sublimed, leaving pores in the dried-clay product. The naphthalene vapor can be cooled and used over again. Although this process has been employed industrially, it is not in common use at the present time.

Still another method of forming the pores in the clayware is to introduce into the clay, which has been made to the consistency of a thick slip, bubbles of gas that are sufficiently stable to remain in the mix until the material is dried and burned. This method seems very attractive, as the cost of the organic material is eliminated; but practically, difficulties are encountered in stabilizing the bubbles so that they are of uniform fine size and evenly distributed throughout the mass. The method patented by Eriksson produces bubbles in the clay mix by the addition of powdered aluminum or zinc and alkali so that each particle of aluminum is decomposed and forms a bubble of hydrogen. Others accomplish the same effect by stirring finely divided lime into the soft mix and then acidifying, with the result that each particle of lime is decomposed, forming a bubble of carbon dioxide. Still another method, similar to those patented by Ericson and Roos, consists in making a preformed foam, which is then stirred into the clay slip. Such a foam, of course, has to have considerable stability in order that this process may be successful.

All the bubble-structure refractories must have some stabilizing device to hold the bubbles as individuals and prevent their coalescing into larger ones. In the Husain and Bole patents, this is accomplished by adding gypsum plaster to the mix and permitting it to set, thus stabilizing the bubbles. In fact, there is probably no successful bubble process that does not use some stabilizing method, either plaster or a flocculent; and for this reason, it is difficult, because of the fluxing elements added, to make high-temperature insulating firebrick by the bubble process.

A high porosity can also be obtained by adding to the mix an inorganic refractory material having itself a high porosity. Such materials include diatomaceous earth, bloated clay pellets, or porous grog.

Molding Methods. The molding of the insulating firebrick mixes is not particularly difficult, as the mixtures containing organic matter can be readily molded by the soft-mud method, whereas the bubble method is best carried out by pouring the slip into molds until the mass has set. Use of the stiff-mud or dry-press method for these highly porous materials is difficult; for if they contain organic matter, they are usually rather elastic and difficulties from pressure cracking are troublesome. On the other hand, if the bubble process is used, any considerable amount of pressure would squeeze the air out of the structure and the density would be too high.

10.8 Abrasion-resisting Metals. In the stiff-mud and dry-press process previously mentioned, numerous metal parts receive considerable wear as a result of the abrasive action of the grog particles in the mix. These parts include the mullers of the wet and dry pan, the blades of pug mills and augers, auger dies, and the die boxes of re-presses and dry presses. Table 10.4 gives a summary of a number of wear-resisting metals

TABLE 10.4   Use of Wear-resisting Metals

	Pug knives	Auger dies	Muller tires	Dry-press pads	Dry-press liners
"Ni-hard"	×				
High-chrome steel	...	...	...	×	
White cast iron	×	×	×		
Tool steel	...	...	...	×	×
Stellited steel	...	...	...	×	×
Tungsten carbide	...	×			
Chrome moly steel (heat-treated)	...	...	×		

that have been used. There is, however, at present no great uniformity in the use of these materials, as some plants will obtain better life with one and some plants with another owing, perhaps, to the differences in their mixes. A great deal of development work is still to be done in this field.

Dry-press and re-press die wear depends greatly on the mix. One hundred million bricks can be produced from plastic and flint clay in a set of liners; but when high-grog clay bricks or highly burned magnesite or chrome are used, the life may drop to as low as 5,000 bricks.

It has been found economical to use heavy cast stellite liners locked into the mold cavity for abrasive mixes. When these wear, they are all removed and trued up on a surface grinder, then replaced with a shim backing equal in thickness to the metal removed. This process may be repeated a number of times before the liners become too thin. For most uses hard surfacing alloys are built up on the surface of the part with a welding torch, as described by Kough.[12] Small pressed parts are often made in tungsten carbide dies, but the cost and brittleness preclude their use to any extent for heavy refractories.

10.9 Grog Sizing. The manufacturer of refractories is coming more and more to realize the importance of maintaining consistency of particle sizes of the grog or flint clay portion of his mix, especially where he has to make a low-porosity or low-shrinking refractory or an unburned refractory with a chemical bond. It has also been found that proper sizing helps the spalling resistance. Some manufacturers are going so far as to divide their particles by screens into a series of size fractions and then are recombining them by definite weights to produce the desired distribution. Others control size by proper adjustments of screens and crushers.

If the one-component system is considered first, i.e., particles having a single diameter, it will be found that the closeness of packing depends on the arrangement of particles and their shapes. Taking the simplest case, i.e., spheres, there are five different methods of packing, which are shown in Fig. 10.38 and the results given in Table 10.5.

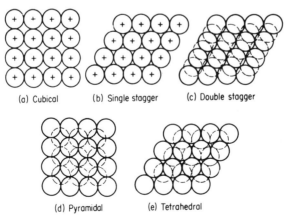

(a) Cubical   (b) Single stagger   (c) Double stagger

(d) Pyramidal   (e) Tetrahedral

FIG. 10.38 Various methods of packing spherical particles.

TABLE 10.5   Voids with Different Methods of Packing[2]

Type of packing of spheres, *in Fig. 10.37*	*Voids,* %
Cubical......................................	47.64
Single stagger (cubical tetrahedral).............	39.55
Double stagger.............................	30.20
Pyramidal.................................	25.95
Tetrahedral................................	25.95

In the case of irregular particles as would be obtained from a crusher, the packing varies between 40 and 50 percent, although theoretically, with perfectly arranged cubes, the voids would be zero.

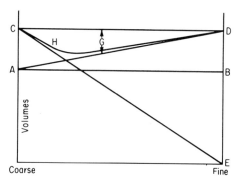

FIG. 10.39   Packing diagram for two sizes of particles.

Consider next the two-component system wherein a coarse grain and a fine grain are mixed and shaken down to the minimum volume; the clearest picture of the situation can be obtained by referring to the packing diagram of Westman[2] shown in Fig. 10.39.   Along the base is plotted the proportion of coarse and fine in the mix, giving a constant total weight. On the vertical scale are plotted volumes.   A line AB represents the true volume of the solids.   Point C represents the apparent volume of the coarse mix, and D the apparent volume of the fine mix, the apparent volume being the true volume plus the volume of voids.   Now suppose that a small proportion of the coarse material is replaced by fine material. It is obvious that up to a certain point, the fines will simply fill the voids between the larger grains and therefore the total apparent volume will decrease in proportion to the amount of fines added.   Accordingly, the theoretical apparent volume will follow the curve CE.   Now starting from the other end of the diagram, if a certain amount of the fine material is replaced by the coarse, the coarse material will displace an equal volume of fine material and the voids existing in the fines.   Therefore, the total volume will decrease owing to an elimination of a portion of the voids in the finer material. Consequently, the line AD will represent the theoretical apparent volumes. Where the lines CE and AD cross will be the minimum volume from a mixture of coarse and fine having an infinite ratio between the diameters. Actually, however, we have a finite ratio, and the actual curve of apparent volume will be indicated by H. The line CD is the apparent volume of the fractions before mixing, and the distance G represents the shrinkage in volume during the mixing process.

Many investigators have studied particle packing in addition to the classic work of Westman and Hugill.[2] McGeary[21] investigated the packing of spherical particle Cooper and Eaton[22] the compacting of ceramic powders, Rose and Robinson[35] the difference between a perfect and a "hyperperfect" mix, and Schwartz and Weinstein[36] a compaction model of ceramic powders.

In order to show the influence of grog sizing from normally crushed material, it will be of interest to examine some of the data from a paper by Gugel and Norton[23] to see how the dry density is influenced, not only by the grog, but also by the bond clay, moisture, and compacting. The grog was a sintered sedimentary kaolin with a true density of 2.69 and very few pores.

The grog particles were vibrated to minimum bulk volume in a graduate. The results for the six single fractions are shown in Table 10.6. The porosities of the coarser fractions are higher than the fine ones because of the side effects.

TABLE 10.6 Porosity of Single Grog Sizes

Fraction No.	Median diameter, mm	Bulk density	Porosities, %
1	3.53	1.40	47.9
2	1.76	1.35	49.7
3	0.79	1.42	47.2
4	0.31	1.48	45.2
5	0.18	1.53	43.1
6	0.08	1.53	43.1

These grog fractions were blended into binary systems, as shown in Table 10.7, with the low porosity values in boxes. The lowest values, of course, occurred when diameter ratios were largest, with a value of 26.7 for 55.6 coarse and 44.4 fine.

The results from ternary systems are shown in Table 10.8 where the minimum porosity is 23 percent, not a large gain from the binary low. This value comes at 44.8 coarse, 13.8 medium, and 41.4 fine.

TABLE 10.7 System with Two Grog Sizes

% coarse	% fine	Coarse fraction													
		4–8					14–35				35–65			65–100	
		Fine fraction													
		8 14	14 35	35 65	65 100	−100	14 35	35 65	65 100	−100	35 65	65 100	−100	65 100	−100
91.0	9.0	47.2	45.0	45.0	44.6	43.1	47.2	47.2	46.1	44.6	45.7	45.0	42.7	43.5	42.0
84.4	15.6	47.2	42.7	41.2	41.2	40.1	46.8	45.0	43.8	41.2	45.0	42.0	40.5	42.4	39.0
77.0	23.0	46.4	42.0	38.6	38.3	35.7	46.5	42.4	42.0	38.3	44.2	40.9	38.3	42.0	37.5
71.5	28.5	46.8	40.9	36.0	34.2	32.0	46.1	40.5	39.0	35.3	43.5	39.0	35.3	42.0	37.2
66.7	33.3	**46.1**	40.9	34.2	32.3	30.1	**45.0**	38.3	37.9	33.1	43.5	37.9	34.2	42.4	36.8
62.5	37.5	46.8	40.9	**32.3**	31.6	27.5	**45.0**	38.3	35.7	30.1	43.1	37.2	33.8	42.4	35.7
55.6	44.4	47.2	**39.0**	33.5	28.2	26.7	45.0	37.9	33.1	**29.4**	**42.7**	**36.8**	33.1	42.0	**35.3**
50.0	50.0	47.2	39.7	34.5	**27.5**	27.1	45.3	37.9	31.6	30.1	**42.7**	**36.8**	**30.8**	**41.6**	36.0
45.5	54.5	47.2	40.1	34.5	30.1	29.4	45.3	**37.2**	**30.8**	30.8	43.1	36.5	**30.8**	42.0	36.0
40.0	60.0	47.2	40.9	36.0	29.7	28.2	45.3	37.2	32.0	30.8	42.7	30.5	31.6	42.7	36.0

TABLE 10.8   System of Three Grog Sizes
Components, %

Coarse	Medium	Fine	Porosity, %
4–8	14–35	−100	
48.5	14.3	37.2	24.1
54.2	12.5	33.3	24.5
48.0	16.0	36.0	24.5
52.0	12.0	36.0	24.9
48.2	11.1	40.7	24.9
4–8	35–65	−100	
44.8	13.8	41.4	23.0
46.6	26.7	26.7	23.2
45.4	18.2	36.4	23.4
40.6	21.9	37.5	23.4
51.1	17.8	31.1	23.6
54.2	12.5	33.3	23.8
56.5	13.0	30.5	24.1
4–8	65–100	−100	
52.8	11.1	36.1	24.5
54.3	11.4	34.3	24.7
54.6	27.2	18.2	24.3
51.9	14.8	33.3	25.1

It was hoped that by adding a well-dispersed kaolin to a low-porosity ternary system the pores might be filled and thus give a lower porosity. This did not occur, as the more kaolin added the higher the porosity when the sample was vibrated to a minimum volume.   Evidently the kaolin increased the friction between the grog particles and prevented close packing, as shown in Table 10.9.   These same mixes were then placed in a steel die and pressed from 0 to 15,000 psi without moisture.   The results are given in Table 10.10.   At the higher pressures, the lowest porosity is 17.3 percent, only 1 percent lower than the grog alone.

The same samples were moistened with 10 percent water and again pressed in the steel die.   The results in Table 10.11 show that the moist kaolin acts as a lubricant to allow the porosity to be reduced to 14.5 percent.   The water was replaced by various organic lubricants, but lower porosities were not obtained.

The value of 14.5 percent porosity is very low for a green body and should be of value in chemically bonded refractories.

TABLE 10.9   Porosity After Vibration with Additions of Kaolin

Kaolin added, %	Composition		Porosity, %
	Grog	Kaolin	
0	100	0	23.6
5	95.2	4.8	27.2
10	90.9	9.1	29.9
15	87.0	13.0	33.6
20	83.3	16.7	35.2

TABLE 10.10   Porosities of Pressed Specimens with Kaolin Additions (without Water)

Pressure, psi	Grog	Grog + 5% kaolin	Grog + 10% kaolin	Grog + 15% kaolin	Grog + 20% kaolin
0	23.6	27.2	29.9	33.6	35.2
1,000	26.0	27.6	28.0	27.8	28.2
5,000	23.0	23.2	22.4	22.7	22.4
10,000	20.3	20.4	19.4	19.4	18.5
15,000	18.7	18.2	17.7	17.3	17.4

TABLE 10.11   Porosities of Moistened Specimens Containing Kaolin (10 Percent of Water by Volume)

Pressure, psi	Amount of kaolin added, %			
	5	10	15	20
5,000	22.7	21.2	21.2	19.6
10,000	20.4	19.2	18.4	16.3
15,000	17.4	17.4	16.5	14.5

10.10   Automatic Batching.   Since the Second World War there has been active interest in mechanizing batching for all types of refractories. The rapid development of electronic control devices has made possible simple and reliable batching systems.   As an example of how they might work, the very simple device in Fig. 10.40 may be considered.   Bins A, B, C, and D contain the four batch ingredients.   Each bin has below it a weighing feeder that will give, let us say, exactly 10 lb per min when it is run by a constant-speed motor (a).   These motors are each connected to a time switch that can be set to any desired interval.   For example, a 19-min setting would mean 190 lb delivered to the batch hopper. There could be a set of cams for each type of batch with any desired proportions of A, B, C, and D, quickly interchangeable. This very simple system can benefit from the addition of a push-button control that will automatically set the time switches for each formula.

In the working setup there are many safeguards, so that any defective unit can be automatically bypassed. In other words, if the batcher runs

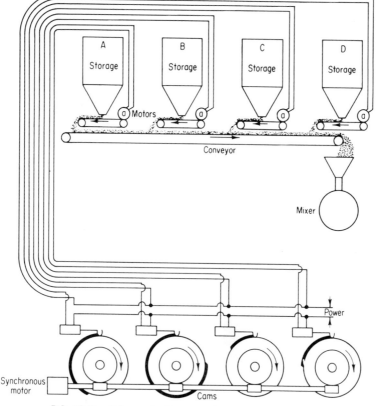

FIG. 10.40 Schematic diagram of automatic batching.

at all, it must give the correct result. Such a system can be readily made to operate in a reliable manner if the raw materials are uniform and will feed out of the bin continuously without hangups. Sometimes vibrators are used on the lower part of the bin. Then, moisture of the feed must be controlled or a correction factor fed into the computer.

Automatic batching is used for making up mixes for various types of refractory in which more than one component is needed. They are particularly valuable for refractory mortars, plastics, concretes, etc., not only saving labor but guarding against mistakes.

10.11 Chemical Bonding. A great deal of interest has been shown recently in refractories that are chemically bonded with a setting material, which gives to the brick sufficient mechanical strength so that it can be used in the furnace without previous burning. In order to do this, it is necessary to have a large proportion of well-fired nonplastic material in the brick and have it sized in such a way as to give a very dense packing. In addition, the bricks must be pressed under high enough pressure to reduce the voids as far as possible. Bricks of this type have been made of magnesite, chrome, and fireclays.

Most of the bonds used today are oxychlorides or oxysulfates. There seems to be very little specific information available on the exact formulas used.

The chemically bonded refractories in some types of service give results superior to the burned refractories, especially in resistance to spalling; consequently, they have definitely taken a place in our refractory production. Of course, the first thought on considering this type of refractory is the great saving in firing cost; but when the problem is analyzed, it is usually found that with the additional cost of the bond and the necessity for carefully sizing the nonplastic, which comprises practically the whole brick, the total cost is little, if any, less than for the burned brick.

BIBLIOGRAPHY

1. Westman, A. E. R.: The Capillary Suction of Some Ceramic Materials, *J. Am. Ceram. Soc.*, **12**, 585, 1929.
2. Westman, A. E. R., and H. R. Hugill: The Packing of Particles, *J. Am. Ceram. Soc.*, **13**, 767, 1930.
3. American Society for Testing Materials, Tentative Standard E 24–37T, 1937.
4. Norton, F. H.: An Instrument for Measuring the Workability of Clays, *J. Am. Ceram. Soc.*, **21**, 33, 1938.
5. Hauser, E. A., and S. Hirshon: Behavior of Colloidal Suspensions with Electrolytes, *J. Phys. Chem.*, **43**, 1015, 1939.
6. Johnson, A. L., and F. H. Norton: Fundamental Study of Clay, III, Casting as a Base Exchange Phenomenon, *J. Am. Ceram. Soc.*, **25**, 337, 1942.
7. Norton, F. H., A. L. Johnson, and W. G. Lawrence: Fundamental Study of Clay, VI, Flow Properties of Kaolinite-Water Suspensions, *J. Am. Ceram. Soc.*, **27**, 149, 1944.
8. Heindl, R. A., G. B. Massengale, and L. G. Cossette: The Slip Casting of Clay Pots for the Manufacture of Optical Glass at the National Bureau of Standards, *Glass Ind.*, **27**, 177, 1946.
9. Norton, F. H.: Fundamental Study of Clay, VIII, A New Theory for the Plasticity of Clay-Water Masses, *J. Am. Ceram. Soc.*, **31**, 236, 1948.
10. Coughanour, L. W., and F. H. Norton: Influence of Particle Shape on Properties of Suspensions, *J. Am. Ceram. Soc.*, **32**, 129, 1949.

11. Kingery, W. D., and J. Francl: Fundamental Study of Clay, XII, Drying Behavior and Plastic Properties, *J. Am. Ceram. Soc.*, **37**, 596, 1954.
12. Kough, H. V.: Hardsurfacing for Structural Clay Products Machinery, *Bull. Am. Ceram. Soc.*, **35**, 228, 1956.
13. Bloor, E. C.: Plasticity: A Critical Survey, *Trans. Brit. Ceram. Soc.*, **56**, 423, 1957. Noble, W., A. N. Williams, and F. H. Clews: Influence of Moisture Content and
14. Forming-pressure on the Properties of Heavy Clay Products, *Trans. Brit. Ceram.*
15. *Soc.*, **57**, 414, 1958. Moore, F.: The Rheology of Ceramic Slips and Bodies, *Trans. Brit. Ceram. Soc.*, **58**, 470, 1959.
16. Baudran, A., and C. Deplus: Rheological Behavior of Clay Bodies, *Trans. Brit. Ceram. Soc.*, **58**, 454, 1959.
17. Bloor, E. C.: Plasticity in Theory and Practice, *Trans. Brit. Ceram. Soc.*, **58**, 429, 1959.
18. Allison, E. B., P. Brock, and J. White: The Rheology of Aggregates Containing a Liquid Phase . . . , *Trans. Brit. Ceram. Soc.*, **58**, 495, 1959.
19. Goodson, F. J.: Experiments in Extrusion, *Trans. Brit. Ceram. Soc.*, **58**, 159, 1959.
20. Parks, J. R., and M. J. Hill: Design of Extrusion Augers and the Characteristic Equation of Ceramic Extrusion Machines, *J. Am. Ceram. Soc.*, **42**, 1, 1959.
21. McGeary, R. K.: Mechanical Packing of Spherical Particles, *J. Am. Ceram. Soc.*, **44**, 513, 1961.
22. Cooper, A. R., Jr., and L. E. Eaton: Compaction Behavior of Several Ceramic Powders, *J. Am. Ceram. Soc.*, **45**, 97, 1962.
23. Gugel, E., and F. H. Norton: High-density Firebrick, *Bull. Am. Ceram. Soc.*, **41**, 8, 1962.
24. Capriz, G.: A Theoretical Analysis of Extrusion Processes, *Trans. Brit. Ceram. Soc.*, **62**, 339, 1963.
25. Levitskii, G. D.: Mathematical Analysis of the Operation of Vacuum Pugmill Auger Blades, *Glass & Ceramics, Moscow*, **20**, 434, 1963.
26. Leuking, W. C., and R. S. Bradley: Effect of Forming-pressure on Properties of a Fireclay Mix, *Am. Refractory Inst. Tech. Bull.* 104, October, 1963.
27. Freebury, L. S.: Presses for Refractories Production, *Refractories J.*, **39**, 254, 1963. Pels Leusden, C. O.: The Rheological Properties of Plastic Bodies in the Heads
28. of Augers, *Trans. Ninth Int. Ceram. Congr.*, **553**, 1964.
29. Stoops, R. F.: 98.2% Density with Vibrating Impact Presses, *Ceram. Ind.*, **82**, 54, 1964.
30. Franke, G.: Influence of Shaping-pressure and Lubricants on the Manufacture of Fireclay Bricks from Non-plastic Batches, *Silikat Tech.*, **15**, 159, 1964.
31. McRichie, F. H.: A Device for Determining Pressure Distribution in Dry-pressing Refractories, *Bull. Am. Ceram. Soc.*, **43**, 501, 1964.
32. Catchpole, C.: Isostatic Pressing, *Trans. Brit. Ceram. Soc.*, **1**, 385, 1964. Schurmann, W.: The Fully-automatic Preparation of Tar-dolomite, *Interceram*
33. *Luebeck Ger.*, (2), 108, 1964.
34. Way, R. E.: Tar-bonded Brick: Future of the Refractories Industry? *Brick Clay Record*, **144**, 72, 1964.
35. Rose, H. E., and D. J. Robinson: The Density of Packing of Two-component Powder Mixtures, *Powder Met.*, **8**, 20, 1965.
36. Schwartz, E. G., and A. S. Weinstein: Model for Compaction of Ceramic Powders, *J. Am. Ceram. Soc.*, **48**, 346, 1965.

CHAPTER ELEVEN

Drying

11.1 Introduction. The refractories industry has been adopting forming methods with low-water-content mixes so rapidly that the drying problems are becoming quite simplified. However, there are a few products, such as large shapes, where drying methods must be carefully controlled to obtain efficiency in the operation.

11.2 The Mechanism of Drying. The work of Sherwood[1,2,3,4] and his collaborators has done much to give a clear picture of the drying process in porous solids. It is logical to divide the drying into two periods: the constant-rate period and the falling-rate period.

The Constant-rate Period. At the start of the drying process, when the clay contains enough water at least to fill the pores, there will be a continuous film of water over the surface particles, as evidenced by the comparatively dark color of the wet clay. It is therefore not unexpected to find that the rate of drying per unit area of surface is substantially the same as for a free water surface. This rate in terms of temperature, humidity, and velocity of the air is shown in Fig. 11.1. As the drying proceeds at this rate, it is obvious that water must travel through the pores toward the surface to supply the evaporation loss. In all but exceptional cases, where the drying rate is very high or the clay very fine grained, the water will travel by capillary forces as fast as it is carried away. In the case of clay, the piece will shrink in volume equivalent to the volume of water lost until the particles touch, and thereafter air must enter the pores to take the place of the water. Soon after this happens, it will be found that the water cannot travel to the surface fast enough to maintain an

unbroken surface film; so the drying rate starts to decrease.   This rather
sharply defined point may be called the "critical point" and corresponds
approximately to cessation of shrinkage.

It might be thought that a nonshrinking material like sand and water
would not show any constant-rate period because the pores would start
to empty at once.   Experiments show this to be true when the particles are
fine; but when the particles are coarse, the resistance to flow is so slight that

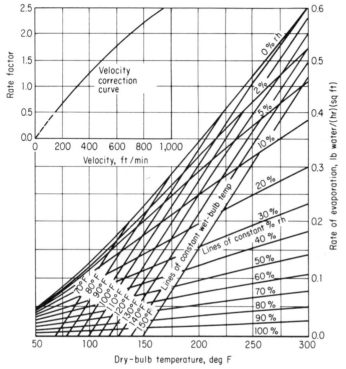

FIG. 11.1   Evaporation chart for free water surface.   (*C. B. Shepherd, C. Hadlock,
and R. C. Brewer, Industrial and Engineering Chemistry,* 1938.)

even with air in the pores, the surface drying rate is the governing factor.
This is shown clearly in Fig. 11.2 for a fine and coarse sand.

The Falling-rate Period.   Beyond the critical point, the drying rate is
governed by the rate of transfer of water from the interior to the surface
and rapidly becomes less as the drying proceeds.   It is believed that
beyond the critical point, the continuous surface film is broken and the
water surface recedes into the capillaries, giving the light color to the clay
at this stage.   As the drying proceeds, more and more of the water is

converted inside the structure into vapor, making it necessary for this to travel out through the nearly empty capillaries to the surface.

In Fig. 11.3, an attempt has been made to show how an enlarged section

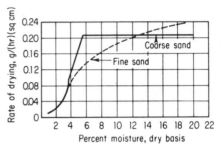

FIG. 11.2   Drying rates of sand-water mixes.   (*After Sherwood.*)

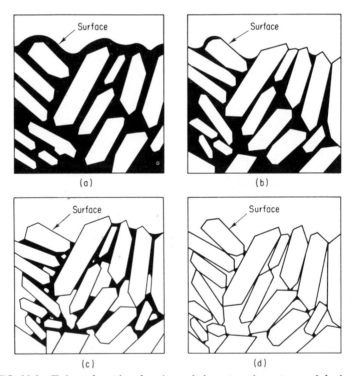

FIG. 11.3   Enlarged section drawings of clay at various stages of drying.

of the clay would look at various stages of the drying process.   In *A*, the clay particles are well separated by a water film, which also runs continuously over the surface.   In *B*, the amount of water has decreased until the particles touch one another, but there is still a continuous surface film.

In *C*, the water has decreased until the surface layer is broken and the level recedes into the capillaries with some air in the structure. Water is brought to the surface by a capillary flow and as vapor. In *D*, the water has still further decreased until it is found only in a few places where the particles come closest together. Here all the transfer of water is in the

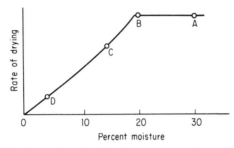

FIG. 11.4   A typical drying-rate curve for a clay.

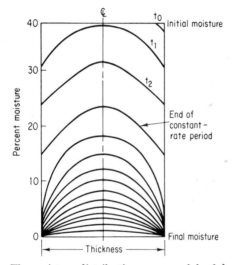

FIG. 11.5   The moisture distribution across a slab of drying clay.

form of vapor. These four conditions are shown on the typical drying-rate curve for a clay in Fig. 11.4.

The last of the water comes off slowly because it is adsorbed on the surface of the dry particles. Practically, however, it is never necessary to remove the last trace of water. In fact, a bone-dry piece is apt to be brittle and hard to handle.

Moisture Gradient in Clay Body. The moisture distribution across a slab of drying clay may be shown to follow the lines in Fig. 11.5. At the critical point, the moisture at the surface drops rapidly toward zero.   In the constant-rate period, however, the curves are parallel and may be expressed mathematically by Newton's law of diffusion

$$\frac{\delta M}{\delta \theta} = K \frac{\delta^2 M}{\delta x^2}$$

in the same way as for heat flow.   Here

M = the moisture concentration per unit volume

θ = the time

x = the distance from the surface

K = a constant

The solution of this equation and its application to the constant-rate period has been carried out by Gilliland and Sherwood,[4] which permits the calculation of the moisture distribution in a slab after any time interval.   Reference should be made to the original papers for a full discussion of this calculation.

The moisture movement in drying clay masses has been studied by several investigators.   Macey[10] considers again his earlier suggestion of an ice structure in the adsorbed water layers.   Moore[9] gives an excellent review of the drying mechanism.   Ford and Noble[8] show that long storage does not always ensure a uniform water distribution.

11.3   Drying Shrinkage.   The drying shrinkage of a clay or body is an important characteristic because it largely governs the maximum safe rate of drying.   The drying-shrinkage curve is also very helpful in studying some phases of the clay-water system.

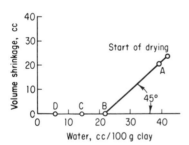

Mechanism of Drying Shrinkage.   The mechanism of shrinkage is a very simple one.   If the volume shrinkage is plotted against the volume of water in the clay, as in Fig. 11.6, it will be seen that the shrinkage curve is a straight line with a 45° slope.   In other words, the volume shrinkage is just equal to the volume of water lost in drying down to the critical point, where the interparticle film becomes so thin that the particles touch one another and shrinkage can go no further.   This will be made clearer by referring back to Fig. 11.3, which shows the condition of the clay and water at the points lettered on the curve of Fig. 11.6.

FIG. 11.6   Theoretical drying-shrinkage curve of clay.

The water lost from the clay in drying between its original condition and the critical point is often referred to as "shrinkage water," and the remainder as "pore water."

The critical point is not usually so sharply defined as shown in Fig. 11.6, nor does the lower end of the shrinkage curve always lie along the axis. When clays of the swelling type, such as bentonite, are present, the curve is above the axis and only shrinks the final amount just before dryness, because of the fact that the last of the water is more firmly held between the crystal layers than on the surface of the particles. Other clays have shrinkage curves falling below the axis as a result, at the end of the drying process, of a slight expansion that has not yet been explained but is perhaps due to the opening up of kaolinite books on losing the last trace of water.

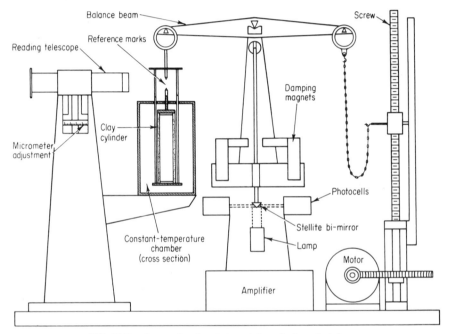

FIG. 11.7   Apparatus for determining drying shrinkage.

Measurement of Drying Shrinkage.   Shrinkage can be measured by either a length change or a volume change.   The former is more useful for studying the type of shrinkage curve, whereas the latter gives more accurately the overall shrinkage.

The apparatus[6] shown in Fig. 11.7 is a convenient one for obtaining shrinkage curves.   The test specimen is in the form of a hollow cylinder $7\!/_8$ in. outside diameter, 2 in. long, with a $1\!/_8$-in. wall.   It is placed on an automatic balance in a chamber with controlled humidity and temperature; accordingly the change in length can be precisely determined with a micrometer telescope.

Volume shrinkage is usually measured by placing the specimen, at various stages of dryness, in a mercury volumeter. Volume shrinkage can be converted to linear shrinkage by the curve in Fig. 11.8.

Water-film Thickness. The ceramist is much interested in the thickness of the water film separating the clay particles. If the total linear shrinkage and the number of films per unit length are known, this factor can be readily calculated. The total shrinkage can be easily determined, but the number of films is more difficult to estimate. The most recent work on this subject is by Norton and Johnson[7] who measured the drying shrinkage of a number of monodisperse size fractions of a kaolin. Provided the size and shape of the particles are known, the number of

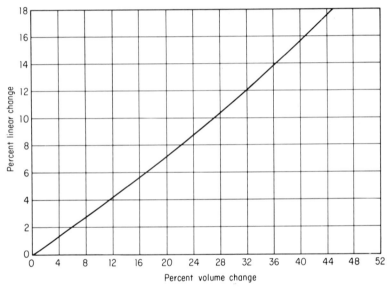

FIG. 11.8 Curve for the conversion of volume to linear shrinkage based on initial dimensions.

films are readily computed. In this way, they found a value of about 0.005 μ as the film thickness in a plastic mass, or a layer of about 50 water molecules. This value can also be computed by determining the thickness of the shrinkage water when spread over an area equal to the surface area of the particles. The result is of the same order of magnitude.

Shrinkage Curves. Figure 11.9 illustrates drying-shrinkage curves for a few typical clays. Curve A for a fireclay gives the typical secondary shrinkage, and curves B and C show a slight expansion.

11.4 Dry Strength. The dry, or "green," strength of clays is a unique and more or less unexplained property. Such theories as postulate molecular cohesion, felting of the particles, or organic colloids are not entirely substantiated as yet. It is known that the dry strength increases as the area of the particles increases (finer particles). Also the strength increases with more of some adsorbed ions. For example, a kaolin having a strength (modulus of rupture) of 400 psi when dialyzed showed an increase to 1,000 psi when saturated with sodium ions. These facts indicate that strength is due, at least in part, to attractive forces between the clay mineral crystals. This is an excellent field for further research.

11.5 Factors Influencing Drying Efficiency. When clayware is dried, the operation must be carried out as rapidly as possible in order to cut down the size and cost of the drying equipment. Great strides have been

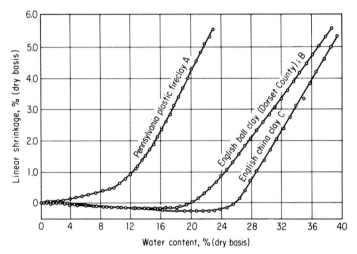

FIG. 11.9 Drying-shrinkage curves of typical clays.

made in the last ten or fifteen years in increasing the speed and efficiency of drying by means of controlled humidity driers.

Maximum Drying Rate. The maximum drying rate is determined by the stresses set up in the ware and the ability of the body to resist them. The same factors apply as will be discussed in Chap. 16 for spalling. The tendency to crack for similar-sized pieces is a function of the moisture gradient, the slope of the shrinkage curve, and the flexibility of the structure. The only variable that can be controlled in the drying operation is the moisture gradient; and for a given rate of drying, this can be decreased only by reducing the viscosity of the water flowing through the capillaries. The viscosity of water decreases quite rapidly with temperature, as shown in Fig. 11.10, which explains the reason why, in modern driers, the ware is heated rapidly to the maximum temperature in a saturated atmosphere before much drying takes place.

The size of the refractory piece influences the time required to dry.

Macey[5] has shown that under constant drying conditions, various sizes of cubes dry in times proportional to the length of the edge of the cube. Also the larger the piece the greater will be the tendency to crack under given drying conditions. It should also be realized that when the water content is sufficiently low to come below the critical point, the shrinkage stops and therefore the drying can be carried on as rapidly as desired, provided the vapor can escape from the interior without setting up a bursting pressure.

Distribution of Heat. It is quite a step from drying a single brick in the laboratory to drying a full car in a production drier. Circulation of air

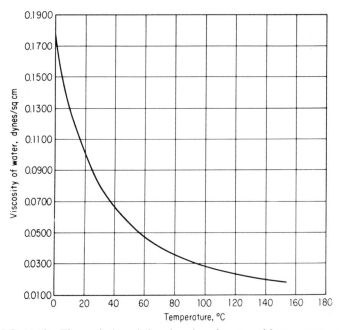

FIG. 11.10 The variation of the viscosity of water with temperature.

must be adjusted both to bring heat to the brick at the required rate and, at the same time, to carry away the water vapor. In order to dry all the bricks alike, a large volume of air must be circulated with the proper distribution of velocity.

Conservation of Heat. A definite amount of heat is required to evaporate the water; but in addition, heat must be supplied for the waste air leaving the drier, for the heat in the bricks, and for the loss through the drier walls. Therefore, an efficient drier will have a minimum volume of exit gases, by means of recirculation, and will have well-insulated walls.

11.6 Drier Calculations. It is not within the scope of this book to consider drier design, as this is well covered in more general treatises.

However, it might be well to point out some of the principles involved. The heat content of the air supplied to the drier, between the entrance and exit temperatures, must be sufficient to provide for the heat required in the drying operation. Also the volume of the air passing through the drier must be so great that the moisture added to it will not bring it above the saturation point at the exit temperature (or temperature of the entering bricks).

11.7 Types of Driers. The hot floor is used to some extent at the present time for drying and tempering soft-mud bricks and for drying large shapes. The heat is supplied by means of steam pipes running in conduits or tiles just under the surface of a concrete floor. The temperature of the floor can be controlled by valves, and the floor is usually divided

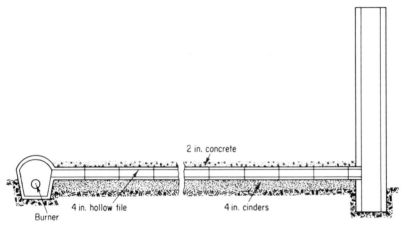

2 in. concrete

4 in. hollow tile   4 in. cinders

Burner

FIG. 11.11   Section showing construction of a direct-fired hot floor.

up into small sections so that the temperature can be regulated according to the material drying at any particular point on the floor. Hot floors are sometimes heated by waste heat from the kilns or by direct heat from coal or oil fires, from which the hot gases pass under the floor in tile flues as shown in Fig. 11.11. These methods, however, are not so satisfactory as steam heating, because it is difficult to control the temperature of the floor and because the temperature will vary considerably from one end of the floor to the other. Sometimes, the thickness of the floor is varied from the hot to the cold end to give a more even surface temperature. From one-half to four 9-in. equivalents can be dried per square foot of floor per 24 hr, depending on the clay, the water content, and the size. In some plants, one-half of the drier capacity is in the hot floor. Overhead unit heaters are often used to circulate air from above in order to increase the drying capacity per square foot.

Tunnel driers are used mainly for bricks and small shapes. They consist of a long tunnel in which rack cars of bricks can be placed.   Heat is supplied by steam coils under the tracks or, in some cases, by waste heat, which passes through the tunnel.   The green bricks are usually pushed into the tunnel at the cooler end and gradually progress to the hotter end until they are completely dry.   The bricks or shapes are placed on shods that are held in the rack cars.   Sometimes perforated shods are used to speed up the drying of the underside of the bricks.   They cannot be used, however, with very soft clay, for they will mark the brick.   Dry-pressed brick are often dried directly on the tunnel-kiln cars with a great saving in handling costs.

The more modern type of drier is the continuous-tunnel type, in which the loaded cars are pushed at a uniform rate through a tunnel having care-

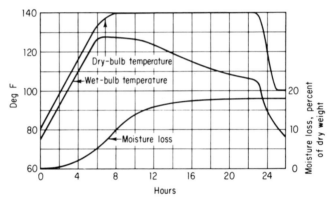

FIG. 11.12   A typical drying schedule for firebrick in a modern humidity drier.   (*J. Am. Ceram. Soc.*)

fully regulated temperature and humidity values at each section.   Fans or injectors are used to circulate the air around the steam coils and through the bricks in order to ensure a rapid and uniform transfer of heat to the drying material.   Usually recirculation is used; i.e., the air is passed around a closed circuit, the only vent being a small escape pipe at the top of the tunnel to allow the moisture from the bricks to pass off.   The tunnel is divided longitudinally into sections, and each section is controlled for temperature and humidity by automatic controllers.   In general, the temperature and humidity are high at the start to heat the brick with little loss of water from the surface.   The humidity is then gradually decreased as the bricks pass through the tunnel.   The regulation, however, must be determined for the particular kind of clay being dried.   Figure 11.12 shows a typical temperature distribution.

The temperature in driers varies according to the nature of the clay and size of the articles.   Generally it is kept as high as possible, 50°C (about

120°F) to 90°C (about 190°F).   The capacity of a drier depends also upon the clay and the size of the articles.   It is an economy to dry as rapidly as possible, as this reduces the investment in the drier, cars, and shods for a given output.   Typical drying conditions are shown in Table 11.1. Infrared driers have been used in the ceramic industry for thin ware, but this principle has not as yet found much use in the refractories field.

Since the design and construction of efficient driers require a great deal of technical knowledge and experience, the amateur drier builder seldom turns out a satisfactory job.   The scientific study of drying problems has lately resulted in great advances in this field.   Reference 11 should be consulted for the details of drier design.

TABLE 11.1   Drier Conditions

Type of brick	Maximum drier temperature		Time to dry 9-in. brick, hr
	°F	°C	
Hand-molded.................	160	71	48
Stiff-mud, re-pressed..........	160	71	24
Silica......................	150–250	66–121	24
Chrome and magnesite........	180	82	16
Insulating firebrick...........	180	82	30

BIBLIOGRAPHY

1. Sherwood, T. K.: Drying Solids, I, *Ind. Eng. Chem.*, **21**, 12, 1929.
2. Sherwood, T. K.: Drying Solids, II, *Ind. Eng. Chem.*, **21**, 976, 1929.
3. Sherwood, T. K.: Drying of Solids: Application of Diffusion Equations, *Ind. Eng. Chem.*, **24**, 307, 1932.
4. Gilliland, E. R., and T. K. Sherwood: Drying of Solids: Diffusion Equations for the Period of Constant Drying Rate, *Ind. Eng. Chem.*, **25**, 1134, 1933.
5. Macey, H. H.: The Effect of Temperature and Humidity on the Rate of Drying of Clay Shapes, *Trans. Brit. Ceram. Soc.*, **37**, 131, 1938.
6. Norton, F. H.: Precise Measurement of Drying Shrinkage, *Ceram. Age*, **33**, 7, 1939.
7. Norton, F. H., and A. L. Johnson: Fundamental Study of Clay, V, Nature of Water Film in Plastic Clay, *J. Am. Ceram. Soc.*, **27**, 77, 1944.
8. Ford, R. W., and W. Noble: Moisture Redistribution in Plastic Clay during Storage, *Trans. Brit. Ceram. Soc.*, **59**, 58, 1960.
9. Moore, F.: The Mechanism of Moisture Movement in Clays with Particular Reference to Drying: A Concise Review, *Trans. Brit. Ceram. Soc.*, **60**, 517, 1961.
10. Macey, H. H.: Moisture Movement in Plastic Clay, *Trans. Brit. Ceram. Soc.*, **62** (1), 67, 1963.
11. West, H. W. H.: The Effect of Structural and Operating Modifications on the Efficiency of a Tunnel Dryer, *Trans. 9th Intern. Ceram. Congr.*, **501**, 1964.

CHAPTER TWELVE

Firing of Clay and Other Refractory Materials

12.1 Introduction. All refractories, with the exception of those chemically bonded, plastics, or castables, are fired to stabilize and strengthen the structure; even these exceptions contain a large percentage of heat-stabilized aggregate. The firing step in refractories manufacture must be closely controlled, as fuel, kiln upkeep, and labor amount to a large portion of the total cost of manufacture. For these reasons, it is essential that there is a good understanding of the reactions taking place during the firing cycle. Therefore, in this chapter the firing reactions occurring in the usual refractory materials are discussed.

There has been a gradual rise in firing temperatures, as it is now generally appreciated that a brick should be stabilized at a temperature as high as or higher than the use temperature. High-fired superduty brick are fired at 1480°C (about 2700°F), kaolin brick at 1700°C (about 3100°F), and some basic brick as high as 1870°C (about 3400°F).

12.2 Effect of Heat on Raw Clay. The reactions occurring in clays during heating are one of the most fundamental subjects in the field of ceramics. For this reason it is well for the producer of refractories to learn all he can about this subject. There is available a large number of papers in this field extending over the past fifty years. These have been well digested by Brindley and Nakahira;[23] so they will not be considered in detail here. However, it will be attempted to summarize the latest thinking on this important subject.

Kaolin. It is logical to start with well-crystallized kaolinite and follow the changes as it is slowly heated. There are many methods available to observe these changes, which may be listed as follows:

1. Weight loss[7]
2. Differential thermal analysis[5,8]
3. X-ray diffraction[10]
4. Rehydration
5. Electron microscope[24]

As these methods are familiar ones today, no attempt will be made to describe them here, but references are given at the end of this chapter.

When the well-crystallized kaolin is slowly heated, nothing happens until 450°C (about 840°F) is reached, at which point there is a loss in weight of 14 percent and a heat absorption of 170 cal per g. Beyond this temperature, x-ray data indicate a breakdown of the kaolinite structure, which is replaced by a material called meta-kaolin. In the past this was called amorphous, but with more refined techniques some order, probably two-dimensional, is indicated which may be the Si-O sheets. At about 950°C (about 1740°F) there is a sharp evolution of heat and a new crystal phase appears, having a spinel structure which is oriented along the crystal axes in the original kaolinite. Above 1050°C (about 1920°F) this spinel structure gradually breaks down into mullite and cristobalite with an amorphous or glassy phase.[14] Again the mullite is oriented in relation to the original kaolinite, as clearly shown in the beautiful electron photomicrographs of Comer.[24] At a temperature of 1350°C (about 2460°F) the composition is 30 percent mullite, 15 percent cristobalite, and 55 percent glass. This sequence of events is represented in Fig. 12.1.

Fireclay. The changes in less perfectly crystalline kaolinite or fireclay are somewhat different, for the transformations are less sharp and there is little or no alignment between the kaolinite starting material and the resulting mullite.

McGee[50] studied the minerals in kaolins and fireclays with various cooling rates. He found the mullite content at about 28 percent for all conditions, but the cristobalite content of 15 percent with slow cooling fell to 2 percent in the fireclays on quenching. With the kaolins the cristobalite remained high even when quenched.

While a great deal of light has been thrown on these changes during the heating of clay in the last few years, there are still many questions to be answered. For example, what is the true structure of mullite? Why does the exothermic peak occur so suddenly? Why will the meta-kaolin rehydrate?

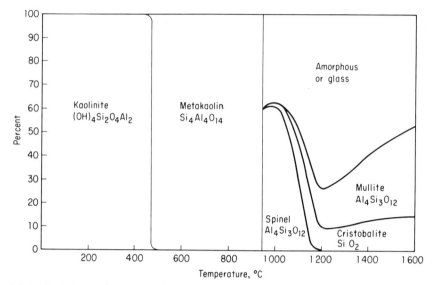

FIG. 12.1   Schematic diagram showing the amount of constituents when well-crystal-lized kaolin is heated.

12.3   Effect of Heat on Other Hydrated Minerals.   Such minerals as gibbsite [Al(OH)$_3$], bayerite [Al(OH)$_3$], boehmite [AlO(OH)], and diaspore [HAlO$_2$] are altered on heating to corundum as the final product.   However, on the way there are many varieties of alumina, as shown by Stirland, Thomas, and Moore.[15]

When pure gibbsite is heated to 150°C (about 300°F), some of the (OH) group are driven off to form the mineral boehmite.   On further heating to 275°C (about 530°F), the remaining (OH) groups are driven off forming γ-Al$_2$O$_3$, a crystal with a spinel-like structure.   Cooper et al.[27] has shown that a parallelism has been found between the crystographic orientations in the original and final crystals.   Further heating results in corundum (Al$_2$O$_3$).

Diaspore is more stable than gibbsite; so it does not lose the (OH) group until a temperature of 350°C (about 660°F) is reached.   The reason for this is evident from its crystal structure where the hydrogen is strongly held between two oxygens.   On dehydration γ-alumina is first formed; then above 550°C (about 1020°F) it slowly inverts to corundum. Again the orientation of the final crystal is related to the starting crystal.

From the practical viewpoint, few of these alumina minerals are found without silica additions, usually in the form of clay.   Therefore, the end point is generally a mixture of mullite, corundum, and silica-alumina glass.

12.4   Influence of Impurities on Heat Effects.   Most fireclays contain impurities that have important effects on the firing properties.   These impurities, generally carbonates or sulfides, break up themselves to form oxides which then react with the clay. Table 12.1 gives some of the common impurities with their decomposition temperatures. Clay ware fired too rapidly to allow the reaction gases to escape from the structure may result in black cores or bloating.

TABLE 12.1   Breakdown Temperatures of Clay Impurities

	°C	°F
$FeS_2 + O_2 \rightarrow FeS + SO_2$	350–450	660–840
$4FeS + 7O_2 \rightarrow 2Fe_2O_3 + 4SO_2$	500–800	930–1470
$Fe_2(SO_4)_3 \rightarrow Fe_2O_3 + 3SO_3$	560–775	1040–1430
$C + O_2 \rightarrow CO_2$	350 →	660 →
$S + O_2 \rightarrow SO_2$	250–920	480–1690
$CaCO_3 \rightarrow CaO + CO_2$	600–1050	1110–1920
$MgCO_3 \rightarrow MgO + CO_2$	400–900	750–1650
$FeCO_3 + 3O_2 \rightarrow 2Fe_2O_3 + 4CO_2$	800 →	1470 →
$CaSO_4 \rightarrow CaO + SO_3$	1250–1300	2280–2370

Fluxing materials that lower the viscosity of the glass phase permit reactions to take place at increased rates, or at the same rate at lower temperatures. Alkalies, iron compounds, fluorides, and alkaline earths often are the cause of this increased reaction rate when heating. Tyrrell[30] showed that if the impurities in the clay were over 5 percent (3 percent TiO_2, 1 percent Fe_2O_3, 1 percent alkali), the resulting mullite was volume-stable.

12.5   Firing Shrinkage and Porosity of Clays. *Firing Shrinkage.*

From the practical viewpoint of use, the firing shrinkage of clay is of great interest, for the greater the shrinkage the greater the difficulty in holding finished tolerances. Also, high shrinkages enhance the danger of cracking in the kiln. In general, the more compact the formed clay, the lower the shrinkage. This compaction may be inherent in the clay, as, for example, flint clay, or it may be produced by high pressure.

The firing-shrinkage curves in Fig. 12.2 illustrate this clearly. Curve *A* is a sedimentary kaolin, pulverized and molded at low pressure into a bar. Shrinkage starts at 550°C (about 1020°F), continuing at a fairly even rate until 950°C (about 1740°F) is reached, where there is a sudden shrinkage corresponding with the change of meta-kaolin to the spinel structure. Then there is little shrinkage until 1100°C (about 2010°F) is reached; while beyond this, shrinkage continues rapidly as mullite and cristobalite form together with an even more fluid glassy phase until zero porosity is reached.

Curve B is for the flint clay which was crushed and molded with water at low pressure. The initial part of the shrinkage is quite similar to the kaolin, but at higher temperatures the shrinkage ceases and even shows a slight expansion due to bloating caused by trapped gases in the glassy phase.

Curve C represents the same kaolin of curve A, but here it was molded by isostatic pressing at 20,000 psi. It will be seen that the shrinkage at the high temperatures is materially reduced. It is also found that the

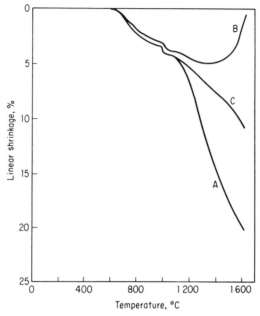

FIG. 12.2   Firing-shrinkage curves.

high pressure allows reaching zero porosity at lower temperatures because of the strain energy built into the clay particles.

A number of theories have been suggested for the cause of firing shrinkage when a glassy phase is present. It has been suggested that the stretched-membrane theory proposed for drying shrinkage (p. 119) may apply here. In this case the shrinkage forces are due to the glassy phase acting in the surface capillaries. It is certainly true that high shrinkage rates occur in clay only when some glassy phase is present.

Porosity. Porosity of fired clays is an important characteristic. Distinction can be made between open pores found at temperatures below maturity and closed pores sometimes found above maturity (overfiring). The percent volume of open pores may be measured by three weighings substituted into the following equation:

$$P = \frac{W - D}{W - A} \times 100$$

where P = percent of pore volume in relation to the bulk volume
$\quad\quad W$ = weight of saturated specimen in air
$\quad\quad D$ = weight of dry specimen in air
$\quad\quad A$ = weight of the saturated specimen submerged in water
The specimen may be saturated by boiling in water for 2 hr.

The determination of the total pores, including the closed ones, is best carried out by grinding the specimen finely enough to open all pores and measure the true density by the pycnometer method. The total pores will then be

$$p = \frac{d_t - d_b}{d_t} \times 100$$

where d_t is the true density, d_b the bulk density, and p the percentage of total pores in relation to the bulk volume. The bulk volume is usually computed from measurement of a cube cut to size with a diamond saw.

In some cases it is of interest to know the size distribution of the open pores, as the size influences permeability and some other properties. The most practical method consists of subjecting the sample to mercury under various pressures and measuring the volume penetration against the surface tension. The pressure required is inversely proportional to pore diameter. Pore sizes from 8.5 to 0.10 μ or less can be determined. Watson, May, and Butterworth[13] describe equipment for this purpose, and Hill[25] gives results on some ceramic specimens.

12.6 Firing Properties of Nonclay Refractories. There are, of course, a number of refractory materials that are not derived from clay.

Silica. Silica, one of the most important of refractory minerals, has received a great deal of study, and consequently more data on its properties are available than on any other mineral. The most complete treatise on silica is that of Sosman,[49] which the reader should consult for many of the details for which space is not available here.

In nature, silica is very common and most often occurs as the mineral quartz. It is found in many different types of mineral deposits, is an essential constituent of such acid igneous rocks as granite and rhyolite, and also occurs in veins, in sedimentary rocks such as sandstones and quartzites, and in unconsolidated river and beach sands.

Quartz can be transformed by heating into a number of polymorphous modifications. These forms, together with their stability ranges, are shown in Fig. 12.3 taken from Sosman.

There are at least six crystalline varieties of silica: quartz, tridymite, and cristobalite and each of these has a high- and a low-temperature modification. All the varieties are characterized by a three-dimensional network of SiO_4 tetrahedrons, each corner being shared with neighboring tetrahedrons. Thus a silicon atom is always shared by two oxygens, which checks the formula SiO_2, long ago determined by chemical analysis. The density and refractive indices of tridymite and cristobalite are very

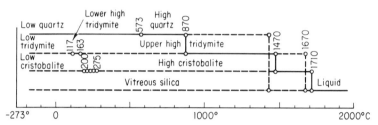

FIG. 12.3 Polymorphous modifications of silica and their stability ranges. (*R. B. Sosman, "The Properties of Silica," Reinhold Publishing Corporation, New York, 1927.*)

similar, but they differ considerably from those of quartz, as is shown in Table 12.2.

TABLE 12.2   Density of Forms of Silica
At Room Temperature

	Index of refraction	Density
Quartz............	$\epsilon = 1.553$ $\omega = 1.544$	2.65–2.66
Tridymite.........	$\alpha = 1.469$ $\beta = 1.473$ $\sigma = 1.470$	2.26–2.33
Cristobalite.......	1.486	2.27–2.32

Quartz is remarkably pure, and consequently the values for its indices of refraction and density are usually very closely alike. There are only meager data on the chemistry of tridymite and cristobalite, but the data that are available indicate that they are usually impure and consequently vary in density and refractive index. On the other hand, it is quite possible that these variations may be due to the little care usually given to the selection of material for such measurements.

In all cases, the high-temperature forms have a higher symmetry than the low-temperature forms. Note that the diagonal groups of tetrahedrons in high-temperature cristobalite form nearly a straight line whereas in low-temperature cristobalite, they form a puckered chain; the same is true in the high- and low-temperature-quartz relationship.

The differences between the linking of the tetrahedrons in the high- and in the low-temperature forms of each mineral are slight, but they are considerable in quartz, tridymite, and cristobalite. When a high-temperature form is transformed to a low-temperature form, the tetrahedrons are shifted and rotated but none of the links is broken, and consequently the high-low transformations of each mineral are rapid and reversible (see Fig. 12.3). These inversions are accompanied by volume changes, given in Table 12.3, together with the heat effects that produce them.

TABLE 12.3   Density Changes Accompanying Inversions

Form	Density change	Heat effect, g-cal per g
Quartz, low-high..................	−0.02	4.5
Cristobalite, low-high..............	−0.04	5.0
Tridymite, low-lower high...........	−0.01	1.0
Tridymite, lower high-high..........	0.00	

The structures of the minerals quartz, tridymite, and cristobalite are quite different from one another, and the transformation of one to the other requires a marked change in the linking of the tetrahedrons. Consequently these transformations are extremely slow, and tridymite and cristobalite can exist indefinitely in the metastable state at room temperature. In fact, it is impossible to form quartz from either tridymite or cristobalite by strictly thermal reactions.

The sluggish inversions that are so important to the manufacturers of silica refractories can be greatly influenced by other materials called "mineralizers." A great deal of work has been done on the effect of mineralizers in changing quartz to cristobalite or tridymite in the kiln, and many of these are noted in the references at the end of the chapter.[18,19,28, 29,31,41,49] The most common one is lime, which is used in most of the commercial silica bricks, partly as a bond and partly as a mineralizer, to hasten the conversion of the quartz to cristobalite and tridymite (Fig. 12.4). Many other mineralizers have been suggested, such as fluorides, titanium dioxide, boric oxide, ferric chloride, and sodium chloride. It should be remembered, however, that in commercial production, it is undesirable to cause the inversion to proceed with too great rapidity; for if the volume change proceeds too fast, it will rupture or make unsound the structure of the brick.

Eaton et al.[20] discuss thermal treatment on properties of silica brick, and Chaklader and Roberts[17] show how the properties are related to the constitution.

Kyanite and Associated Minerals. Kyanite (triclinic), andalusite (orthorhombic), and sillimanite (orthorhombic) all have the formula Al_2SiO_5. The structural relations of these minerals are complex, and for a general discussion the reader should see Greig.[1] On heating, they change over to mullite and glass or to mullite, glass, and cristobalite. The

FIG. 12.4 Pennsylvania ganister from Mt. Union quarry, 50×, transmitted light with crossed nicols. (*Harbison-Walker Refractories Co.*)

crystalline breakup of these minerals has been thoroughly studied by Greig, and the temperature ranges are shown in Table 12.4.

TABLE 12.4 Decomposition Temperatures of Silica-Alumina Minerals

Mineral	Kyanite	Andalusite	Sillimanite
Decomposition starts..............	1100°C (2010°F)	1410°C (2570°F)	1550°C (2820°F)
Complete decomposition*..........	1410°C (2570°F)	1500°C (2730°F)	1625°C (2960°F)
Peak on thermal curve............	1420°C (2590°F)	1510°C (2750°F)	1560°C (2840°F)
Density before firing..............	3.6–3.7	3.1–3.3	3.3

* This temperature varies with rate of heating and size of grains.

As the density of the mullite is 3.16 and the density of the silica glass and cristobalite is less, it is obvious that there must be an expansion in volume during the conversion. This expansion is particularly evident in the case of kyanite, which has the highest original density, but shows only to a very small extent with the other two minerals. The expansion of kyanite on heating may be a valuable property when combined with clays, as it offsets the shrinkage of the clay. Unfortunately, however, the expansion of the kyanite takes place through a rather small temperature interval, whereas the shrinkage of the clay continues over a much longer range; consequently the two effects do not exactly balance each other. As will be shown later, however, a refractory can be produced that will have exactly the same fired volume as the initial volume. The expansion characteristics of kyanite are greatly dependent on the size of the grains, a fact that should be taken into account when considering the total expansion.

The difficulty in getting kyanite during the last war has led companies in this country and Germany to produce mullite-type brick from other materials. The general method seems to be the calcination of bauxite and clay in a rotary kiln to form grog and then bond this together with raw material and fire to high temperatures.

A production plant for synthetic mullite using various raw materials is described by Hawkes.[32] It is now believed that mullite is not of a specific composition $3Al_2O_3 \cdot 2SiO_2$ (72 percent Al_2O_3) but may have alumina in solid solution, perhaps as high as 78 percent Al_2O_3, and some evidence indicates that it may approach the 1:1 ratio of kyanite. An equilibrium diagram for $SiO_2 \cdot Al_2O_3$, based on the latest findings, is shown in Fig. 12.5.

Richardson and Lester[33] studied the expansion effects associated with mullite formation in fireclay-bauxite mixtures. Löcsei[42,43] finds aluminum fluoride (AlF) causes mullitization at lower temperatures. Konopicky et al.[36] discuss the commercial production of mullite. Ďurovič[34] explains the structural relation between sillimanite and mullite. Of course, some mullite is made in the electric furnace and crushed as grog.

Magnesite. The firing properties of magnesite depend largely on its composition. For refractory purposes, a dense, low-porosity material is desired. On heating the raw magnesite, the CO_2 is driven off at temperatures between 400 and 900°C (about 750 and 1650°F), forming immediately the only known form of MgO, periclase. At first these crystals are so minute that they can be detected only by means of x-rays; but with increasing time and temperature, they grow until they form the usual honeycomb structure. It should be remembered, however, that the periclase crystal is cubic and not hexagonal.

The burned magnesite has a true density of 3.55 to 3.63 and a porosity of 15 to 27 percent. The greater the impurities present and the higher the firing temperature the lower the porosity.   It is desirable to fire the grain well so that no appreciable shrinkage will occur on firing the brick.

The bond in magnesite refractories is mainly glass, but crystalline material such as forsterite does occur.   Iron oxide, in the form of finely divided magnesioferrite, $MgFe_2O_3$, occurs to the extent of 7 or 8 percent in some types of magnesite, which allows the production of a strong and dense product at temperatures very much lower than would be possible with the pure magnesia.   Other impurities such as lime, silica, and alumina form liquid, which acts as a bond.   The burning process is discussed in the references.

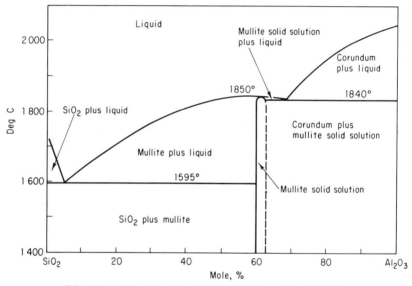

FIG. 12.5   Silica-alumina diagram.   (*Aramaki and Roy.*)

Modern magnesite refractories are made of purer materials and are fired at higher temperatures.   Often the magnesia is combined with a high-purity chromite to form direct-bonded brick.   Firing temperatures as high as 1870°C (3400°F) are employed.   Varma and Roberts[35] studied the burning of natural magnesites and found the crystal change to periclase between 1700 and 1800°C (about 3090 and 3270°F).   Allison et al.[11] studied the sintering properties with high-purity MgO.   Treffner[44] gives excellent pictures of the microstructure of periclase, while Kriek et al.[21] and Van Haeften[37] studied the sintering mechanism.

Since the use of basic refractories in open-hearth roofs became common after the Second World War, a great amount of effort has gone into making magnesite-chrome refractories with much better hot strength than was possible with the older glassy bond.   In spite of the fact that open hearths have been superseded by the oxygen converter, this basic study is helpful in other uses. The hot strength was increased by the rather obvious step of reducing the glass-forming oxides in the raw materials and firing high enough to obtain a direct bond between the crystalline elements. The literature on this subject is voluminous, but a few of the more important references will be given.

Van Dreser et al.[38] showed that by proper particle sizing, selection of chromite and magnesite, and firing at 1700°C (about 3092°F) a highly stable brick resulted. Houseman and White[22] studied the bonding in high-fired basic brick. Richardson et al.[26] also studied the bonding and properties of basic brick at high temperatures. Davies and Walther[45] discussed direct bonding in basic brick. Other references include El-Shahat and White,[46] Hayhurst and Laming,[47] Rigby et al.,[39] Scheerer et al.,[48] and Dodge.[16]

Fired basic brick, particularly magnesite and dolomite, may be impregnated with hot tar or pitch to fill the pores. In use the organics are carbonized, or the brick may be fired in a reducing atmosphere. In order to fill the pores completely, the heated brick is placed in vacuum to evacuate the pores and then subjected to the liquid tar at high pressure.

Lime. Calcium carbonate can be decomposed readily to calcium oxide; and on heating at temperatures as low as 1400°C (about 2550°F), crystals of calcium oxide can be formed. Such crystals are chemically much more stable than the low-calcined lime and undoubtedly would serve as an excellent basic refractory if they could be properly protected from moisture.

Zirconia. There are a number of polymorphous modifications of zirconium dioxide. The natural mineral baddeleyite is monoclinic. Pure monoclinic ZrO_2, often called the "*C* form," is stable up to 1000°C (about 1830°F) where it is transformed to the β form, which is tetragonal; this transformation is reversible. The tetragonal form is transformed irreversibly at 1900°C (about 3450°F) to the α_2 form, which is trigonal. The α_2 form is transformed to the α_1 form, which is also trigonal, at 625°C (about 1160°F), an inversion that is rapid and reversible.

It should be remembered that the inversion temperature and the crystallographic modifications obtained at various temperatures are dependent to a large extent on the composition and impurities that are present. This is again true in the ZrO_2 system. If the nitrate, oxychlorate, oxylate, or certain hydroxides of zirconium are heated, the tetragonal form is obtained at temperatures around 600°C (about 1110°F). This tetragonal form is metastable and is converted to the monoclinic form on further heating. The tetragonal form is also obtained if pure zirconia is heated in the presence of silica at about the same temperature; but whereas the tetragonal form obtained by heating the above-mentioned compounds of zirconium is transformed to the monoclinic form on further heating, the silica-bearing form remains tetragonal up to 1460°C (about 2660°F), where it is transformed to zircon, $ZrSiO_4$. If certain oxides such as MgO and CaO are present, a cubic form is obtained at about 1700°C (about 3090°F). This isometric form does not transform on cooling and is obtained only in the presence of the oxide impurities (other types of impurities might act in the same way). Less than 3 percent magnesia in solid solution will give rise to the isometric form, and the maximum amount of magnesia that exists in solid solution in ZrO_2 is about 28 percent, which corresponds to the formula $Mg_2Zr_3O_8$.

The most useful compound of zirconium, other than the oxide, is zirconium silicate, $ZrSiO_4$, which corresponds to the natural mineral zircon. This mineral is stable up to 1540°C (about 2800°F), where it decomposes to ZrO_2 and cristobalite or ZrO_2 and glass, according to Curtis and Sowman.[9] Reassociation takes place slowly between 1260 and 1540°C (about 2300 and 2800°F).

Much research has been carried out on the stabilization of zirconia in order to make a useful product. Weber et al.[12] studied stabilization with CaO, MgO, and CeO_2, measuring the phases present at a series of temperatures. They showed $CaO \cdot ZrO_2$ solid solutions were stable at all temperatures. Hamano et al.[40] studied the advantageous effects of partial stabilization to give best heat-shock resistance. It is not entirely clear whether or not stabilized zirconia remains stable over long periods of heating.

Chromium Minerals. Chromite is the natural spinel-type mineral of theoretical formula $FeCr_2O_4$. It is thermally stable and apparently undergoes no polymorphous modifications up to the fusion point. However, the natural mineral always contains impurities in solid solution, such as Ca, Mg, and Al, and admixed mineral impurities such as the silicates. Silica, if in large quantity, is objectionable, as it forms considerable quantities of glass on heating, which lowers the mechanical strength of the chromite refractories at high temperatures. The addition of MgO in the right amount to form the compound forsterite will diminish the amount of glass and provide a superior refractory (see the patents of Seil[51]). Considerable study has been given to bricks of mixtures of magnesite and chromite, which have given trouble because of a gradual growth due to absorption of iron oxide.

Magnesium Silicate Minerals. There are a number of minerals of this class such as olivine, talc, and serpentine, which may form the basis of forsterite refractories. In most cases, sufficient magnesite must be added to the natural mineral to convert the low-melting metasilicate to the high-melting orthosilicate. The amount of magnesia added as well as the grain size of the various minerals is quite important and is covered by patents such as that of Goldschmidt.

Carbon and Graphite. These refractories are used extensively in aluminum electrolysis cells, the lower linings of blast furnaces, some hot-blast cupolas, molds for fusion-cast refractories, and of course for electrodes in electric-arc furnaces. Carbon and graphite have many advantages for high-temperature use:

1. Very high hot strength
2. Not wet or attacked by most metals or slags
3. Excellent thermal-shock resistance
4. Easily machined to size
5. Wide range of thermal and electrical conductivity

The flow sheet in Fig. 12.6 gives an idea of the manufacturing methods. Petroleum coke is the usual starting point for a low-ash product such as electrodes; but for blocks, coke from coal may be used. The carefully sized coke is hot-mixed with tar or pitch and then formed by extrusion or pressing in a die. The cooled pieces are then set in a baking furnace,

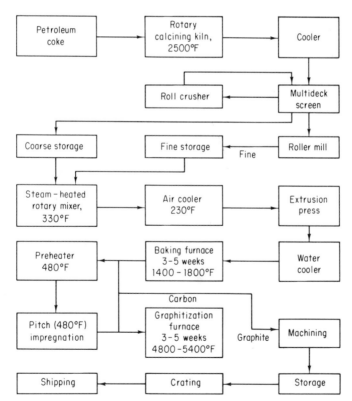

FIG. 12.6 Flow sheet for carbon and graphite electrodes.

bedded in coke, and held at a temperature of 760 to 980°C (about 1400 to 1800°F) for 4 to 6 weeks. Pieces to be graphitized are placed in a graphitizing furnace at 2650 to 2990°C (about 4800 to 5400°F) for 3 to 5 weeks. It is interesting to note that this is one of the highest-temperature industrial operations carried out on a large scale. The finished pieces, if necessary, can be machined to accurate sizes.

Natural graphite flakes are blended with fireclay and formed into crucibles and retorts. The clay protects the graphite from oxidation, and the oriented graphite plates give hot strength and thermal-shock resistance. This refractory is not new, as there is a crucible of this type (for melting gold) in the Cairo Museum, which looks as if it were just off our warehouse shelf, so alike in shape and texture is it to our modern product. Yet this was made about 3,000 years ago.

12.7   Solid Reactions.   *Importance in Ceramic Processes.*   A great many ceramic firing processes take place with little or no liquid phase present; accordingly many believe that the reactions go on entirely between solids. In spite of the great importance of these types of reactions, it was not until recently that a thorough study was made of this phenomenon, although as early as the year 1909, Wright pointed out the fact that calcium oxide crystals would grow rapidly at least 1000°C (about 1830°F) below their fusion point. Most of the early work in this field has been carried out in Germany by G. Tammann and his coworkers and by Goldschmidt. Heavy refractories seldom are sufficiently free from silica to make solid-state reactions a major process. Nevertheless, as we are using purer and purer materials for some of our refractories, solid-state reactions become of more interest.

Mechanism of Solid Reactions.   Solid reactions apparently take place because of loosening of the atomic bonds by thermal agitation, which permits mutual diffusion of the various atoms into the adjacent parts of the structure. The solid reactions can be divided logically into two groups; the first in which we simply grow large crystals from small ones and the second group where we produce a new type of crystal from two dissimilar ones.

The process of diffusion is greatly influenced by many factors in addition to the composition of the original crystals. In the first place, it has been found that the solid reaction progresses much more rapidly at a temperature where one of the crystals involved passes through an inversion point. For example, if a solid reaction is going on between quartz and another mineral, the reaction would be greatly accelerated at a temperature of 573°C (about 1060°F) where the low-temperature quartz changes to the high-temperature form. This is probably due to the loosening of the bonds at the inversion point and permitting of much greater rates of diffusion. It has also been found that gases, such as water vapor and fluorine, act as mineralizers and greatly accelerate the solid reaction. Again the reactions proceed much more rapidly, as we should expect, when the reacting particles are fine and in intimate contact. It has also been found that the reaction can be accelerated by seeding the mass of reacting crystals with a few fragments of the new crystal to be formed.

One example of a solid reaction is the production of periclase crystals of considerable size at temperatures far below the softening point. This is the type of reaction where fine crystals are changed into large crystals of the same kind. Another example is the production of spinel from magnesium oxide and alumina. Here an entirely new crystal is formed, and this can be carried out at temperatures far below the point where any liquid could be formed in the system.

The problem of sintering, that is, causing a packed powder to change into a dense solid, is receiving much attention in the field of technical ceramics. As heavy refractories seldom are glass-free, solid-state sintering will not be covered here; however, much information may be found in the book "Kinetics of High-temperature Processes," W. D. Kingery, editor, John Wiley and Sons, Inc., New York, 1959.

12.8 Hot Pressing. *Principle.* If a powdered material is heated to a temperature somewhat below its melting point, there is a tendency for the powder to sinter together and gradually approach a nonporous mass. If glass is present, this sintering goes on over a considerable temperature range as with the ordinary refractory or whiteware body; but if there is no glass present, the sintering occurs in a narrower temperature range as discussed in the preceding section on solid reactions. It has been known for a long time that if external pressure is applied to the powder while it is being heated, this sintering is greatly accelerated and the particles coalesce into a solid mass at a very much more rapid rate and at a lower temperature level.

Sintering under pressure has certain obvious advantages, such as the production of low-porosity refractory bodies, which would be difficult to form in the usual firing operation. It is also possible to produce articles very accurately shaped, as there is little shrinkage or distortion during the operation. Another advantage is the fact that the operation can be carried out very rapidly, saving considerable time and labor. Again the raw material does not have to be carefully sized, and binders or lubricants are not needed.

On the other hand, there are certain disadvantages, as a temperature cannot generally be used above the point where the powder will react vigorously with the mold. Then again, at the high temperatures the life of the mold is short. Also, means must be provided for slowly cooling the finished piece to prevent cracking if it is a nonmetal.

Methods. Hot pressing has received much attention for forming dense refractories of various kinds, but most of the work has been of an experimental nature.   Boron carbide seems to be the only material hot-pressed in any quantity.   For the high temperature needed, carbon dies are used which have a rather short life.   The only way hot pressing could be used for heavy refractories is by an impact compacting of a hot slug in a cool metal mold, that is, coining.   While experiments have been carried out along these lines, production of heavy refractories has not developed.

12.9   Firing Properties of Refractory Bodies.   It is seldom desirable to produce a refractory directly from a raw plastic clay because of the large

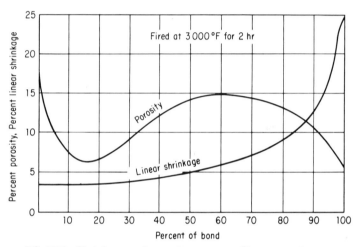

FIG. 12.7   Shrinkage and porosity curves.   Clay-grog mixtures.

shrinkage and distortion that would occur in firing.   A reduction in firing shrinkage is a very desirable accomplishment for the following reasons:

1. It permits the production of a fired refractory that will be close to the desired dimensions.

2. It prevents warping of the refractory on firing.

3. It permits a greater weight of finished bricks to be fired in a given kiln.

4. It makes it unnecessary to control the maximum temperature of firing as carefully as when large shrinkages occur.

Use of Grog.   One of the simplest methods of reducing the firing shrinkage of a refractory is by introducing a nonplastic material such as grog, an ancient expedient.   As this material has already been fired and has reached a stable volume, it acts somewhat as a skeleton to hold the brick together in a stable form during the firing operation.   Figure 12.7 shows a curve of firing shrinkage and porosity of a mixture of kaolin and grog burned to a temperature of 3000°F (about 1650°C) for 2 hr.   The grog was sized for dense packing, bonded with raw Georgia kaolin, and formed in a mold by hand.   It is interesting to note that the grog alone has a 3 percent linear shrinkage when fired at this temperature.   Careful microscopic studies of individual pieces of the same grog fired at the particular temperature showed no volume change and no rounding of the corners; but apparently when the grog is formed into a mass of particles in contact, the surface tension or pressure effects are such as to cause rounding of the sharp points and edges sufficient to produce welding and shrinkage.

It is also interesting to note that bond clay can be added up to approximately 25 percent without any change in the burning shrinkage.   As the volume of voids in this grog is approximately 25 percent of the bulk volume, it will be seen that the bond is simply acting as a filler in the pore spaces and that when it shrinks, it does not affect the volume of the whole piece, since this is governed by the grog.   However, when the amount of bond is increased beyond 25 percent, the total shrinkage begins to increase more and more rapidly until 100 percent of bond is reached.   This is due to the fact that thicker and thicker films of shrinking bond are built up between the grog grains.

The porosity, which is 17 percent for the all-grog body, falls, as should be expected, with the addition of bond clay to 6 percent when the bond reaches 20 percent.   With more bond, the porosity increases to a maximum of 15 percent and then falls back to 6 percent for all bond.

The close-packed moistened kaolin specimens pressed at 15,000 psi shown on p. 147 were fired at 1600°C (about 2900°F) and 1700°C (about 3100°F) with the results shown in Table 12.5.   At the higher temperature the total porosity was 5 percent and open pores as low as zero, with the modest linear shrinkage of 4 to 5 percent.   The closed pores are shown on polished sections as fine cracks due to bond shrinkage around the large

TABLE 12.5   Properties of Fired Specimens Pressed at 15,000 psi

Firing tempera- ture, °C	Composition	Linear shrink- age	Bulk density		Total porosity		Open pores fired
			Green	Fired	Green	Fired	
1600	Grog + 5% kaolin	1.8	2.20	2.31	19.4	14.1	12.2
	Grog + 10% kaolin	2.0	2.22	2.35	17.4	12.8	10.3
	Grog + 15% kaolin	2.6	2.24	2.39	16.5	11.1	9.6
	Grog + 20% kaolin	3.5	2.28	2.50	14.5	7.0	3.8
1700	Grog + 5% kaolin	5.0	2.20	2.55	18.2	5.2	0.7
	Grog + 10% kaolin	5.0	2.22	2.59	17.4	3.7	0.4
	Grog + 15% kaolin	4.3	2.24	2.54	16.5	5.5	0.0
	Grog + 20% kaolin	4.2	2.28	2.55	14.5	5.4	0.6

grog grains.   This investigation carried out on kaolin could undoubtedly
be repeated on fireclay.   Thus there is no doubt that nonporous brick
can be made by conventional methods.

Expanding Minerals.   The shrinkage of a refractory clay can also be
controlled by adding a mineral that expands on firing.   Such a mineral
may be kyanite, which, mixed with clay in the raw state or with clay and
grog, will give a very definite expansion.

To illustrate this, a few mixes were made by Lane[3] using the proportions given in Table 12.6.

TABLE 12.6   Kyanite-Clay Mixes Illustrated
in Fig. 12.8

Mix	Dense kaolin grog, %	Raw N. C. kyanite, %	Raw kaolin, %
A	85	0	15
B	81	4	15
C	77	8	15
D	73	12	15

Grog sizing		Kyanite sizing	
Screen	%	Screen	%
4–8	35	14–35	50
8–14	15	35–60	30
14–35	6	60–100	10
35–60	4	100	10
60–100	20		
100	20		

These mixes were molded in a dry press and fired to various temperatures at which they were held for 3 hr.   The shrinkage curves are shown in
Fig. 12.8.   It will be seen that the kyanite gives a real expansion over the
range from 2350 to 2600°F (about 1290 to 1425°C) but beyond this point
shrinks faster than the grog and clay mix—a disadvantage when it is
attempted to make a refractory to be volume constant at temperatures
higher than the original firing.   However, the 12 percent kyanite mixture
has the same volume at a firing temperature of 2975°F (about 1635°C) as
when molded.   Kyanite having too coarse grains ruptures and weakens
the structure, whereas too fine grains give little or no expansion.   Grains

of all one size break down over a very small temperature range, giving an undesirable peak in the expansion curve.

One of the chief advantages of a low-firing shrinkage is the ability to hold close size tolerances without extreme stabilization of the maturing temperature. However, it will be seen from Fig. 12.8 that the kyanite additions do not accomplish this, for the shrinkage-temperature curves are even steeper with kyanite than without it.

The minerals sillimanite and andalusite were also tried out in similar mixtures; and although they showed a small amount of expansion, they were not nearly so effective as the kyanite.

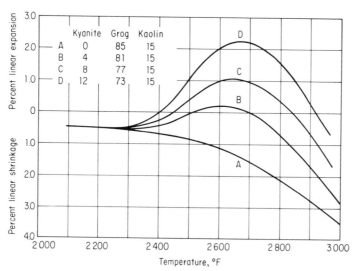

FIG. 12.8 Expansion caused by kyanite.

It is theoretically possible to add a high-density mineral like fused alumina to a mixture of grog and bond, which after long heat-treatment will dissolve in the glass and form mullite of lower density. Practically, however, this solution goes on with extreme slowness and seems difficult to carry out in a practical way, although such a method is suggested in a patent of Lambie and Ross.[2] The secondary expansion of high-alumina clays is probably caused by this reaction.

Bloating. Reduction of shrinkage can also take place if the clay bloats or forms bubbles in the glass owing to breaking up of sulfides, carbonates, etc. This increase in size by bloating may account for the expansion of some of the high-alumina clays and flint clays in firing, but a great deal more work needs to be done on this question before we shall know the exact mechanism causing it. Bloating, if it occurs to any great extent, is not particularly desirable, as it is hard to control and will not occur under any appreciable load.   Therefore, a shape being fired may show no change in horizontal dimensions but may shrink materially in a vertical direction.

12.10   Specific Heat. *Values.*   Actually the specific-heat values for the common refractory materials are closely alike.   Therefore, unless refined work is being carried out, all the fired materials will approximate those of Al_2O_3 given in Table 12.7.

TABLE 12.7   Specific Heat of Alumina

Temperature range, °C	*Mean specific heat*
30–100	0.206
30–300	0.226
30–500	0.240
30–700	0.250
30–900	0.258
30–1100	0.265
30–1300	0.271
30–1500	0.276
30–1700	0.280

12.11   Fusion-cast Refractories. *Composition.*   The whole subject of fusion-cast glass-tank refractories is discussed in an able manner in the book "Fused Cast Refractories," A. A. Litvakovskii, National Science Foundation, Washington, D.C., 1961, translated from the Russian. The author develops the steps taken to arrive at the present-day compositions.   Refractory fusions of alumina have been used since the beginning of the century for abrasive grain, but it was not until the year 1921 that Dr. Gordon Fulcher of the Corning Glass Works actually cast blocks for glass refractories.

The compositions generally used at first were alumina, chrome-alumina, and mullite; but a zirconia-containing block, particularly resistant to glass, was patented by Field.[4]   This composition, largely used in the melting and refining zones, consists of 32 to 40 percent ZrO_2, 45 to 50 percent Al_2O_3, and the remainder SiO_2, TiO_2, Fe_2O_3, and alkalies; 99 percent α-alumina blocks are used in the superstructure, and 90 percent alumina (partly α- and partly β-alumina) are used in the refining zones; 60 to 80 percent α-alumina plus 20 to 30 percent Cr_2O_3 are used in certain places, such as fiber-glass tanks.

Manufacture.   The manufacture of these blocks requires a process quite foreign to the conventional ceramist.   The flow sheet in Fig. 12.9 shows the steps in producing glass-tank blocks.   Figure 12.10 shows a block being poured from the electric furnace into a mold, where the pouring temperature for the zirconia composition is 1800°C (about 3270°F).   The mold is made from slabs of resin-bonded sand assembled to give the exact size and held together by packing in cans with a refractory or insulating

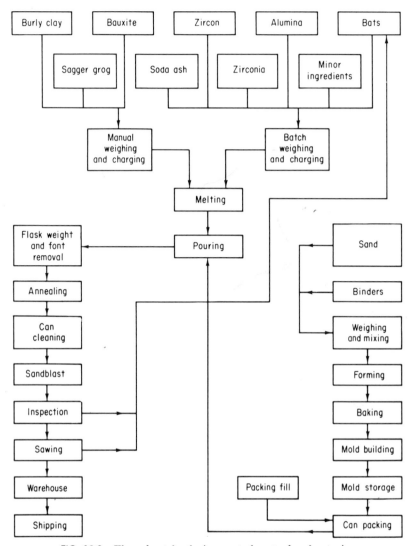

FIG. 12.9   Flow sheet for fusion-cast glass-tank refractories.

filler.   The cooling is slow, taking several days.   There is bound to be some shrinkage cavity, but it is planned to have this cavity come in a place where it does no harm (or in the portion of the block where the cavity may be cut off).   The blocks are then sandblasted, the sprue removed, and stored ready for shipment.   Figure 12.11 shows an installation of these blocks in the side wall of a glass tank.

Recently, fused-cast blocks of magnesia-chromite composition have found extensive use in the steel industry for areas of very severe service

FIG. 12.10   Pouring a tank block.   (*Corhart Refractories Co.*)

FIG. 12.11   Blocks installed in a glass tank.   (*Corhart Refractories Co.*)

in electric arc, steel melting furnaces, and parts of oxygen converters. The composition generally used is 58 percent MgO, 19 percent Cr_2O_3, 6 percent Al_2O_3, 3 percent SiO_2, and 13 percent Fe_2O_3. A flow sheet (Fig. 12.12) shows the production methods. The batch is melted and poured at 2500°C (about 4500°F), at which time basic grog is added to the

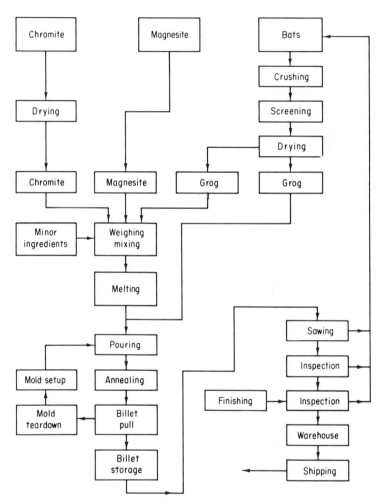

FIG. 12.12 Flow sheet for fusion-cast basic refractories.

stream to control the shrinkage. The molds are graphite slabs packed in steel cans. After the billet, which is 72 in. long, 6 in. thick, and 12 to 18 in. wide, has cooled, it is cut up with a diamond saw into precisely sized keys, wedges, or straights. At present, more dollar value goes into fusion-cast refractories for the steel industry than for the glass industry.

BIBLIOGRAPHY

1. Greig, J. W.: Formation of Mullite from Cyanite, Andalusite, and Sillimanite, *J. Am. Ceram. Soc.*, **8**, 465, 1925.
2. Lambie, J. M., and D. W. Ross: Refractory Mixture, U.S. Patent 1,769,297, 1930.
3. Lane, R. O.: S.M. Thesis, Massachusetts Institute of Technology, 1936.
4. Field, T. E.: Cast Refractory Product, U.S. Patent 2,271,366, 1939.
5. Norton, F. H.: Critical Study of the Differential Thermal Method for the Identification of the Clay Minerals, *J. Am. Ceram. Soc.*, **22**, 54, 1939.
6. Grim, R. E., and W. F. Bradley: Investigation of the Effect of Heat on the Clay Minerals Illite and Montmorillonite, *J. Am. Ceram. Soc.*, **23**, 242, 1940.
7. Nutting, P. G.: Some Standard Thermal Dehydration Curves of Minerals, *U.S. Geol. Survey Profess. Paper* 197-E, 1943.
8. Speil, S., et al.: Differential Thermal Analysis—Its Application to Clays and Other Aluminous Materials, *U.S. Bur. Mines Tech. Paper* 664, 1945.
9. Curtis, C. E., and H. G. Sowman: Investigation of the Thermal Dissociation, Reassociation, and Synthesis of Zircon, *J. Am. Ceram. Soc.*, **36**, 190, 1953.
10. Brindley, G. W.: X-ray Identification of Clay Minerals, *Mineral. Soc. London*, 1951.
11. Allison, A. G., et al.: Sintering of High-purity Magnesia, *J. Am. Ceram. Soc.*, **39**, 151, 1956.
12. Weber, B. C., et al.: Observations on the Stabilization of Zirconia, *J. Am. Ceram. Soc.*, **39**, 197, 1956.
13. Watson, A., J. O. May, and B. Butterworth: Studies of Pore Size Distribution. *Trans. Brit. Ceram. Soc.*, **56**, 37, 1957.
14. Vaughan, F.: Effect of Heat on Kandite-containing Clays, *Trans. Brit. Ceram. Soc.*, **57**, 39, 1958.
15. Stirland, D. J., A. G. Thomas, and N. C. Moore: Observations on Thermal Transformations in Alumina, *Trans. Brit. Ceram. Soc.*, **57**, 69, 1958.
16. Dodge, N. B.: Mineral Placement of the Constituents in Five Types of Basic Brick, *Bull. Am. Ceram. Soc.*, **37**, 139, 1958.
17. Chaklader, A. C. D., and A. L. Roberts: Relationships between Constitution and Properties of Silica Refractories, *Trans. Brit. Ceram. Soc.*, **57**, 115, 1958.
18. Hill, V. G., and R. Roy: On Tridymites, *Trans. Brit. Ceram. Soc.*, **57**, 496, 1958.
19. Chaklader, A. C. D., and A. L. Roberts: The Constitution of Silica Refractories before and after Heat Treatment, *Trans. Brit. Ceram. Soc.*, **57**, 115, 1958.
20. Eaton, N. F., T. J. G. Glinn, and R. Higgins: Effect of Thermal Treatment on the Properties of Silica Refractories, *Trans. Brit. Ceram. Soc.*, **58**, 92, 1959.
21. Kriek, H. J. S., W. F. Ford, and J. White: The Effect of Additions on the Sintering and Dead-burning of Magnesia, *Trans. Brit. Ceram. Soc.*, **58**, 1, 1959.
22. Houseman, D. H., and J. White: Development of Bond Strength during Firing—A New Approach and Its Technical Implications, *Trans. Brit. Ceram. Soc.*, **58**, 231, 1959.
23. Brindley, G. W., and M. Nakahira: The Kaolin-Mullite Reaction Series, I, II, and III, *J. Am. Ceram. Soc.*, **42**, 311, 1959.
24. Comer, J. J.: Electron Microscope Studies of Mullite Development in Fired Kaolinites, *Bull. Am. Ceram. Soc.*, **43**, 378, 1960.
25. Hill, R. D.: A Study of Pore-size Distribution of Fired Clay Bodies, *Trans. Brit. Ceram. Soc.*, **59**, 189, 1960.
26. Richardson, H. M., K. Fitchett, and M. Lester: Bond Structure and the Behaviour of Basic Bricks at High Temperatures, *Trans. Brit. Ceram. Soc.*, **59**, 483, 1960.
27. Cooper, A. C., D. A. R. Kay, and J. Taylor: The Free Energy of Formation of Mullite, *Trans. Brit. Ceram. Soc.*, **60**, 124, 1961.

28. Konopicky, K., L. Patzak, and K. Wohlleben: The Glass Content of Silica Bricks, *Ber. Deut. Keram. Ges.*, **38**, 403, 1961.
29. Chaklader, A. C. D., and A. L. Roberts: Transformation of Quartz to Cristobalite, *J. Am. Ceram. Soc.*, **44**, 35, 1961.
30. Tyrrell, M. E.: Effect of Impurities on Sintered Mullite, *U.S. Bur. Mines Rept. Invest.* 5957, 1962.
31 Konopicky, K., and C. Zaminer: The Mineralogical Structure of Silica Bricks, *Ber. Deut. Keram. Ges.*, **39**, 44, 1962.
32. Hawkes, W. H.: The Production of Synthetic Mullite, *Trans. Brit. Ceram. Soc.*, **61**, 689, 1962.
33. Richardson, H. M., and M. Lester: Laboratory Experiments on High-alumina Refractories, *Trans. Brit. Ceram. Soc.*, **61**, 773, 1962.
34. Ďurovič, S.: Isomorphism between Sillimanite and Mullite, *J. Am. Ceram. Soc.*, **45**, 157, 1962.
35. Prasada Varma, C. N. S., and A. L. Roberts: Effect of Heat on Natural Magnesites, *Trans. Brit. Ceram. Soc.*, **61**, 212, 1962.
36. Konopicky, K., W. Lohre, and G. Routschka: The Problem of Synthetic Mullite, *Ber. Deut. Keram. Ges.*, **40**, 337, 1963.
37. Van Haeften, A. W.: Sintered Magnesite, *Sprechsaal*, **96**, 343, 1963.
38. Van Dreser, M. L., and W. H. Boyer: High-temperature Firing of Basic Refractories, *J. Am. Ceram. Soc.*, **46**, 257, 1963.
39. Rigby, G. R., R. F. Hutton, and B. G. Hamilton: Reactions Occurring in Basic Brick, *J. Am. Ceram. Soc.*, **46**, 332, 1963.
40. Hamano, K., H. Yamada, and T. Ohba: Microstructure of Stabilized Zirconia and Some Considerations about Partial Stabilization of Zirconia, *J. Ceram. Assoc. Japan*, **71**, 224, 1963.
41. Sosman, R. B.: Phases of Silica, *Bull. Am. Ceram. Soc.*, **43**, 213, 1964.
42. Löcsei, B. P.: Topaz and Mullite Formation at Low Temperatures in the Kaolinite-Aluminum Fluoride System, *Keram. Z.*, **16**, 350, 1964.
43. Löcsei, B. P.: The Influence of Small Amounts of Aluminum Fluoride on the Properties of Ceramic Products, *Epitoanyag*, **16**, 363, 1964.
44. Treffner, W. S.: Microstructure of Periclase, *J. Am. Ceram. Soc.*, **47**, 401, 1964.
45. Davies, B., and F. H. Walther: Direct Bonding of Basic Brick, *J. Am. Ceram. Soc.*, **47**, 116, 1964.
46. El-Shahat, R. M., and J. White: The Systems $MgAl_2O_4$–$MgCr_2O_4$–Ca_2SiO_4 and $MgFe_2O_4$–$MgCr_2O_4$–Ca_2SiO_4, *Trans. Brit. Ceram. Soc.*, **63**, 313, 1964.
47. Hayhurst, A., and J. Laming: Modifications of Microstructure of Chrome-Magnesite Refractories by Heat Treatment, *Trans. Brit. Ceram. Soc.*, **63**, 135, 1964.
48. Scheerer, P. E., H. M. Mikami, and J. A. Tauber: Microstructure of Chromite-Periclase at 1650–2310°C., *J. Am. Ceram. Soc.*, **47**, 297, 1964.
49. Sosman, R. B.: "The Phases of Silica," Rutgers University Press, New Brunswick, N.J., 1965.
50. McGee, T. D.: Constitution of Fireclays at High Temperatures, *J. Am. Ceram. Soc.*, **49**, 83, 1966.
51. Seil, G. E.: Petrography and Heat Treatment of Chromite Refractories, reprinted for E. J. Lavino and Company from *Proc. Ceram. Conf. Bull.* 14, Pennsylvania State College.

CHAPTER THIRTEEN

Kilns for
Burning Refractories

FIG. 13.1 The inside of a Korean kiln running up a hillside.

13.1 Introduction. We know that kilns have been used by man for many thousands of years, as their remains are frequently found in the excavations of ancient-sites of early civilization. These primitive kilns were of the updraft type and, in most cases, consisted only of a firing chamber, a perforated floor, and a wall. The construction of a permanent kiln roof was too much of a structural problem for most of the early potters, and recourse was made to a temporary roof of green poles and raw clay, which would hold in place long enough for the very low temperature of firing used at that time. There is, however, one exception, as a kiln has been found in Tepe Gawra that was believed to have had a permanent roof at a period as early as 4500 to 4000 B.C.

The early Chinese and Korean kilns, which consisted of a series of chambers extending up the side of a hill, were remarkably similar to our modern chamber kilns and permitted a considerable degree of recuperation as the fire advanced from one chamber to another. An interesting kiln of this type is shown in Fig. 13.1.

It should be remembered that in dealing with these early kilns as well as our modern ones, four essentials must always be present: (1) a means for producing the heat, (2) a support for the ware, (3) a container such as the walls and crown to confine the heat in the working space, and (4) a means to transfer the heat from the source to the ware.

13.2 General Principles. *Production of Heat.* As some of the refractories are still fired with coal as a fuel, this method of heating will be discussed first. Coal is used because of its low cost and because most of

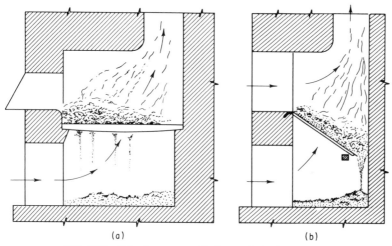

(a)   (b)

FIG. 13.2 Kiln furnaces with flat and sloping grates.

the refractories are not particularly sensitive to impurities in the kiln gases that might arise from using coal as a fuel. The coal is generally burned in furnaces with sloping or flat grates as shown in Fig. 13.2. The air for combustion passes partly through the fuel bed and partly over the top, the hot gases generally going upward into the kiln over a bag wall, as shown in Fig. 13.3. In the round or rectangular kilns, a considerable number of furnaces are spaced around the wall of the kiln in order to distribute the heat as evenly as possible.

The manufacturers of refractories generally prefer for their kilns a good grade of bituminous coal giving a long flame; also a low ash content and a high fusion point of the ash is desirable. In most cases, it is necessary to have a low sulfur content so that sulfates will not deposit on the surface of the bricks during the water smoking period.

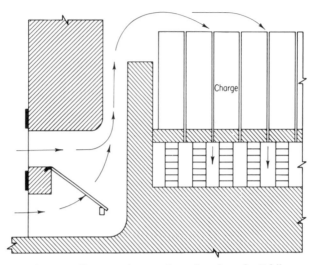

FIG. 13.3   Passage of gases through a downdraft kiln.

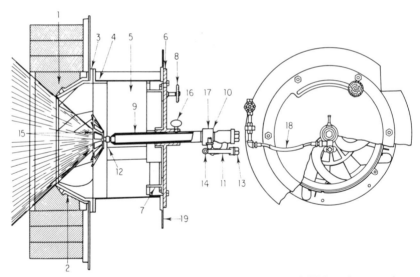

FIG. 13.4   A mechanical atomizing burner.   (*Babcock & Wilcox Company.*)

It was formerly the practice to fire the coal by hand, although now mechanical stokers are being used to a considerable extent.

Many of the coal-fired kilns waste a good deal of heat as a result of radiation from the more or less open fire mouths and escaping gases from the fire bed that are not pulled into the kiln by the draft.   These conditions have been improved in some of the modern installations by tighter doors and forced draft.

Oil has many advantages as a fuel for firing refractories; and in some localities, the cost compares favorably with coal. Advantages of oil are a more perfect temperature control, the possibility of a higher maximum temperature, a generally cleaner kiln atmosphere, and a smaller labor requirement for the firing operation. Oil is, of course, obtainable in a number of different grades; but from the cost standpoint, the heavier grades of oil, such as Bunker C, are generally used. A number of different types of burner are available for firing with oil. In principle, they all break the oil up into a spray of fine particles and intimately mix them with

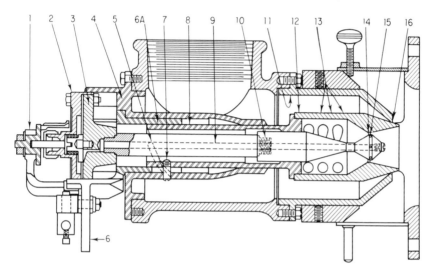

FIG. 13.5 An air-atomizing oil burner. (1) Oil-control valve lever. (2) Oil-control valve cam. (3) Oil-control valve consisting of a V-groove in a flat surface. (4) Burner backplate. (5) Curved slot. (6) Operating lever. (7) Inner-nozzle operating pin. (8) Inner air-nozzle operating tube. (9) Oil tube. (10) Oil nozzle. (11) Outer air nozzle. (12) Inner air nozzle. (13) Primary air-supply openings. (14) Primary air discharge. (15) Oil nozzle. (16) Secondary air-discharge opening. (*Hauck Manufacturing Company.*)

the combustion air. In burners of the mechanical atomizing type, as shown in Fig. 13.4, the oil is atomized by forcing it through a fine orifice at relatively high pressure. This method is particularly suitable for burners of high capacity, as the orifice becomes very tiny for small burners and there is apt to be danger of particles getting through the filter and plugging the orifice.

Oil is also fired in air- or steam-atomizing burners where high-pressure air or steam is used to inject the oil into the kiln and at the same time break it up into a fine mist. Secondary air is generally supplied from a low-pressure source. A burner of this type is shown in Fig. 13.5. In setting up an oil-burning system, care should be taken that it be properly

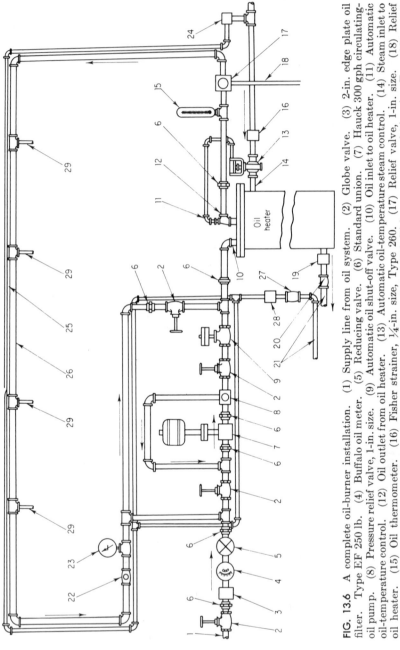

FIG. 13.6 A complete oil-burner installation. (1) Supply line from oil system. (2) Globe valve. (3) 2-in. edge plate oil filter. Type EF 250 lb. (4) Buffalo oil meter. (5) Reducing valve. (6) Standard union. (7) Hauck 300 gph circulating-oil pump. (8) Pressure relief valve, 1-in. size. (9) Automatic oil shut-off valve. (10) Oil inlet to oil heater. (11) Automatic oil-temperature control. (12) Oil outlet from oil heater. (13) Automatic oil-temperature steam control. (14) Steam inlet to oil heater. (15) Oil thermometer. (16) Fisher strainer, ¼-in. size, Type 260. (17) Relief valve, 1-in. size. (18) Relief oil line to sump or barrel. (19) Fisher strainer, ⅜-in. size, Type 260. (20) Fisher steam trap, ⅜-in. size, Type ST. (21) Condensate discharge to sump. (22) Check valve. (23) Oil pressure gage. (24) Steam-pressure reducing valve, 50 to 5 lb. (25) Steam-tracer line, ½-in. iron pipe or copper tubing. (26) Loop oil line to burners. (27) Steam trap, ½-in. size, Type ST. (28) Fisher strainer, ½-in. size, Type 260. (29) Leads to burners. (*Hauck Manufacturing Company.*)

laid out to ensure reliable operation. The heavier oils have to be heated and must be kept hot until they actually reach the burner; otherwise the viscosity will become so great that proper atomization is impossible. A good filter that will permit cleaning without shutting down the oil flow is necessary, and often a complete recirculating system is desirable. In

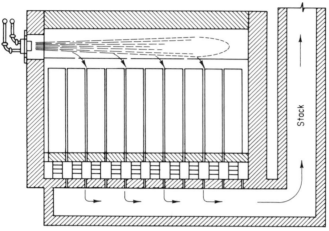

FIG. 13.7 An oil-fired kiln in which the flame passes above the charge.

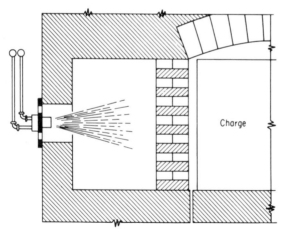

FIG. 13.8 A kiln in which the oil flame passes through a checkerwork.

Fig. 13.6 is shown a layout for an oil-fired kiln that has been found quite satisfactory.

In regard to the arrangement of the burners in the kiln, they may fire over the charge in a downdraft kiln, as shown in Fig. 13.7, or they may fire against the side of the charge, preferably through a checker brick to distribute the heat, as shown in Fig. 13.8. In setting up the burner, care must be taken to allow sufficient combustion space for the particular burner used; otherwise carbon will tend to build up on the surfaces around the flame. The recommendations of the burner manufacturers should be followed when the combustion chamber is laid out.

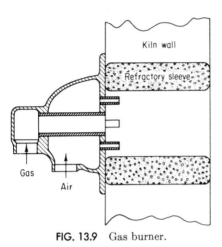

Natural gas has become more generally available since the Second World War. For this reason, many kilns firing refractories now use this fuel, often with standby supplies of propane[7] for periods when the gas supply is inadequate. In other cases kilns are equipped with oil burners for use interchangeably with gas burners.

Natural gas is an ideal fuel, as it produces a clean, easily controlled heat. There are many types of natural-gas burners available for the kiln. In the older installations, only the primary air was mixed with the gas, while secondary air was drawn in around the burner. This method did not allow a satisfactory control of temperature and atmosphere. Today, all the air is introduced into the burner, giving excellent combustion control. A typical gas

FIG. 13.9   Gas burner.

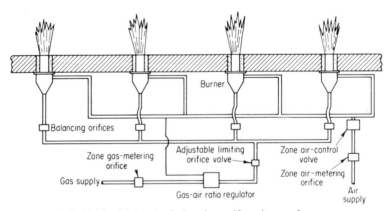

FIG. 13.10   Piping for balancing orifices for gas burners.

burner[2] is shown in Fig. 13.9 and a piping scheme for balanced orifices in Fig. 13.10.

Producer gas is used to some extent in firing kilns, especially in Europe. In most cases, the gas is used hot and without scrubbing; consequently provision must be made for draining and cleaning the connecting pipes between the producer and the burner, as they periodically fill up with tar and soot. In regions where only low-grade coal is available, the producer-gas firing is found to be quite successful. The producer gas gives a comparatively low-temperature flame, so there is little danger of overheating any part of the charge; and as the volume of gas is large because of this low heat value, it tends to give an even temperature throughout the kiln. The small-unit gas producer has been steadily improved and is a much more reliable and efficient piece of equipment than it was in the past.

There is not space here to go into the principles of combustion, but it should be remembered that the amount of excess air used is of the greatest importance in all kiln firing. The greatest combustion efficiency is reached with very low amounts of excess air; but on the other hand, a large volume flow of gases is needed to carry out the moisture in the water-smoking stage of firing, to provide uniformity of temperature, and to supply enough oxygen to burn away carbonaceous matter.

There is a considerable saving of fuel when preheated air is used. Preheat is seldom used, however, in periodic kilns. In the case of chamber and tunnel kilns, preheat temperatures as high as 2000°F (about 1100°C) are reached with a real saving in fuel. Preheating is used in some cases to attain higher temperatures than would be possible with cold air. The reduced price of commercial oxygen in the last few years also makes possible its addition to the combustion air to aid in reaching high temperatures.

Transfer of Heat. One of the most difficult problems in kiln design is to transfer the heat efficiently from its source to the ware. The problem is unusually difficult in the kiln because the ware is in small units stacked into a considerable volume and therefore radiant-heat transfer, so important in furnaces such as those used in the metallurgical industry, can play little part. For the same reason, conduction of heat into a bulky charge of small units is of little consequence. Therefore, we must rely almost entirely upon convection for heating of the charge.

The transfer of heat from a moving stream of gas to a solid surface is well understood. The transfer rate depends mainly on the temperature difference between the gas and solid and the velocity with which the gas is passing by the solid. The curves in Fig. 13.11 give an idea of the rate of heat transfer under different conditions between a gas and a solid. The transfer of heat from the moving gas by radiation from the gas itself and the incandescent particles that it contains are quite important in furnaces of large volume and great flame thickness, such as would be found in the glass-tank or open-hearth furnace; however, in the kiln, where the gas stream is divided into relatively thin layers, this radiation cannot play a very important part, although attempts are made in practice to use long-flame coal to bring the heat down to the bottom of the charge of a downdraft kiln.

Equalization of Temperature. One of the most important problems in kiln design is to produce a uniform temperature throughout the volume of

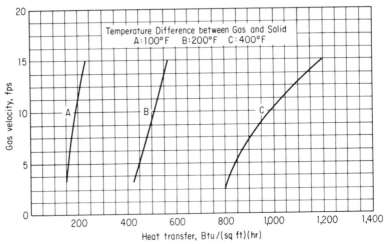

FIG. 13.11   Rate of heat transfer between a gas and a solid.

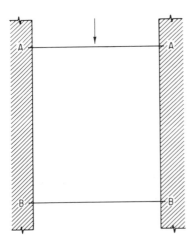

FIG. 13.12   The effect of flow velocity on temperature distribution.

the charge. This is by no means an easy matter, as few large periodic kilns now operating show a difference of less than 50°F from top to bottom of the charge. In tunnel kilns, with their small cross section, the temperature uniformity is better, but here again it is apt to vary from place to place in the car by an appreciable amount.

Considering first the case of a downdraft kiln under equilibrium conditions, it may be readily shown that the temperature difference between top and bottom is related to the mass flow of gases. In Fig. 13.12, the section AA may represent the top of the charge and BB the bottom. If M equals the weight of gases flowing through the channel in a unit time and S the heat lost between AA and BB by transfer laterally through the walls per unit time, then

$$S = (t_1 - t_2)c_p M$$
$$t_1 - t_2 = \frac{S}{c_p M}$$

where c_p = the specific heat of the gas

t_1 and t_2 = the temperatures of AA and BB, respectively

We may therefore conclude that a more even temperature is obtained by a large volume flow, which may be realized by the addition of excess air to the products of combustion. This equation also shows that insulation of the kiln side walls will give a more even vertical temperature distribution.

Another method of equalizing the temperature and one that is theoretically very attractive is the recirculation of the combustion gases through the charge. In this way, a large volume flow is obtained, thus decreasing the temperature difference; while at the same time, the volume of gases leaving the kiln is relatively small. The only difficulty with the recirculation method is the practical problem of handling the high-temperature gases. At the lower temperatures encountered in heat-treating furnaces, gases can be recirculated with water-cooled, alloy fans at temperatures to 2000°F (about 1100°C); but above this, the only possible means is by injector action, a method quite successful in driers but one that has not been developed for kilns at the present time. Recently, recirculation has been used successfully in some tunnel kilns at the preheating end.

The equalization of temperature at right angles to the path of flow is readily accomplished by adjustments in the floor openings, so that the hotter areas have the openings reduced and the cooler areas have them enlarged. By trial and error, a very uniform temperature can be attained over the horizontal cross section of the charge.

One often hears the kiln fireman state that he uses a damper or a close-set floor to back up the heat in the kiln and give a more uniform temperature. It is not generally understood, however, just what the effect of dampers and restrictions to the flow accomplish. The curves in Fig. 13.11 show clearly the great increase in heat transfer from the gas to the charge with increasing velocities. The point in the kiln where the flow is restricted has an increased velocity, and thereby more heat is transferred from the gas to the charge. As the usual problem is to raise the temperature at the bottom of the charge, it is possible to have a relatively open setting in the charge itself and as small floor openings as are permitted by the available draft. Owing to the high velocity through the floor openings, the heat from the gas will be given up to a much greater extent at the floor than in the other parts of the charge and will therefore tend to compensate for the gradual cooling of the gas as it goes down through the charge.

Temperature uniformity can be attained in a muffle kiln by varying the area of contact of the gas with the muffle wall from the burner to the exit point; i.e., at the burner, the passage for the gas would only have a small width in contact with the muffle, whereas in the exit, it would have a large width. By properly adjusting these areas, a muffle kiln with absolute uniformity of temperature can be constructed. Laboratory kilns that have been made in this way, though bottom-fired, show a temperature at the top of the muffle within a few degrees of the bottom temperature.

Insulation of the kiln tends to increase the uniformity of inside temperature, because (1) it tends to keep the inside walls of the kiln at a more uniform temperature, and because (2) an insulated kiln requires less fuel and therefore the gases entering the kiln can be at a lower temperature. Also, as explained from the diagram in Fig. 13.12, the insulation of the walls of the kiln increases the uniformity of temperature by minimizing the heat loss as the gases pass through the charge.

Kiln Efficiency. The efficiency of the kiln is usually defined as the ratio of the heat required to bring up the ware to its maximum temperature divided by the amount of heat supplied by the fuel. This definition is perfectly logical when applied to the periodic kiln; but when it is used under conditions where recuperation occurs, then it may become absurd, as efficiencies of over 100 percent have been determined on actual kilns. This can perhaps be made clear by the set of heat-balance diagrams shown in Fig. 13.13 beginning with a nonrecuperative kiln and ending with one having a high recuperative effect. It will be seen that the heat required to bring the ware up to temperature is the same in each case but the external heat supplied to the kiln becomes progressively smaller as the recuperative action increases. It would seem more reasonable to give fuel consumption per thousand standard-sized bricks rather than to express efficiency in percentage.

It should be stated here that a high degree of kiln efficiency is often incompatible with good uniformity of temperature because high efficiency demands a low temperature of exit gases and these, in turn, mean a low temperature of the charge in the bottom of the kiln. Only by recuperative action or recirculation can good temperature uniformity be obtained together with high efficiency.

It is obvious that the insulation of kilns will increase their efficiency, but unfortunately very few data are at hand to show just what the saving might be, as there is seldom an opportunity to compare similar kilns, one of which is insulated and the other not. It should be brought out, however, that the effect of insulation is not only to reduce the heat passing through the walls and the crown of the kiln but also to reduce the volume of exit gases in proportion. This fact is little understood, and it may be well to emphasize it with an example, under steady temperature conditions, in Table 13.1. This follows because, for a constant exit temperature, the heat lost in the flue gases is proportional to the heat input.

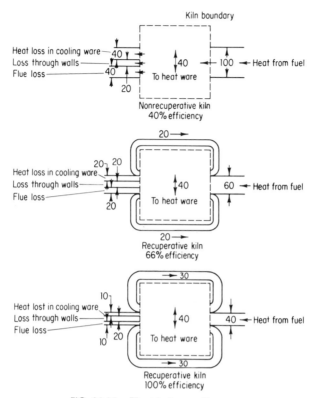

FIG. 13.13 Heat-balance diagrams.

TABLE 13.1 Saving Due to Insulation[4]

	Uninsulated kiln	Insulated kiln	
		As usually computed	As it occurs
Heat passing through walls, Btu.........	50,000	25,000	25,000
Heat lost in flue gases, Btu..............	50,000	50,000	25,000
Heat supplied to the kiln, Btu...........	100,000	75,000	50,000
Heat saved, %.......................	0	25	50

13.3 Periodic Kilns. The periodic kiln has been much used in the refractories industry. Its advantages are a comparatively low first cost and flexibility in operation. The disadvantages are the poor fuel economy secured, the unevenness of temperature when the charge is piled high, and the strain on the brickwork due to the repeated heating and cooling. Although the periodic kiln is being gradually displaced by the tunnel kiln,

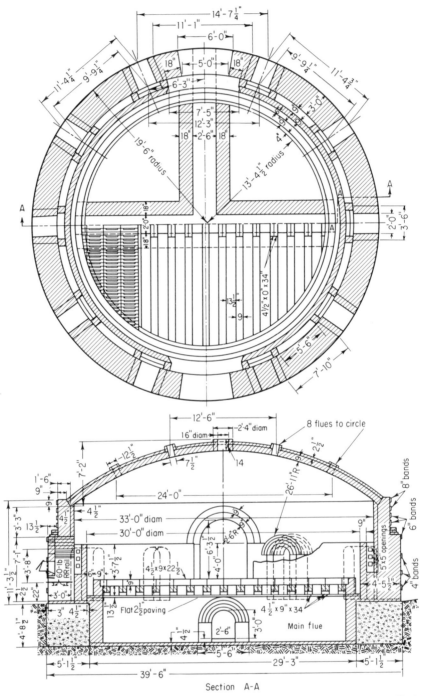

FIG. 13.14 A round downdraft kiln for firing refractories. (*J. Am. Ceram. Soc.*)

it will always be used to some extent for burning large shapes and special bricks.

Common Types of Kilns. The periodic kiln generally used to fire refractories is either of the circular downdraft type or of the rectangular type. The former is usually preferred because of the low cost for a given capacity and the even temperature distribution possible. The round kilns are made in sizes of 26 to 42 ft inside diameter, but sizes from 30 to 36 ft are generally preferred. The capacity ranges from 30,000 to 140,000 nine-inch equivalents. The fireboxes are arranged around the kiln and vary in number from 8 to 18 according to the size and type. The construction of the walls and crown is practically the same in all cases, but there is a great diversity of arrangements for the floor and flues.

In Fig. 13.14 is shown a well-designed, round, downdraft kiln for burning refractories. It will be noted that the flues are arranged in such a way as to give an even draft at all parts of the floor.

Although there is considerable variation in the dimensions of this type of kiln, the figures in Table 13.2 are generally used. It is generally considered good practice to connect two or four kilns to a single stack having separate flues for each kiln. This method gives a hot stack under all conditions, so that a good draft is produced when starting a kiln. The height of stack generally varies from 30 to 40 ft.

TABLE 13.2   Kiln Areas

Floor to top of crown, ft.........................	11–19
Height of bag wall above floor, ft..............	3–5
Area of floor openings, % of floor area..........	2.8–10
Area of grates, % of floor area.................	8–12.6
Area of main flue, % of floor area..............	1.4–3
Area of stack, % of floor area.................	1–1.6

Usually the turnover time in this type of kiln, burning clay refractories, is approximately 14 days: 4½ days for firing, 5½ days for cooling, 2 days for drawing, and 2 days for setting. When burning silica brick, the kiln is held at the maximum temperature for some length of time, so the turnover time is somewhat longer.

The modern tendency seems to be toward the use of blowers and suction fans to accelerate the burning and cooling. In this way, as much as 50 percent can be taken off the burning and cooling time, which means a marked increase in the output of the kiln. There is, however, a limit to the rate of firing, caused by the burning properties of the clay used. On the other hand, the cooling rate is limited because of the strains set up in the kiln itself by a too rapid change in temperature.

There are a number of other types of periodic kilns for burning refractories, but they are not so generally used. Descriptions of them can be found in the references given at the end of this chapter.

Fuel Consumption. The fuel consumption of periodic kilns is comparatively large and varies with the burning temperature and the amount of insulation used on the kiln. In Table 13.3 are given some figures on the fuel used in a number of periodic kilns.

TABLE 13.3   Fuel Consumption of Periodic Kilns

Type of kiln	Capacity, 9-in. equivalents	Fuel	Maximum temperature °C	°F	Firing time, hr	Fuel per 1,000 bricks (9-in. equivalents)
Round downdraft......	65,000	Coal	1270	2318	192	2,200 lb
	32,000	Coal	1290	2350	150	1,450 lb
	60,000	Coal	1320	2408	156	1,200 lb
	48,000	Coal	1170	2138	132	1,500 lb
		Coal		2600		3,000 lb
		Coal	1150	2100		2,100 lb
		Coal	1250	2280		2,500 lb
		Coal	1316	2400		2,400 lb
Rectangular downdraft .	42,000	Coal	1280	2336	163	1,770 lb
	60,000	Coal	1270	2318	168	1,800 lb
Round downdraft......	76,000	Oil	1270	2318	110	125 gal
		Oil	1400	2550		115 gal
Rectangular (insulated) downdraft	10,000	Oil	1650	3000	115	400 gal
	5,000	Oil	1650	3000	100	500 gal
Round downdraft......	50,000	Natural gas	1250	2280	216	18,000 cu ft

The heat losses in periodic kilns are mainly in the flue gases, as shown in Table 13.4, which gives average figures for periodic kilns burning fireclay bricks with coal.

TABLE 13.4   Heat Balance of a Typical Coal-fired Periodic Kiln

	%
Heat to raise temperature of bricks, or efficiency..............	15–45
Heat to drive off moisture.................................	0–3
Heat in combustion of ash.................................	4–10
Heat in combustion in flue gases...........................	0–5
Heat in dry flue gases.....................................	20–45
Heat in water vapor in gases..............................	4–7
Heat to raise temperature of kiln walls.....................	4–15
Heat to raise temperature of kiln crown....................	2–4
Heat lost by walls and crown.............................	3–9
Heat lost in bottom and flues.............................	4–12

Setting Methods. The method of setting the brick in the periodic kiln depends on their shape, the type of refractory, and individual plant practice. The setting is usually in a checkerwork pattern, with about ½ in. between bricks, making up a bench extending across the kiln. For straight bricks, this means three set across three others. To prevent the bricks from sticking to one another, silica sand is sprinkled on the top of the bench before placing the next pieces. Brick such as magnesite and chrome cannot support themselves in a high setting and therefore are commonly boxed with silica brick, which carry the load.

FIG. 13.15 A shuttle kiln for rapid firing. (*Denver Fire Clay Co.*)

Shuttle Kilns. These kilns are used for shapes and large pieces. The ware is set on cars similar to those of the tunnel kiln. and in fact the cars may be interchangeable. The kilns[3] are often made to take two rows of cars with burners in the lower side walls and the exhaust flues in the center between the cars. A modern shuttle kiln is shown in Fig. 13.15. These kilns have an advantage over periodic kilns in ease of setting and drawing the ware.

Car-bell Kilns. For special refractories this type of kiln has come into extensive use because of the ease of setting and drawing and the rapid firing cycle. There is often a considerable saving in fuel and labor over the older type of periodic kiln. They are used for small shapes, such as

FIG. 13.16   Car-bell kiln for firing refractories.   (*Bickley Furnaces, Inc.*)

FIG. 13.17   A group of car-bell kilns for high-temperature firing of refractories. (*Bickley Furnaces, Inc.*)

glass-tank feeder parts, and often are fired above 1700°C (about 3100°F). Figures 13.16 and 13.17 show typical installations in a refractory plant. In one type the charge is lifted up into the kiln chamber, while in the other the kiln itself is lifted off the loaded car.

13.4 Continuous Compartment Kilns. The compartment kiln is designed to provide more economical operation than can be given by the periodic kiln. A great number of methods are used, but the general principle is the same in all cases. In Fig. 13.18 is shown a portion of a typical compartment kiln. Each chamber is similar to a downdraft periodic kiln; but instead of the gases passing out of the stack, they pass from one chamber to the bag wall of the next, so that the heat from the combustion gases is used to preheat the bricks in the chambers ahead of the fire. In the same way, the heat from the cooling bricks is used to

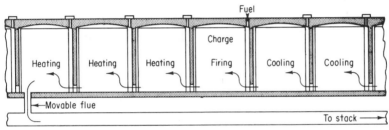

FIG. 13.18 Longitudinal section of a compartment kiln.

preheat the air for combustion. In some cases, this combustion air is drawn off by blowers and passed through the grates or burners; in other cases, the fuel, usually in the form of fine coal, is introduced through ports directly into the chambers under fire.

This type of kiln has not been used at all for refractories in this country, but abroad, where economy of fuel is more important, the compartment kiln has been extensively employed. The main trouble with the older type of chamber kiln was the slow rate of fire travel, which amounted to only about 6 in. per hr. This was due mainly to the considerable resistance to the flow of gases through the long passages. In the more modern chamber kilns, blowers and suction fans are being used to increase the gas velocity, which gives a fire travel as high as 3 to 5 ft per hr, with an immense increase in the capacity of a given kiln.

In a well-designed chamber kiln, practically the only heat loss is by transfer through the walls and crown, for nearly all the heat is abstracted from the cooling bricks and from the combustion gases. Although it is difficult to get strictly comparable figures, it may be said that the fuel consumption runs between one-half and one-third of that of a periodic kiln as shown in Table 13.5. This, of course, is an important saving.

TABLE 13.5   Fuel Consumption of Chamber Kilns

Ware	Fuel	Maximum temperature °F	°C	Fuel per M standard bricks, lb
Fireclay.............	Coal	2450	1340	750
	Coal	2325	1273	1,000
	Coal	2570	1410	1,100
	Producer gas	2590	1420	1,500 (coal)
Silica..............	Coal	2640	1450	900
Magnesite...........	Coal	2680	1470	1,500
	Coal	3000	1650	3,500
	Brown coal	2730	1500	4,000

However, the initial cost for chamber kilns is somewhat higher than for periodic kilns of the same capacity; and in general, the firing of a chamber kiln requires more skill to produce an evenly burned product.   In regard to the upkeep of the chamber kiln, different operators hold rather conflicting views, but some believe it to be considerably higher than for a periodic kiln.

The capacity of chamber kilns depends entirely upon the size and number of chambers, as well as rate of fire travel.   Kilns with an output of over 100,000 bricks per day have been constructed.

13.5   Tunnel Kilns.   The use of tunnel kilns in burning refractories is extending rapidly, and it may be said that practically all the modern refractory plant units are built around a tunnel kiln.   As the design has been improved and experience gained in the operation of tunnel kilns, it is generally recognized that they will produce better bricks at a lower cost than the periodic-kiln unit.   This does not mean that the periodic kiln will not be used, for it still offers advantages in burning large or intricate shapes, where the burning schedule must be different from that of the standard brick.   Because of the great importance of the tunnel kiln, some space will be devoted to a consideration of the principles of operation.

In principle, the tunnel kiln consists of an elongated chamber, which is maintained at a steady temperature, graded in the desired manner, from end to end.   The charge, mounted on cars, is moved continuously through the chamber and encounters in its passage the desired variations in heat.   The economy of the tunnel kiln is obtained by regaining the heat from the combustion gases to heat the incoming charge and by using the heat from the cooling bricks to preheat the combustion air or, in some cases, to dry the bricks.

The advantages of the tunnel kiln are as follows:

1. The tunnel kiln lends itself well to a continuous production process, which minimizes handling.

2. The setting and drawing of the kiln are simple and regular; and with dry-pressed brick, setting can be done directly from the press, eliminating the shods, drier cars, etc.

3. The kiln structure itself, with the exception of the cars, is always at a uniform temperature, so with proper design, upkeep of the refractory of the kiln itself is very low.

4. It is possible by proper design to heat and cool the ware according to any desired schedule, which enables the brick to be burned properly in the shortest possible time.

5. Owing to the relatively small cross section of the charge, the heat is able to penetrate to the center rapidly, thus allowing much more rapid burning than is possible in the bulky charge of the periodic kiln. This rapid burning is an advantage when special orders must be put through rapidly.

6. The tunnel kiln, when properly operated, shows marked economy of fuel in comparison with the periodic kiln.

There are a number of disadvantages to the tunnel kiln, but many of these are gradually being eliminated as more experience is gained with this method of burning.

1. The construction of a tunnel kiln necessitates a considerable capital outlay, as it is impossible to increase the capacity gradually as can be done with a number of periodic kilns. However, the cost per thousand bricks produced is not necessarily higher than that with the periodic kiln.

2. A shutdown on a tunnel kiln is serious, as it will tie up the whole unit for perhaps two to four weeks. With proper operation, a shutdown is very rare, but it does happen occasionally.

3. The upkeep on the car tops is high, because of the fact that they are periodically heated and cooled during each passage through the kiln. Considerable study is being given the problem, and it seems probable that this expense can be greatly reduced by careful design.

4. Some difficulty is experienced in getting a uniform temperature across the section of the tunnel kiln. The top and sides of the charge are often hotter than the bottom. This can be corrected, however, for some tunnel kilns now running give a very uniform temperature distribution.

5. The tunnel kiln requires a considerable length of floor space, and some plants are so situated that this is not readily available.

6. Another objection to the tunnel kiln is that for most efficient operation, it must be run with a uniform charge, i.e., with straight bricks or 9-in. shapes. It is, of course, possible to vary the burning schedule

to accommodate larger shapes or different types of clay, but these changes have a bad effect on the structure of the kiln, and some time is lost in getting the kiln running uniformly after such a change is made.

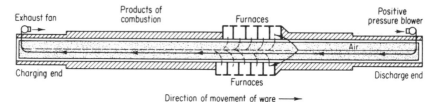

FIG. 13.19   A plan section of a direct-fired tunnel kiln. (*American Refractories Institute.*)

7. The tunnel kiln cannot be run efficiently at a low capacity and therefore lacks the flexibility of the periodic kiln.

Tunnel kilns are made in the direct-fired and muffle types. Because the average refractory is not very sensitive to impurities in the atmosphere, the muffle type, which is more expensive in first cost and in fuel, would rarely be selected for burning refractories; therefore, we shall confine ourselves to the direct-fired type. In Fig. 13.19 is shown a plan section of a typical direct-fired tunnel kiln. The charge enters the kiln at the left-hand end and is gradually heated, reaching a maximum temperature in the hot zone. It is then cooled as it passes out of the kiln. Cold air is forced into the exit end of the kiln with a blower, and this passes through the charge, cooling it and at the same time picking up heat. This air then passes into the combustion zone, where it is mixed with the combustion gases, and then passes through the incoming charge. In this portion of the kiln, it loses heat to the charge; then it is drawn out at the entrance end of the kiln with a suction fan. The primary air used for combustion in the burners is usually drawn out from the cooling section by a fan, but this is small in volume compared with the total gas flow.

FIG. 13.20   Tunnel-kiln recirculation.

At the present time reliable blowers are available for moving gases at temperatures up to 1100°C (about 2000°F). With this equipment either gases can be recirculated in the preheating zone in one cross-sectional area, as shown in Fig. 13.20, to equalize the temperature over the charge, or the gas may be pulled out of the cooler part of the preheating section and reinserted at a hotter portion, as shown in Fig. 13.21. The latter plan increases the flow velocity through the charge and thereby speeds up the heat transfer from the gases to the charge. Figure 13.22 shows one method of indirect cooling.

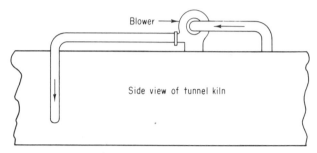

FIG. 13.21   Recirculation in tunnel kiln.

In order to regulate the heating curve of the bricks, the gases are usually not drawn off at one point at the kiln entrance but are taken out at a number of points along the preheating section so that the desired temperature curve may be obtained.

The longer the preheating zone of the kiln the cooler will be the exit gases and the higher the efficiency of the kiln. However, for constructional reasons, the length must be limited; consequently the temperature of the exit gases usually runs between 100°C (about 212°F) and 400°C (about 750°F). In the same way, the outgoing bricks may be cooled to any desired temperature by sufficient length of the cooling zone. It is generally desired to cool the charge sufficiently for easy handling. The bricks usually emerge at temperatures of 50°C (about 120°F) to 150°C (about 300°F).

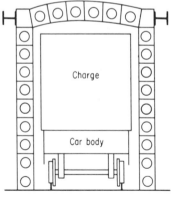

FIG. 13.22 A tunnel kiln with indirect cooling.

The setting of the charge itself should be so proportioned that a comparatively high gas velocity is maintained throughout the length of the preheating zone, and yet excessive power should not be required to pull the desired volume of gases through. In general, the velocity of the gases through the preheating zone will vary from 500 to 2,000 fpm, and the draft at the entrance end needed to supply this velocity will vary from ¼ to 1 in. of water.

A great many methods are used in setting the charge to provide longitudinal gas channels. As the sides of the charge have a tendency to heat before the center, it is common practice to provide a gas channel down the center of the charge and to reduce the clearance at the top and sides to a minimum. The width of the charge usually ranges from 4 to 7 ft, and the height from 4 to 6 ft. Rather flat arches are used in order to prevent gas from flowing over the charge. It is quite necessary in a kiln of this length to make the sections as airtight as possible. This is accomplished by keeping the wall joints well cemented; and in some cases, airtight coatings are applied to the outside of the insulating material.

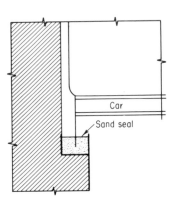

FIG. 13.23   A sand seal under a tunnel-kiln car.

There would tend to be a very large leakage between the car bottom and the kiln wall, because the clearance here cannot be made small on account of the change in dimensions of the car top. This leakage is prevented in most kilns by attaching to each car a blade that runs in a

FIG. 13.24   Tunnel kiln for burning refractories. (*Harbison-Walker Refractories Co.*)

trough filled with sand along each side of the kiln, as shown in Fig. 13.23. This method appears to be quite satisfactory, as sand can be supplied to the entrance end of the trough from which it is gradually carried through to the exit end, where it can be removed and returned.

The furnaces for firing tunnel kilns are arranged along the sides in the firing zone as shown in Fig. 13.24. The number of furnaces usually varies between two and seven on each side, and usually they are staggered. For the use of coal, furnaces and grates similar to those used in periodic kilns are employed; but in this case, they may be designed to work at their maximum efficiency with a constant load. Although a number of kilns have been fired directly with coal, in general it is more difficult to maintain a uniform temperature, and better control is obtained with other fuels. Many tunnel kilns are fired with oil. In most cases, the burners are placed in small chambers at the side of the firing zone, and the oil flame is directed into the charge. Because of the necessarily small distance between the burner and the charge, there is danger of overheating the surface of the brick next to the burner. For this reason, a checkerwork, as shown in Fig. 13.8, is sometimes placed at the mouth of the chamber to distribute the heat more evenly over the surface of the charge.

Kilns fired with natural gas need only a small firing chamber, as the flame is comparatively short. As in the case of periodic kilns, natural gas is an ideal fuel, and perfect control of temperature and atmosphere can be maintained with it.

Producer gas is often used in firing tunnel kilns. In most cases, the flame can be directed immediately into the charge because of the low temperature of the producer-gas flame. Because of the low heat value of this gas, it is not necessary to use a large amount of secondary air to provide a sufficient volume of gases to carry the heat down the kiln. As previously stated, raw producer gas has the disadvantage of depositing soot and tar in the connecting flues, and provision must be made for cleaning them out periodically.

The position of the burners is usually somewhat below the center of the charge in order to bring up the bottom temperature, which tends to lag behind that of the top. Sometimes burners in the hot zone are varied in height from one point to another in order to give a more equal temperature distribution to the charge. Many kilns have been constructed with hollow car bottoms with burners firing under the charge. In this case, the car tops must be made of excellent material to withstand the high temperatures and loads experienced under these conditions.

The tunnel-kiln cars vary from 5 to 7 ft in length and are built of structural steel or made of castings. Generally the cars roll on wheels mounted on roller bearings as in track cars. The track, however, is laid with great care, and there is little side clearance between the flanges and the rails. A number of other methods have been suggested and tried out for supporting the cars, such as balls running in grooves, floats on a canal, and stationary rollers in the kiln.

Tunnel-kiln car tops require much more upkeep than any other portions of the kiln because of the periodic heating and cooling. The top is usually made of large blocks laid on insulating firebrick. Setting sand has a tendency to work into the joints when the car is cool; then when the joints try to close on heating again, the blocks are forced apart so that in time the car "grows" until it cannot go through the kiln. This condition can be helped by blowing the sand out of the cool cars each trip. In some cases refractory concrete[10] has shown good results and can be installed with low labor cost.

The cars are pushed through the kiln generally by a hydraulic pusher operated by a small variable-stroke pump. After one car has been pushed in, the plunger draws back, a second car is inserted by means of a cross transfer, and the process is repeated. At the exit end, the car is taken off on a transfer in the same way. Kilns have been built where the cars are moved continuously rather than intermittently, in which case they are usually driven by a heavy chain with dogs. Means are provided to transfer the load gradually from the driving dog on the forward car to that on the following car to prevent shock and jars. As it is more expensive to construct a mechanism for continuous operation, it is not generally used, because the wait of four or five minutes while changing cars is not serious.

Nearly all tunnel kilns employ air locks at both ends so that the flow conditions in the kiln will not be disturbed by the entrance or exit of cars. However, kilns can be run with one lock and, in the smaller sizes, without any.

The cars are moved at the rate of from 3 to 8 ft per hr, depending upon the length of kiln and the burning properties of the clay, thus giving a total time in the kiln of from 50 to 100 hr; however, kilns are operating successfully at speeds as high as 20 ft per hr and with a total time in the kiln of 12 hr. If the kiln is properly designed so that the combustion gases are evenly distributed around each brick, the limit in rate of heating for many clays is the time taken for the heat to penetrate from the surface to the center of the brick. In the future, we may see clay refractories burned in a much shorter time than is considered possible at present. Figure 13.25 shows a high-temperature tunnel kiln for firing basic brick.

Production and Efficiency. The production of tunnel kilns generally used for refractories varies between 20,000 and 35,000 bricks per day. It may be said that the faster the bricks can be run through a given kiln and still be burned properly the more efficient will be the burning conditions. It takes but little more fuel to burn 30,000 bricks per day in a given kiln than it does 20,000 bricks.

Some manufacturers burn all shapes in their tunnel kilns, even up to 45 in. in length. They also change the burning temperature to accommodate different types of clay that may be run through the kiln. Although this procedure can undoubtedly be followed if necessary, a tunnel kiln can be run efficiently only with one type of setting and one temperature adjustment. Any changes in the setting or in the temperature disturb the balance of the kiln and require a long period for readjustment. Again, if the kiln is running slowly enough to burn large shapes properly, it is running slower than necessary to burn standard bricks. Probably the most desirable condition would be to have one tunnel kiln running on 9-in. sizes and a second kiln on a slower schedule burning only large shapes.

FIG. 13.25 Tunnel-kiln firing basic refractories at 3100°F.

A well-designed tunnel kiln will use from one-half to one-third of the fuel per thousand bricks required for a periodic kiln.[5] In Table 13.6 are given some figures on the fuel consumption of tunnel kilns.

A tunnel kiln reduces the setting and handling cost in a unit by as much as 20 to 35 percent, through the continuous flow of materials. In the case of dry-pressed bricks, it is general practice to set the tunnel-kiln cars directly from the dry press and then to pass them either through a drier or, in some cases, directly to the kiln. This eliminates drier cars, shods, and storage space for the drying bricks.

It is difficult to state definitely the relative qualities of a tunnel-kiln and a periodic-kiln product because the conditions are so varied. It may be said in general, however, that the tunnel-kiln product is more uniform than that of the periodic kiln, both as to size and as to degree of burning. This is due to a more even temperature in the small cross section of the tunnel kiln and to the fact that the stresses resulting from the height of

TABLE 13.6   The Fuel Consumption of Tunnel Kilns

Type of refractory fired	Fuel used	Maximum temperature		Fuel per 1,000 bricks
		°C	°F	
Fireclay..........	Coal	1270	2320	700 lb
		1335	2435	900 lb
		1470	2680	1,250 lb (stokers)
Fireclay..........	Oil	1400	2550	80 gal
		1670	3040	150 gal
Fireclay..........	Producer gas	1150	2100	850 lb (coal)
		1250	2280	970 lb (coal)
		1420	2590	1,150 lb (coal)
		1400	2552	630 lb (coal)
Fireclay..........	Natural gas	1140	2080	6,000 cu ft
		1100	2000	7,000 cu ft
Chrome..........	Oil	1470	2680	500 gal
		1525	2775	600 gal

the setting and the shrinking of the charge as a whole are much smaller in the tunnel kiln.   Kiln marking is rarely found.   The tunnel-kiln product is probably no better than the best of the periodic charge, but a large portion of the charge from the periodic kiln will undoubtedly be inferior to any of the tunnel-kiln product.

Undoubtedly, a modern tunnel-kiln unit can turn out bricks at a lower cost than the old periodic unit, because of both the saving of fuel and the saving of labor.   This fact and the general belief that the product is superior to the periodic product have led many manufacturers to change over their plants to tunnel-kiln operation.

Tunnel kilns are used successfully to fire refractories other than fireclay. Silica bricks have been burned in tunnel kilns here and abroad for some time.   A kiln for burning silica bricks at the rate of 35,000 bricks per day would be approximately 500 ft long to give a comparatively long soak at the maximum temperature, which would run around 1500°C (about 2730°F).   Magnesite and chrome bricks are being successfully burned in tunnel kilns (Fig. 13.25).   These bricks are not able to support a great load while they are being burned; and in periodic kilns, it is necessary to support them more or less completely with silica bricks.   In a tunnel kiln, however, the setting can be made so low that the bricks are well supported during the burning operation, which gives a lower kiln loss and a superior product.

Automation of Setting and Drawing.   A great deal of effort has gone into developing automatic devices for setting and drawing brick, especially in connection with the tunnel kiln.   The greatest progress has been made in the heavy-clay-products field where there is great unity in sizes.[11]   In the case of refractories, many shapes and sizes must be handled, which makes the problem more difficult.   However, some progress has been made in using electronic-tape input control to take care of setting patterns.[6]   It is certain that automatic setting will come into use, but the fact that inspection is needed for a high-grade product like refractories will probably make automatic drawing less attractive.

13.6   Shaft Kilns.   Shaft kilns are used to a considerable extent in burning lime, dolomite, and magnesite, although they have been displaced by the rotary kiln in many instances.

Shaft kilns for burning magnesite are described by Seil.[1]   The coal-fired kiln is 6.5 ft in diameter and 40 ft high.   The fuel weighs 35 percent of the clinker.   A blast pressure of 1 lb is used, which produces a temperature at the base of the stack of 1500°C (about 2730°F) and 250°C (about 480°F) at the top.   The lower lining is magnesite brick, and the remainder firebrick.   The producer-gas-fired kilns are 8 ft in diameter and 52 ft high.

An oil-fired shaft kiln for firing grog is used in Germany[8] at a temperature of 1450°C (about 2640°F).

13.7   Rotary Kilns.   As the demand for high-alumina firebrick has become larger and the availability of suitable raw materials such as diaspore and kyanite minerals has become smaller, the industry has been forced into making grog from kaolin, bauxite, etc.   This was first done by firing bodies in the periodic or tunnel kiln, but it was found less expensive to carry out the operation in a rotary kiln very similar to that used in the cement industry.   The Babcock and Wilcox Company pioneered in making dense kaolin grog in the rotary kiln in the early 1920s.   Now there are a considerable number of refractory plants using this process.

The rotary kiln is structurally like a cement kiln.   It is lined with high-alumina blocks in the hot end and high-duty or medium-duty firebrick in the cooler sections.   The kiln may be fed with lump clay from the mine (such as kaolin) or a mixture may be extruded from an auger into the water-cooled feed tube.   The kilns are usually oil-fired with a long flame.   The clinker is cooled by air which is sometimes used as preheat. Figure 13.26 shows a rotary kiln in a refractory plant for producing grog, and Fig. 13.27 a rotary kiln for firing periclase at 3300°F.

Rotary kilns are slowly replacing shaft kilns for dead-burning magnesite or dolomite.   These kilns are similar to the grog kilns except that they are lined with basic blocks and run at higher temperatures, sometimes at 1900°C (about 3450°F).

13.8   Temperature Control.   Contrary to the conditions of a generation ago, excellent equipment is now available from a number of manufacturers to enable reliable temperature control to be maintained.   This equipment

FIG. 13.26   A rotary kiln for firing grog.   (*Harbison-Walker Refractories Co.*)

FIG. 13.27   Rotary kiln for sintering periclase at 3300°F.   (*Kaiser Refractories Co.*)

is so standardized at present that there seems to be no need to describe it here.   It is still general practice to place pyrometric cones with the charge as a check on the operation, but the best check, of course, is the physical properties of the brick coming out of the kiln.   A control system is described in *Industrial Heating*.[9]

13.9 Handling and Storage.   Products from the kiln are either shipped or put in storage directly.   In recent years, nearly every plant employs lift trucks and wooden pallets for handling as shown in Fig. 13.28. This allows high piling with a minimum of labor. In some cases, the bricks are loaded in the freight cars on the pallets, which are later returned when a carload accumulates.

FIG. 13.28 Drawing refractories fired on kiln cars and loading on pallets. (*Harbison-Walker Refractories Co.*)

Insulating firebricks are generally ground to sizes on automatic grinders and then packed in cartons for storage or shipment.

BIBLIOGRAPHY

1. Seil, et al.: The Manufacture of Refractories and Information Concerning Their Use in the Iron and Steel Industry of West Germany, P. B. 37804, 1946.
2. Wittmer, A. L.: Combustion System Improvements for Oxidation Section of Tunnel Kiln, *Bull. Am. Ceram. Soc.*, **40**, 122, 1961.
3. Garve, T. W.: Shuttle Kiln, *Bull. Am. Ceram. Soc.*, **40**, 134, 1961.
4. Anon.: How New Concept in Kilns Saves $35,000, *Brick Clay Record*, **141**, 46, 1962.
5. Jeschar, R.: Energy Balance of a Tunnel Kiln, *Ber. Deut. Keram. Ges.*, **40**, 596, 1963.
6. Matchett, W. M.: Automatic Brick-setting System, *Claycraft*, **36**, 12, 462, 1963.
7. Skerhoske, E. J.: Propane Air-standby Fuel for Industry, *Bull. Am. Ceram. Soc.*, **43**, 630, 1964.
8. Anon.: Economy, Simplicity Earmark German Oil-fired Shaft Kiln, *Brick Clay Record*, **144**, 57, 1964.
9. Anon.: Large Tunnel Kilns for Refractories Provide Precise Combustion Control, *Ind. Heating*, **31**, 1534, 1964.
10. Caignan, A.: Tunnel-kiln Cars of Refractory Concrete, *Ziegelindustrie*, **18**, 201, 1965.
11. Keller, W.: Automatic Brick-setting Machine, *Ziegelindustrie*, **18**, 451, 1965; *Brick Clay Record*, **147**, 63, 1965.

PART FOUR

Properties

CHAPTER FOURTEEN

Fusion Point

14.1 Introduction. When revising this chapter for the Fourth Edition, it was thought desirable to omit the phase diagrams of refractory materials that have become so numerous in the past twenty years, as these are readily available in "Phase Diagrams for Ceramists" by Levin, Robbins, and McMurdie, published by The American Ceramic Society (1964). Also, fusion points will be confined to the materials used in heavy refractories.

14.2 Fusion Process. At the present time, there is very little understanding of the mechanism concerning the fusing of nonmetallic materials. If a curve of viscosity is plotted against temperature, it will be found that for some materials, the change from a solid to a liquid occurs within a very small temperature interval whereas in other materials, especially silicates, the melting process is a more gradual one and occurs over a large temperature range. As shown earlier, it will be found, in the case of materials having a long softening range, that there is a great difference in the bond strength between the various atoms. For example, in the silicates, the bond strength that holds the silicon in the center of an oxygen tetrahedron is very strong and probably will not break up until much higher temperatures are reached than are necessary to break the weaker bonds. Therefore, we may conceive that when a silicate fuses, at first the weaker bonds are broken and a very viscous liquid results, which contains aggregates of the silicon-oxygen groups. As the temperature goes higher, these groups are broken up by the thermal agitation until the viscosity markedly decreases. Unfortunately, we have no experimental evidence to show just what happens in this process and can merely conjecture a mechanism that most nearly fits the facts.

In the literature, it will be found that the terms "melting point," "softening point," and "fusion point" are used more or less indiscriminately in discussing refractories. The best usage would seem to indicate that the term melting point should be reserved for pure materials and fusion or softening point for more complex or impure materials. It is, of course, difficult to draw a sharp boundary between the two; so it was thought best to use the term fusion point throughout this chapter for everything except the pure materials, as the majority of the refractories show a considerable softening range.

14.3 Method of Obtaining Fusion Points. *Physical Significance of the Fusion Point.* There has been a great deal of confusion in considering just what is meant by the fusion point of the average refractory. We can do no better than refer to the method used by the glass technologist in determining the fusion or softening point of glass. Here, the fusion point is specified as the temperature at which the particular specimen under a definite stress becomes sufficiently fluid to flow at a specified rate. No other meaning will have any significance in discussing materials that have no sharp boundary between the solid and liquid state.

This definition of fusion point can be made clear by the curve in Fig. 14.1 in which the deformation in millimeters of the specimen is plotted against the time, under the conditions where the specimen is heated up at a uniform rate. If we are arbitrarily assuming the fusion point as that temperature at which the rate of flow equals 0.01 mm per sec, a tangent can be drawn to the flow curve with this slope. The point where the curve and tangent meet will represent the fusion temperature, which, in the illustration, is 3143°F. Owing to experimental difficulties, the usual methods of measuring the fusion points do not follow such an exact procedure; consequently, only an approximate value is obtained.

Determination of Fusion Points by Comparison with the Pyrometric Cones. A common method of determining the fusion point is by comparing the bending characteristics of the sample with those of a series of standard pyrometric cones all run in the same furnace. As it can be carried out by persons not skilled in pyrometric practice and with inexpensive equipment, this method has the advantage of simplicity. It has the disadvantage, however, of not giving the highest degree of precision, because of variations in the furnace atmosphere or from other causes that are not easy to control. Under good conditions, however, a precision of $\pm 15°C$ (27°F) should be obtained.

The pyrometric cone equivalent test is described in ASTM Standard C 24–56. A number of furnaces are commercially available in the United States for this test. Smalley and Sosman[3] have compared several of them, and point out good and bad features. One of these furnaces is shown in Fig. 14.2 to give an idea of its construction.

In the year 1956, the National Bureau of Standards made a reevaluation of the end points of the pyrometric cones produced by the Edward Orton, Jr., Ceramic Foundation. Added to the previous list were cones 31½ and 32½ to fill gaps on the temperature scale. The end points determined from this study are listed in Table 14.1.

Determination of Fusion Points with the Optical Pyrometer. A more precise determination of the fusion point can be made with the optical pyrometer, which is sighted on the test specimen heated in a furnace under blackbody conditions. With ordinary care, a precision of $\pm 7°C$ (about 12.6°F) should be obtained; and with especial care in calibration, a precision of $\pm 4°C$ (about 7.2°F) is possible.

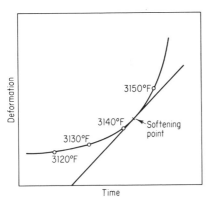

FIG. 14.1 Method of determining the softening point.

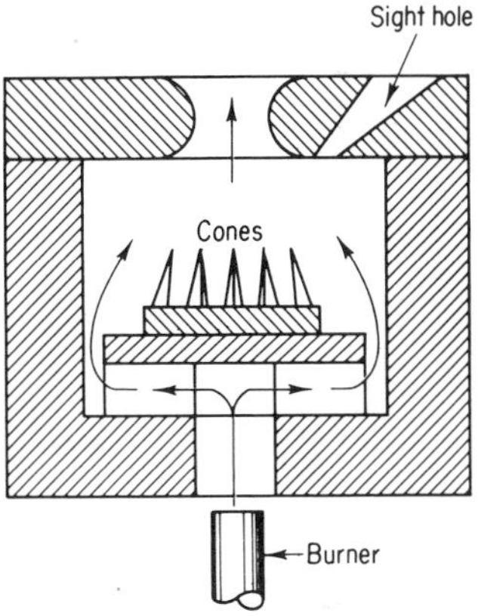

FIG. 14.2 A gas-fired fusion-point furnace.

Most of the fusion points have been made on specimens formed into a slender tetrahedron similar in shape to the standard pyrometric cone. This particular form of specimen has commonly been used because the bending can readily be seen without special measuring apparatus, but it does have certain disadvantages for precise work because (1) the size of the specimen varies, depending on the shrinkage of the refractory material; (2) the density of the specimen varies with the material so that the bending forces are variable; and (3) the bending forces vary with the position of the cone.

Apparently a much better form of specimen for fusion-point tests would be a small bar supported at its ends. On heating, this bar will soften and sag in the middle. By sighting on the center of the bar with a reading telescope, it is possible to plot the amount of settling against time and thus obtain a curve capable of giving a precise value of fusion point as shown in Fig. 14.1. Certain precautions are necessary, however,

TABLE 14.1   End Points of Orton
Pyrometric Cones
Large cones, 150°C per hr. 1956 values

Cone number	End point, °C	End point, °F
12	1337	2439
13	1349	2460
14	1398	2548
15	1430	2606
16	1491	2716
17	1512	2736
18	1522	2772
19	1541	2806
20	1564	2847
23	1590	2894
26	1605	2921
27	1627	2960
28	1638	2980
29	1645	2993
30	1654	3009
31	1679	3054
31½	1699	3090
32	1717	3125
32½	1730	3146
33	1741	3166
34	1759	3198
35	1784	3243
36	1796	3265
37*	1830	3326
38*	1850	3362
39*	1865	3389
40*	1885	3425
41*	1970	3578
42*	2015	3659

* 1926 values, 600°C per hr.

in this type of test: (1) The material should be precalcined in order to shrink it to a stable volume, and (2) the distance between the supports should be so arranged as to give a definite stress in the specimen based on its density and cross-sectional area.

There are many furnaces available for melting-point determinations. One often used for high-temperature work is a vacuum chamber with a resistance-heated tungsten cylinder surrounded with radiation shields. Also, inductively heated vacuum furnaces are used in either a vacuum or an inert atmosphere. Temperatures up to 2500°C (about 4500°F) are easily obtained.

Several excellent makes of optical pyrometer are on the market, but for highly precise work it may be necessary to provide a better optical system. Pyrometers may be calibrated[4] from the freezing point of pure metals or a comparison may be made with a standardized instrument. The U.S. Bureau of Standards is equipped to calibrate pyrometers. The classic paper by Foote, Fairchild, and Harrison[2] still serves as an excellent reference.

Effect of Atmosphere on Fusion Point. The fusion point of pure silica or alumina is little, if at all, influenced by the furnace atmosphere; however, when any impurity is present, especially iron oxide, the atmosphere does

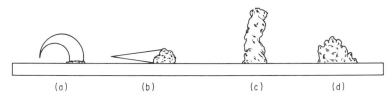

(a)   (b)   (c)   (d)

FIG. 14.3   Behavior of specimen cones on firing.

have a marked effect on the fusion point. Just why the reduced oxide should combine more readily with the other constituents to form a low-melting compound is not known. Fieldner, Hall and Feild[1] found that when the iron oxide in the mixture was reduced to the metallic condition below the softening temperature, little lowering of the fusion point resulted. Apparently the act of reduction makes the iron oxide more active in combining with the other constituents.

Test cones in a reducing atmosphere seldom bend over smoothly. Often a hard coating forms, allowing the molten center to run out at the base. Fusion points should be taken only when the test cone bends down gradually until the tip touches the base, as in (*a*) of Fig. 14.3. When a cone bends only at the base, as in (*b*), it is probable that interaction has occurred between the cone and the plaque. Some materials even under the best conditions will not give good readings but will either bloat, (*c*), or slump down, (*d*). Unless for some special purpose, fusion points should always be made in an oxidizing or neutral atmosphere.

14.4   Fusion Points of Oxides.   The fusion points of the oxides generally found in heavy refractories are given in Table 14.2. Obtaining the higher fusing points is a difficult problem, and many values will see some change as time goes on.

TABLE 14.2   Fusion Points of Some Refractory Materials

Material	Mineral name	Formula	Fusion point °C	Fusion point °F
Alumina.............	Corundum	Al_2O_3	2050	3720
Lime................	Calcite	CaO	2570	4660
Chromic oxide.......	....	Cr_2O_3	2270	4120
Magnesia............	Periclase	MgO	2830	5130
Silica...............	Quartz	SiO_2	1713	3116
Zirconia.............	Zircite	ZrO_2	2710	4910

14.5   Value of Fusion Points.   The value of a fusion point in determining the possibility of a material for specific purposes has often been over-estimated.   What the fusion point does show is whether or not a material is unsuitable above a certain temperature.   For example, if it should be desired to make a firebrick to be used with furnace temperatures of 1600°C (about 2910°F) and various samples of clays were selected, the fusion-point test would eliminate at once all those with a fusion point equal to or below that temperature.   However, a clay softening at 1850°C (about 3360°F) would not necessarily be better for this brick than another clay softening at 1750°C (about 3180°F).

The fusion point of a material is by no means a criterion of its ability to carry a load at high temperatures, which in actual use is of considerable importance in determining the maximum temperature at which it can be successfully used.   For example, commercial magnesite has a fusion point of over 2000°C (about 3630°F); yet it will not carry a load at temperatures over 1400°C (about 2550°F) to 1500°C (about 2730°F).   On the other hand, a mullite brick that has a fusion point of 1780°C (about 3230°F) will show no deformation at temperatures of 1727°C (about 3140°F).

BIBLIOGRAPHY

1. Fieldner, A. C., A. E. Hall, and A. L. Feild: The Fusibility of Coal Ash and the Determination of the Softening Temperature, *U.S. Bur. Mines Bull.* 129, 1918.
2. Foote, Fairchild, and Harrison: Pyrometric Practice, *Natl. Bur. Std. (U.S.) Tech. Paper* 170, 1921.
3. Smalley, H. F., and R. B. Sosman: A Comparison of Gas-fired Pyrometric Cone Equivalent Furnaces, *Bull. Am. Ceram. Soc.*, **32,** 5, 1953.
4. Beerman, H. P.: Calibration of Pyrometric Cones, *J. Am. Ceram. Soc.*, **39,** 47, 1956.

CHAPTER FIFTEEN

Load-bearing Capacity

15.1 Introduction. The importance of the load-bearing capacity of refractories has been overestimated in the past perhaps, for actual load failures are confined to only a small proportion of the furnace construction. However, such areas as roofs, piers, and walls heated from both sides depend on refractories that will resist deformation under long heating; thus, an understanding of the flow properties of heated refractories is essential in the overall picture.

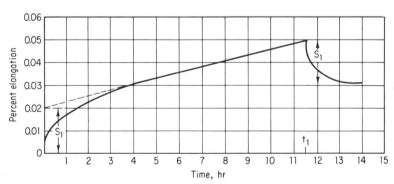

FIG. 15.1　Flow of heated glass.

15.2 Mechanism of Load Failure. *Flow Characteristics of Glass.* The flow properties of a heated glass fiber are shown in Fig. 15.1, in which the elongation is plotted against time for a single load. It will be noted that the flow starts in rapidly but soon decreases until it reaches a constant rate. If the straight portion of the curve is projected back to the zero time axis, the value S_1 is obtained, which represents the elastic strain. If, at time t_1, the load is removed, the glass will contract rapidly at first and then more slowly until it reaches a fixed length. The amount of contraction is equal to S_1, or the original elastic effect. The heated glass is a truly viscous material; for if the steady rate of flow is plotted against stress as shown in A of Fig. 15.2, a straight line passing through the origin results.

The mechanism of flow in glass can perhaps be made clearer by considering a mechanical model, the motion of which reproduces the glass flow with considerable exactness. This model, shown in Fig. 15.3, consists of the spring A, which represents the elastic effect in the glass but is restrained from reaching instant equilibrium by the dashpots B, representing the restraining force or internal friction of the glass. The steady flow of the glass is represented

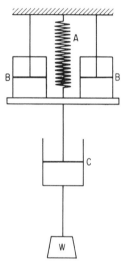

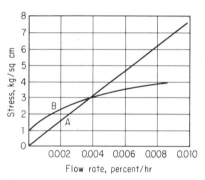

FIG. 15.2 Flow characteristics of a viscous material (A) and an elastico-plastic material (B).

FIG. 15.3 Model simulating the flow in glass.

by the motion of the dashpot C; the weight W provides the extension force.

Flow of Crystals. The flow of single crystals of nonmetals at elevated temperatures has now been studied with great care. It is believed that the single crystal has a definite yield point but deforms plastically below the melting temperature. Research on single crystals is complicated by the fact that some distortion occurs, probably along the slip planes, until there is no longer a single crystal and we must consider the system as many similarly oriented small crystals. A still greater departure from a single crystal is a system of small crystals with random orientation, which gives, in effect, the conditions of an isotropic medium.

Flow of Mixture of Crystals and Glass. This mixture, which is a normal characteristic of most refractory materials, has been studied to a considerable extent. Such mixtures are elastico-plastic materials, similar in flow characteristics to plastic metals; i.e., if the steady rate of flow is plotted against stress, a line results that does not pass through the origin but indicates a certain minimum stress necessary to initiate flow, as shown in curve *B* of Fig. 15.2, a characteristic typical of all plastic materials. The time-elongation curve, however, is very similar to that for a glass, as indicated in Fig. 15.4 for a porcelain. This shows the initial portion of rapid flow and later the gradual conformity to a uniform rate. It is characteristic of almost all refractory materials that the rate of flow soon after the load is applied is much greater than after a uniform

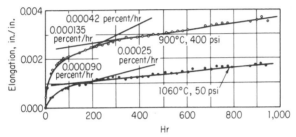

FIG. 15.4 Plastic flow of heated porcelain.

condition has been reached. Also it should be remembered that in plastic flow, the rate of flow is roughly proportional to the fourth power of the stress and not the first power as in viscous flow.

In Fig. 15.5 is shown a more or less idealized type of flow diagram for a refractory material at various temperatures.

The flow relations in an elasticoviscous material may be clarified by a mathematical discussion.

Let S_t = the total flow of the specimen

S_e = the elastic portion of the flow

S_1 = the total elastic flow under the external force F

Now the force acting at any instant to cause elastic flow is

$$F_1 = F(S_1 - S_e)$$

That is, at the start of flow, $F_1 = F$, whereas at the condition of equilibrium flow $F_1 = 0$.

The retarding force restraining the elastic flow is given by

$$F_2 = k \frac{dS_e}{dt}$$

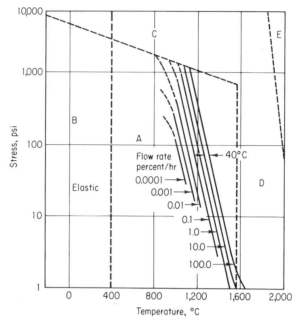

FIG. 15.5 The flow field of a porcelain. A = field of plastic flow. B = field of elastic flow. C = field of instant breaking. D = field of viscous flow. E = field of turbulent flow.

But any time

$$F_1 = F_2$$

or

$$F(S_1 - S_e) = k \frac{dS_e}{dt}$$

Solving for S_e we obtain

$$S_e = S_1(1 - e^{Ft/k})$$

The viscous flow is

$$S_v = k_1 t$$

Then

$$S_t = S_v + S_e$$
$$S_t = k_1 t + S_1(1 - e^{Ft/k})$$

This expression represents the flow relation of Fig. 15.1 and the motion of the model of Fig. 15.3.   After a period of time, the second term becomes zero and the flow is steady.

In the case of elastico-plastic flow, we can set up the expression for the elastic flow as

$$F(S_1 - S_e) = k_2 \left(\frac{dS_e}{dt}\right)^n$$

and for the plastic flow as

$$S_p = k_3 t$$

where S_p = the plastic flow

$n = \frac{1}{4}$ (approximately)

The solution of these equations gives an expression for the total flow of

$$S_t = k_3 t + S_1 \left(1 - \sqrt[3]{\frac{1}{3F^4 S_1^3 + 1}}\right)$$

Reference should be made to the excellent series of papers on the flow of refractories by Clews and his coworkers.[1,2,4]

15.3 Method of Measurement. *ASTM Load Test.* The load resistances of refractories are almost invariably measured with a compression specimen, which is heated under a definite schedule and the amount of deformation determined at the end of the test. However, tension and torsion[3] tests have been made. A typical example of the compression tests is the ASTM standard test, Designation C 16–62. Two standard brick are tested by compression on the end. This is not a constant-temperature test except for the last $1\frac{1}{2}$ hr which merely attempts to bring the center of the brick up to the surface temperature. The measurement of deformation is taken from the beam movement, which includes changes of the holders, furnace frame, and underpinning.

The ASTM load test has been developed over a long period to fit best the needs of the manufacturer and the user of refractories and is particularly useful in control testing. As discussed later, it does not give information particularly suited to the need of the designer.

Other Load Tests. In Europe, the load test is generally carried out on smaller specimens, i.e., cylinders approximately 1 in. in diameter and 1 in. high. This method is preferred because the temperature can be made more uniform over the specimen and the furnace is smaller and more easily controlled. On the other hand, such a small specimen may not represent accurately the structure of a coarse-grained brick, and the specimens are so short that the precision of measuring the elongation is somewhat less than with the larger specimens. However, there are certain advantages with each type of specimen; and with carefully made tests, either should give reliable results. An excellent discussion of the load test is given in the references.

Lythe, Padgett, and Woodhouse[15] find good agreement between compression, tension, torsion, and bending, and suggest the latter as the most convenient type.

Long-time Creep Tests. Recently, some work has been carried out on the load testing of refractories in the same way that creep tests are made

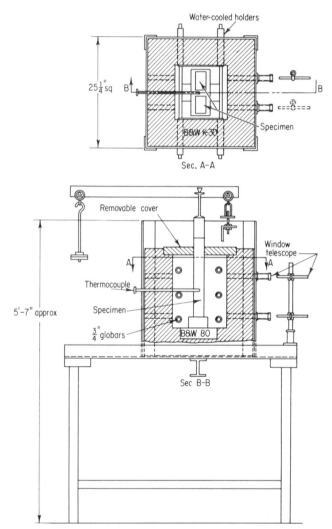

FIG. 15.6 Electrically heated load test furnace. (*J. Am. Ceram. Soc.*)

with metals; i.e., the refractory is heated to a uniform temperature and then loaded for a sufficient length of time to establish a uniform rate of flow. A furnace that is heated by electric globars to the necessary temperature for carrying out this test is shown in Fig. 15.6. The length of the specimen not only is measured at the end of the run but is measured every day with a pair of telescopes sighting on the ends of the specimen in order to determine a time-elongation curve. Figure 15.7 shows such a series of curves for a firebrick run in this way; and in Fig. 15.8, the flow rates are plotted against load on logarithmic paper. It is felt that for design pur-

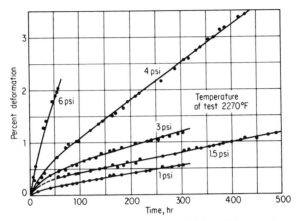

FIG. 15.7 Time-elongation curve for a firebrick. (*J. Am. Ceram. Soc.*)

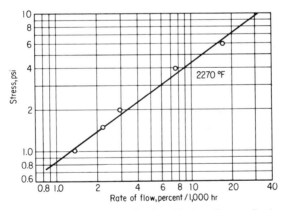

FIG. 15.8 Flow rate vs. load. (*J. Am. Ceram. Soc.*)

poses, the long-time tests are much preferred to the short-time tests previously described. Whether or not the shorter tests give accurate comparisons between various bricks cannot be determined until more long-time tests have been made to reproduce more nearly actual conditions.

A few years ago, there was developed in the Ceramic Laboratory of the Massachusetts Institute of Technology a torsion creep test for refractories of a unique character,[5] as the optical system is in the furnace. The

general plan is shown in Fig. 15.9.   The specimen *a* is put under a steady
torsional stress by the torque wheel *f* and the weight *g* through the refractory holders *e*.   At the end of the gage length of the specimen are cemented
two concave sapphire mirrors, *b*, ⅛ in. in diameter.   Light from lamps *c* is
focused by these mirrors on the ground-glass scale *d*, so that any twist in
the gage length shows up as a separation of the light spots.   The screw *i*
in the nut *j* moves the arm *h* by the micrometer dial *k* to calibrate the
scale *d*.   This method can be used up to at least 1300°C (about 2370°F)

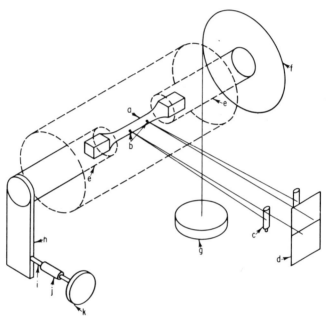

FIG. 15.9   A torsion creep furnace with a sapphire optical system.

with a precision of ± 0.001°.   It should be noted that the holders do not
have to be rigidly attached to the specimen, as slippage does not affect the
optical readings.   This method has many advantages over the tensile test
because of its much greater precision in determining low flow rates.

15.4 Load-test Values for Brick.   There seems little reason to list
many load-test values for various brands of brick tested by the ASTM
method, for these values would be of little use for furnace design; however,
a few are given in Table 15.1.   For those interested, good data are given
by Eusner and Schaefer[8] on superduty and high-duty brick.   More
informative values are given by Crookston and Torgeson[6] for long-
time tests on a few types of brick.   Ford and White,[12] Padfield,[16] and

TABLE 15.1   ASTM Load Tests on Fireclay Brick (25 psi)*

Brand and type	Subsidence %, 2460°F, 1½ hr	Subsidence %, 2640°F, 1½ hr
Superduty (Mo.) high fire...........	0.1	1.2
Standard superduty (Mo.)...........	0.7	3.6
High duty (Tex.)....................	0.3	4.7
High duty (N.J.)....................	1.1	10.3
High duty (Pa.)....................	1.1	5.4
High duty (Mo.)....................	0.7	4.2
High duty (Pa.)....................	1.7	7.3

* From Crookston and Torgeson.[6]

Ford et al.[14] give values for basic brick.   Lahr and Hardy[9] list underload values for silica brick.

Load values for insulating firebrick are given in Tables 15.2 and 15.3.

TABLE 15.2   Test Data on Commercial Insulating Firebrick

Brand	Use limit, °F	Weight, lb per 9 in. straight	Modulus of rupture, psi	Cold rushing strength, psi
A	2000	1.72–1.82	95	120
B	2000	2.21–2.75	197	172
C	2000	1.64	71	104
D	2000	2.27–2.32	107	559
O	2000	1.88	106	176
E	2200	1.87–1.96	165	193
F	2200	2.71–3.07	224	377
G	2200	2.27	129	208
N	2200	2.44	66	235
I	2500	2.40–2.63	196	248
J	2500	2.76	251	408
K	2600	2.30–2.50	133	128
L	2600	2.68–2.80	203	372
S	2600	2.40	127	196
M	2800	2.42–2.48	140	150
T	2900	2.90	235	220

15.5   Factors Influencing the Load-bearing Capacity of Refractories.
In general, it may be said that the load-bearing capacity of a refractory is directly influenced by the amount and viscosity of the glassy phase.

TABLE 15.3 Load-test Results for Insulating Firebrick

Brand	Percentage at 1900°F		Percentage at 2000°F			Percentage at 2100°F			Percentage at 2200°F			Percentage at 2300°F		
	10 lb	25 lb	10 lb	25 lb	40 lb	10 lb	25 lb	40 lb	10 lb	25 lb	40 lb	10 lb	25 lb	40 lb
A	…	…	0.1	0.1	0.1	0.0	0.2	0.3	0.4	0.7	1.1	1.1	2.7	Failed
B	0.1	1.6	0.6	8.3	2.5	2.5	2.9	4.9	3.8	Failed				
C	0.1	Failed	1.0	Failed										
D	…	…	0.9	1.3										
O	…	…	Failed											
E	…	…	0.1	0.0	0.1	0.0	0.1	0.2	0.1	0.1	0.4	1.6	2.6	Failed
F	…	…	0.1	0.4	1.8	0.2	2.5	7.6	2.5	Failed				
G	…	…	0.1	1.8	7.6	1.5	Failed							
N	…	…	Failed											
I	…	…	0.1	0.8	1.2	0.2	2.0	Failed	1.9	Failed				
J	…	…	…	0.2	0.2	0.0	0.2	0.7	0.1	1.7	Failed			
K	…	…	…	…	…	0.3	0.4	0.6	0.3	1.0	2.9	1.4	Failed	
L	…	…	0.0	0.2	0.4	0.1	0.9	1.6	0.9	1.2	5.5			
S	…	…	0.1	0.1	0.2	0.2	0.3	0.8	0.4	2.9	8.2			
M	…	…	…	…	…	0.0	0.4	0.9	0.1	1.0	3.9	0.4	1.3	Failed
T	…	…	…	0.1	0.2	0.2	0.1	0.4	0.2	0.4	0.6	0.3	0.8	…
1	…	…	…	…	…	…	…	…	…	…	0.4	…	…	0.9
2	…	…	…	…	…	…	…	…	…	…	0.2	…	…	Failed

The theoretical basis for the flow of aggregates containing a liquid phase is gone into thoroughly by Allison, Brock, and White.[13]   However, they admit that there are cases where there is a solid-to-solid bond with the liquid in isolated pools. This condition gives greatly increased hot strength, as shown by the direct-bonded basic brick.   It also should be remembered that the crystalline material, free from glass, may deform, but only at high temperatures and loads.

Anything that can be done to diminish the amount of glass or raise its viscosity will increase the resistance to flow.   This can be accomplished by adding to the refractory a material that will form from the glass a more viscous material or a material containing more crystals.   For example, a chrome brick containing silica as an impurity may have a comparatively low load value; whereas if the proper amount of magnesite is added to the brick, it will form the crystalline magnesium orthosilicate, forsterite, which will greatly increase the load-bearing value of the brick.   Also, an increase in firing temperature will generally increase the load-bearing capacity because it distributes the glassy phase more thoroughly and permits it to absorb crystallizing elements.   This is shown in Table 15.4,

TABLE 15.4   Effect of Reburning on the Load-bearing Properties of Firebricks

	As received				After reburning to 1620°C*			
	Shrinkage starts		10% shrinkage		Shrinkage starts		10% shrinkage	
Brand	°C	°F	°C	°F	°C	°F	°C	°F
A	1200	2192	1510	2750	1480	2696	1590	2894 S
B	1260	2300	1440	2624	1460	2660	1550	2822
C	1200	2192	1490	2714	1460	2660	1570	2858
D	1170	2138	1460	2660	1490	2714	1620	2948
E	1090	1994	1460	2660	1430	2606	1580	2876
F	1250	2282	1500	2732	1450	2642	1580	2876 S
G	1200	2192	1500	2732	1490	2714	1590	2894 S
H	1260	2300	1520	2768	1490	2714	1650	3002

* S indicates failure by shear.   Load, 25 psi.

which lists a number of commercial firebricks that were tested as they came from the kiln and then after refiring at a higher temperature.

The structure of the brick also has some influence on the load-bearing capacity.   For example, a brick of high porosity will have less load-bearing capacity than one of the same material with a lower porosity, as there is less material in the brick to carry the load in the latter case. It must be remembered, however, that the porous bricks are lighter and therefore do not need to carry a heavy load.

15.6 Application of Load-test Results to Design. It will be clear from the preceding discussion that the results of the usual short-time load test are of little value to the designer of furnaces. Only with the long-time test that has come to equilibrium flow are values available for the designer to use in computing allowable stresses in the structure. From the curve in Fig. 15.8, the data may be utilized in the following example: A pier under the hearth of a heat-treating furnace has an allowable deformation of 10 percent in its estimated working life of 4,000 hr, which amounts to a rate of 2.5 percent in 1,000 hr. On the curve this corresponds to a loading of 1.7 psi.

It is probably safe to say that very few load failures occur in actual furnace construction and then only in cases where the refractory part is heated all the way through, such as in piers, partition walls, and highly insulated crowns. Nearly all other parts of furnaces have a considerable temperature gradient from the hot to the cool side, and therefore the load is supported by a relatively cool portion of the refractory, which is well able to carry the load. It is difficult with the data at hand at the present time to compute the load-bearing capacity of a wall with a variable temperature gradient, but an approximation can be made by dividing the wall into a number of sections parallel to the face and considering the mean temperature of each section as a temperature for obtaining the load-bearing value. If these sections are all summed up, the load-bearing value of the whole wall can be roughly computed. It will be found, however, that in most cases, the cooler section will be responsible for supporting almost the whole load.

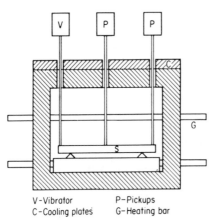

V-Vibrator   P-Pickups
C-Cooling plates   G-Heating bar

FIG. 15.10 Resonant frequency apparatus.

15.7 Modulus of Elasticity at Elevated Temperatures. The elastic properties of refractories can shed much light on the rather complex aggregate of crystals and glass that they consist of. They will show the strength of bonding and pick up discontinuities and internal stresses in the structure. The dynamic methods of measurement are discussed by Lakin,[10] and Lakin and West[11] describe a test useful at elevated temperatures. This consists of a furnace, as shown in Fig. 15.10, in which a brick may be forced into sonic oscillation and the resulting vibration picked up. Ault and Ueltz[7] present some values of elastic modulus for refractories at high temperatures.

BIBLIOGRAPHY

1. Clews, F. H., H. H. Macey, and A. T. Green: Behavior of Refractory Materials under Stress at High Temperatures, I, Introductory Experiments with Porcelains under Torsion, *Bull. Brit. Refractories Res. Assoc.*, **60**, 31, 1941.
2. Clews, F. H., H. M. Richardson, and A. T. Green: Behavior of Refractory Materials under Stress at High Temperatures, II, Compression Experiments on a Refractory Insulating Product, *Bull. Brit. Refractories Res. Assoc.*, **61**, 51, 1941.
3. Swallow, H. T. S.: Furnace and Apparatus for Carrying Out Refractoriness-underload Tests on Firebricks, *Trans. Brit. Ceram. Soc.*, **44**, 93, 1945.
4. Clews, F. H., H. M. Richardson, and A. T. Green: The Behavior of Refractory Materials under Stress at High Temperatures, III and IV, *Trans. Brit. Ceram. Soc.*, **45**, 161, 1946.
5. Stavrolakis, J. A., and F. H. Norton: Measurement of the Torsion Properties of Alumina and Zirconia at Elevated Temperatures, *J. Am. Ceram. Soc.*, **33**, 263, 1950.
6. Crookston, J. A., and D. R. Torgeson: Long-time Load Tests on Commercial Classes of Fire-clay Brick, *J. Am. Ceram. Soc.*, **35**, 265, 1952.
7. Ault, N. N., and H. F. G. Ueltz: Sonic Analysis for Solid Bodies, *J. Am. Ceram. Soc.*, **36**, 199, 1953.
8. Eusner, G. R., and W. H. Schaefer, Jr.: Fifty-hour Load Test for Measuring the Refractoriness of Super-duty and High-duty Fireclay Brick, *Bull. Am. Ceram. Soc.*, **35**, 265, 1956.
9. Lahr, H. R., and C. W. Hardy: Refractoriness-underload Properties of Silica Bricks, *Trans. Brit. Ceram. Soc.*, **56**, 369, 1957.
10. Lakin, J. R.: Determination of the Elastic Constants of Refractories by a Dynamic Method, *Trans. Brit. Ceram. Soc.*, **56**, 1, 1957.
11. Lakin, J. R., and C. S. West: Determination of Modulus of Elasticity of Refractories at High Temperatures, *Trans. Brit. Ceram. Soc.*, **56**, 8, 1957.
12. Ford, W. F., and J. White: The Mechanical Properties of Basic Refractories at High Temperatures, *Trans. Brit. Ceram. Soc.*, **56**, 309, 1957.
13. Allison, E. B., P. Brock, and J. White: The Rheology of Aggregates Containing a Liquid Phase with Special Reference to the Mechanical Properties of Refractories at High Temperatures, *Trans. Brit. Ceram. Soc.*, **58**, 495, 1959.
14. Ford, W. F., A. Hayhurst, and J. White: The Effect of Bond Structure on the High Temperature Tensile Behavior of Basic Bricks, *Trans. Brit. Ceram. Soc.*, **60**, 581, 1961.
15. Lythe, T. W., G. C. Padgett, and D. Woodhouse: High-temperature Mechanical Testing of Refractory Materials, *Trans. Brit. Ceram Soc.*, **62**, 19, 1963.
16. Padfield, R. C.: Hot Strength of Basic Brick, *Bull. Am. Ceram. Soc.*, **44**, 537, 1965.

CHAPTER SIXTEEN

Spalling

16.1 Introduction. Spalling is usually defined as a fracture of the refractory brick or block resulting from any of the following causes:

1. A temperature gradient in the brick, due to uneven heating or cooling, that is sufficient to set up stresses of such magnitude as to cause failure

2. Compression in a structure of refractories, due to expansion of the whole from a rise of temperature, sufficient to cause shear failures

3. Variation in coefficient of thermal expansion between the surface layer and the body of the brick, due to surface slag penetration or to a structural change in service, great enough to shear off the surface layer

Many consider only the first type as true spalling. It is undoubtedly the predominating effect in refractories such as silica, magnesite, and chrome. As this type is caused by the inherent properties of the refractory, it offers a good opportunity for careful analysis. The second type of spalling is almost always due to poor furnace design, and the refractory itself can seldom be blamed. The third type can be minimized by using bricks burned higher than their subsequent operating temperature and by preventing slag penetration insofar as is possible.

The spalling properties of refractories cannot be consistently improved until we know the true mechanism of spalling. Although the study of stresses and fractures in brittle material is by no means well understood, light has been thrown on the matter by Booze,[2] and Phelps,[3] and Preston.[4] Later work by such authors as White[8] and Jorgen[10] has advanced the theory.

16.2 Temperature Stresses in Solids. It may be of interest to examine first the distribution of stress in an elastic solid when its surface is heated or cooled. It is rather surprising that there has been so little recognition of the difference in stresses resulting from heating as differentiated from cooling. As will be shown later, the stresses as well as the fracture are different in the two cases. An investigation, therefore, has been carried out by the photoelastic method to determine the stresses occurring in various objects when suddenly heated and cooled at the surface.

Photoelastic Method. The method of studying stresses in various structures by photoelasticity has been used to a considerable extent in determining the distribution of stress in objects where a direct computation is impossible. For example, valuable studies have been made of the stresses in hooks, eyebolts, ships' hulls, and gear teeth, although it is believed that very little work has been previously carried out on the measurement of temperature stresses by this method.

Photoelastic models must be made from a transparent isotropic medium. In this case, clear bakelite was selected because of its high stress–optical coefficient.

Polarized light is passed through a plane specimen, which becomes double refracting under stress and retards the two wave systems according to the values of the principal stresses p and q. The two wave systems interfere when passed through an analyzer and give isochromatic bands proportional to the difference $p - q$. The isochromatic lines are also lines of principal shear, for the principal shear stress at any point is equal to one-half the difference of the two principal stresses. The separate values of p and q can be determined by repeating the observations with circularly polarized light, which shows the isoclinic lines, or lines of equal direction of principal stresses. The values of p and q can be calculated from the isoclinic and isochromatic lines, the boundary conditions, and the constants of the material. It should be emphasized that this method applies only to a plane object.

Temperature Stress Tests. The first model to be studied was a disk. This disk was carefully maintained at 50°C (122°F); then the circumference was suddenly heated to 100°C (212°F) by steam, and the resulting development of stresses watched until they had reached a maximum value, which required about 45 sec. The experiment was then repeated by suddenly chilling the circumference from 50°C (122°F) to 0°C (32°F) by means of ice water, and the stresses were similarly studied. In Fig. 16.1 are given the magnitudes and directions of the shear and resultant stresses under these conditions.

It will be observed that when the circumference is suddenly heated, tangential compression of considerable magnitude is developed near the surface but the center of the disk is under a slight tension having an equal magnitude in all directions.   The radial stresses are zero at the circumference but increase slowly toward the center.   On the other hand, when the circumference is cooled, the surface is under considerable tension and the center of the disk is under light compression.   In both cases, the shear

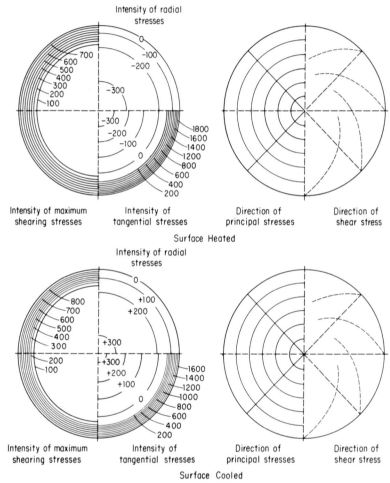

FIG. 16.1   Stresses in a disk.   (*J. Am. Ceram. Soc.*)

stresses are the same and reach a maximum at the circumference.   Their direction is everywhere at 45° to the radii.

The next model tested was a brick heated on the end.   The stresses are shown in Fig. 16.2 in the same way as for the disk.   On heating, a compression maximum occurred at the center of the end; but on cooling, this was changed to a tension.   It will be noted that the directions of the shear stresses are approximately 45° to the surface everywhere and are a maximum at the center of the end.

In the case shown in Fig. 16.3, the model brick was heated at the end and about one-third of the way up the side. A somewhat different type of stress distribution resulted, for the increase in stresses at the sides causes the stresses at the end to decrease. It will be noted that at the corners, the shear lines are curved and follow the isothermals closely.

Figure 16.4 shows a model representing a brick in a wall when this wall is heated and cooled at the surface. The isothermals will then be parallel to the heated surface. It will be noted that the resultant normal stresses are nearly parallel and perpendicular to the surface of the wall and that the shear stresses are at 45° to them.

Conclusions. We may conclude from the preceding results that

1. The stresses set up in a solid by suddenly heating the surface through a certain temperature interval are of the same order but of opposite sign from those set up by cooling the surface through the same temperature range.

2. No high tensile stresses are developed on sudden heating.

3. The maximum shearing stresses are equal to one-half the maximum resultant stresses.

4. The stresses are generally low at the corners and interior of the specimen but reach their highest value along the heated sides at a distance from the corners.

16.3 Study of Cracks Developed in Spalling. In order to correlate the types of cracks formed in the material when heated and cooled, with the stress distribution previously determined, a number of spheres and bricks were made up with a fine-grained, porcelain-like structure in order to make spalling occur readily. These objects were gradually heated to 900°C (about 1650°F) for the first condition and were placed in still air to cool. Under the second condition, the objects were heated by placing suddenly in a furnace that had a temperature of 900°C (about 1650°F).

Spheres. Let us first consider the sphere that was suddenly cooled (Fig. 16.5). It will be noted that in general, the fractures follow great circles. When the portions of the sphere were separated, it was noted that the cracks entered the surface in every case at approximately 90°. On the other hand, when the sphere is suddenly heated, a quite different type of fracture is obtained, as shown in Fig. 16.6. In this case, the fractures entered the surface at 45° and tended to split off circular segments of the sphere having a height of approximately one-quarter the diameter.

Bricks. Standard-size bricks were made up of the same material as the spheres. They were also heated and cooled in the same way. In Fig. 16.7 are shown the fractures resulting from rapid cooling. It will be noted that the fractures enter, in nearly every case, at right angles to the surface

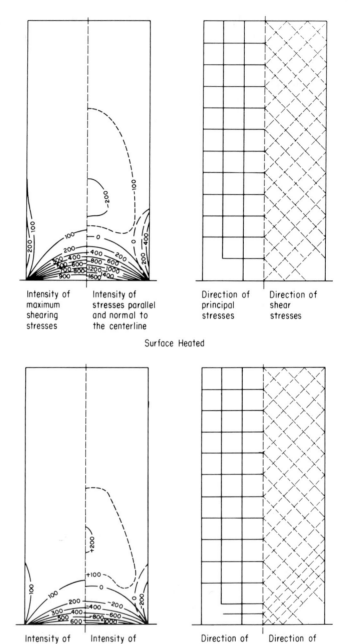

Intensity of maximum shearing stresses | Intensity of stresses parallel and normal to the centerline | Direction of principal stresses | Direction of shear stresses

Surface Heated

Intensity of maximum shearing stresses | Intensity of stresses parallel and normal to the centerline | Direction of principal stresses | Direction of shear stresses

Surface Cooled

FIG. 16.2 Stresses in a brick heated at one end.   (*J. Am. Ceram. Soc.*)

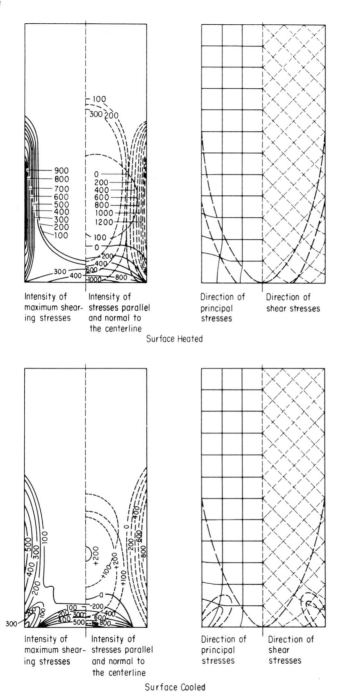

FIG. 16.3 Stresses in a brick heated on one end and one-third of the distance up the side. (*J. Am. Ceram. Soc.*)

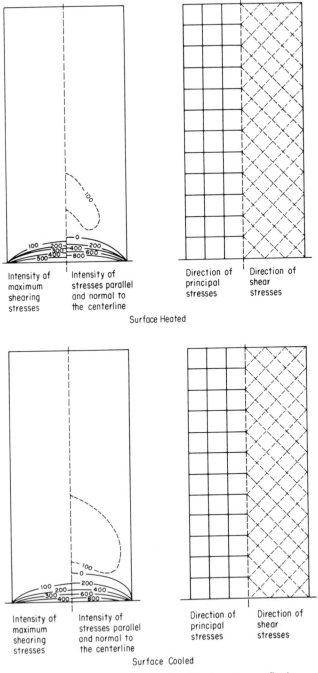

FIG. 16.4   Stresses in a wall.   (*J. Am. Ceram. Soc.*)

and tend to break the brick into cubes.   Figure 16.8 shows the same brick when heated suddenly.   The typical spalling failure is shown at the edges and corners; and in every case, cracks enter the surface at approximately 45°.

As a brick does not consist of a homogeneous material, it is possible to have a weakening or disintegration of the structure aside from actual fractures.   This effect is usually due to a difference between the coefficient of

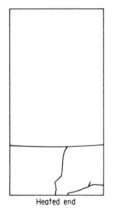 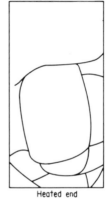

FIG. 16.5   Cracks in a sphere suddenly cooled. (*J. Am. Ceram. Soc.*)

FIG. 16.6   Cracks in a sphere suddenly heated. (*J. Am. Ceram. Soc.*)

FIG. 16.7   Cracks in a brick suddenly cooled. (*J. Am. Ceram. Soc.*)

FIG. 16.8   Cracks in a brick suddenly heated. (*J. Am. Ceram. Soc.*)

thermal expansion of the bond and that of the grog particles.   This may be due to the use of different materials for the grog and bond or to a low-burned bond when the material is the same throughout.   Goodrich[5] has examined the loss in strength due to repeated heatings and finds a distinct weakening of the brick structure.

Conclusions.   We may conclude from the preceding studies that spalling is due to both shear and tension failures.   On sudden heating, spalling is due almost entirely to shear stresses and, on sudden cooling, to tension stresses except at the corners and edges, where shear stresses may still cause failure.   The shear values are about equal on heating and cooling, but there are no large tension stresses on rapid heating.

Shear failure occurs suddenly; i.e., complete pieces are split off at each cycle of heating, and they begin to split at the edges and corners, so that a brick tends to approach a spherical shape on the end.   The shear cracks follow in the lines of shear determined by the photoelastic method very closely, as is indicated by the slight curvature of the shear fractures at the corners.

Tension cracks seldom occur near the corners but usually first appear halfway up the brick where the tensile stresses are a maximum.   The tension cracks enter the surface at 90°.   They do not occur as suddenly as do shear fractures but penetrate deeper and deeper at each cycle until finally the cracks meet at the center.   In every case, the tension failures follow approximately the lines of principal stress, although the deviation was considerably more than for the shear stress, owing undoubtedly to the length of the cracks and the greater influence of discontinuities.

It should be noted at this time that the stress distribution is much altered after a crack starts, as was pointed out by Preston.[4]   Generally, however, the tension cracks enter and leave at 90° and shear cracks at 45° to the surface.   The fracture need not follow the direction of stress everywhere but will travel in such a way as to release the maximum strain energy.   A great deal of confusion has been caused in the past by failure to differentiate between the effects of heating and cooling.

16.4   Theory of Spalling Due to Shear Stresses.   As shown in the preceding pages, a spalling failure may occur either by a shear or by a tension fracture.   If we first assume a shear failure, which appears to be the most common type, it is possible to analyze the mechanism of spalling in terms of the physical properties of the material.

Temperature Distribution.   It is necessary at first to consider the distribution of temperature in a homogeneous solid.   We shall first make the assumption that this solid has one infinite plane surface.   Originally this solid is at a uniform temperature of t_0, but at time τ_0 the surface is suddenly cooled to a temperature of t_2.   The problem is to find the temperature distribution in the solid for different values of τ.   For the solution, it will be necessary to apply Fourier's equation of heat flow.   As the surface is infinite, the isothermal planes will be everywhere parallel to the surface of the solid.   Putting $t_0 = t_1 - t_2$ or simply shifting the temperature scale, the temperature t at any distance below the surface x and at any time τ will be given by

$$t = \frac{2t_0}{\sqrt{\pi}} \int_0^{x/2h\sqrt{\tau}} e^{-\beta^2} d\beta \tag{16.1}$$

where h = a constant of the material

The diffusivity of a material, or the rate at which a point in a hot body will cool under definite surface conditions, is known as h^2 and is equivalent to $k/\rho c_p$, where k is the conductivity, ρ the density, and c_p the specific heat. It should be noted that the preceding equation is in the form of the probability integral and can therefore be readily evaluated from tables.

As will be shown later, the stress set up in a homogeneous material as a result of a change in temperature is proportional to the temperature gradient at any point. It will therefore be necessary to differentiate the preceding equation with respect to x.

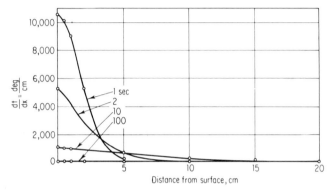

FIG. 16.9 Temperature gradients at different levels below the surface of a cooling body. (*J. Am. Ceram. Soc.*)

We have

$$\frac{\partial t}{\partial x} = \frac{2t_0}{\sqrt{\pi}} e^{-x^2/4h^2\tau} \frac{1}{(-2h\sqrt{\tau})}$$

$$= \frac{-t_0}{h\sqrt{\pi\tau}} e^{-x^2/4h^2\tau} = \frac{dt}{dx} \tag{16.2}$$

To show the practical application of this equation, curves have been plotted in Fig. 16.9 to give the temperature gradient $h = 0.080$ when a body is cooled suddenly from 850°C (about 1560°F) to room temperature. The gradient is, of course, highest at the surface but falls rapidly toward the interior. Near the surface, the gradient falls rapidly with increasing time; but at a certain depth, the gradient increases with the time. This is one of the reasons why spalling does not usually take place immediately after the specimen is chilled.

In an actual case, the surface is not instantaneously cooled (1) because a cooling medium cannot be instantaneously introduced and (2) because perfect thermal contact cannot be made between the cooling medium and the surface of the brick. The equations can be modified to take care of this effect, as suggested by Ingersoll and Zobel,[1] by adding a fictitious distance beyond the true surface. In this instance, however, as we are not interested in the region very close to the surface, it will be more convenient to use a constant k to modify h.

Only the case of an infinite plane surface has been considered. At the edge of a solid, the isothermal planes will not be parallel to the surface but will be rounded off as in Fig. 16.10. This can be explained by considering that point A is cooled by the area abc in addition to the cooling received from an infinite surface such as a point at B. As the isothermals recede further from the surface, the area abc increases and corresponding departures from parallel conditions exist at corners.

It is important to notice that the temperature gradient is slightly less at the edges and the corners than at the faces of a cooling brick; hence, corners spall off because of mechanical weakness rather than because of greater temperature gradients.

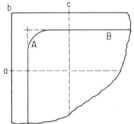

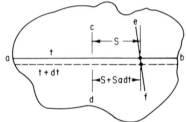

FIG. 16.10 The rounding of isothermal planes at the edge of a solid. (*J. Am. Ceram. Soc.*)

FIG. 16.11 A portion of a heating or cooling brick. (*J. Am. Ceram. Soc.*)

Temperature Stresses. Now that we have the temperature distribution in a solid, it is possible to study the distribution of stresses. We shall first assume that the material is an elastic, homogeneous medium. It will be shown later that in the temperature range at which spalling occurs, the brick is perfectly elastic. The homogeneity will depend upon the structure of the brick and upon the process of manufacture.

The exact evaluation of the stresses in a brick under a given temperature distribution cannot be solved mathematically except in a few simple cases; however, it is required to find the tendency to spall only in terms of the physical properties of the material; i.e., comparative but quantitative results are desired among various types of brick tested in the same manner.

Let us consider an isothermal plane ab in a portion of a heating or cooling brick at a given instant of time (Fig 16.11). Let the temperature of the plane ab be t and that of another isothermal plane dx distant be $t + dt$, which gives the temperature gradient dt/dx. Let the plane cd cut the isothermal surfaces at a neutral point of shear. Now measure along ab a distance s, and erect the plane ef. If the temperature throughout the brick were t, this plane would be normal to ab; but under the existing conditions, it will cut the $t + dt$ isothermal at a distance $s + s\alpha\, dt$ from cd. The coefficient of expansion is α.

The shear strain is measured by the deflection of one plane with respect to another divided by the distance between them. In this case,

$$\phi = \frac{s\alpha\, dt}{dx} = s\alpha \frac{dt}{dx} \tag{16.3}$$

This means that the strain set up in a heating or cooling body is proportional to the size, the coefficient of thermal expansion, and the temperature gradient. This expains why large glassware cracks on cooling or heating more readily than small ware. In large pieces of refractory ware, the temperature distribution may be so uneven that strict proportionality to s may well be doubted. This does not greatly concern us, for we shall compare specimens of the same shape and size and with the same temperature gradient. The shear fracture theoretically should occur at 45° to the isothermal but often is parallel to it. However, the direction of fracture does not affect the reasoning. As pointed out by Clark, the maximum stress or strain is not strictly a constant of the material at a given temperature but varies to some extent with the duration of stress.

From Eq. (16)

$$\frac{dt}{dx} = \frac{k_1}{h} \tag{16.4}$$

where $k_1 = $ a constant
This follows, for dt/dx is closely proportional to $1/h$ because small changes of h in the term $e^{-x^2/4h^2\tau}$ have little effect when τ is reasonably large.

Then the tendency to spall S_s is

$$\frac{\phi}{\phi_b} = \frac{s\alpha(dt/dx)}{\phi_b} = \frac{s\alpha(k_1/h)}{\phi_b} = k_1 \frac{s\alpha}{h\phi_b} = k_2 \frac{\alpha}{h\phi_b} \tag{16.5}$$

where $\phi_b = $ the maximum shearing strain or flexibility

This shows that the spalling tendency of a brick is proportional to the coefficient of thermal expansion divided by the square root of the diffusivity and maximum shearing strain.

16.5 Theory of Spalling Due to Tensile Stresses.

The preceding theory has been based entirely on the shear failure of a material under shear stresses. It was seen that this failure would occur on sudden heating of the brick. The sudden heating produces local compressive stresses in the material, since the portion at a higher temperature tends to try to expand, but this being prevented by the cooler portions places the hotter portion of the brick in compression. It is known that brittle materials fail in shear when subjected to compressive stresses; hence, we get a shear failure under these conditions.

If the brick is suddenly cooled, a different phenomenon is involved. When one portion of the brick is suddenly cooled below the temperature of the adjacent material, the part of the brick at the lower temperature tends to contract, but this contraction is prevented by the adjacent portions, which are at a higher temperature. In this case, tensile stresses are set up in the cooler portion; and since the shearing strength of a brittle material is usually considerably greater than the tensile strength, the brick will fail in tension, provided sufficiently rapid local cooling is present.

Let us then visualize a portion of the brick as shown in Fig. 16.12, which we will assume to be at the uniform temperature t_0. This means that at

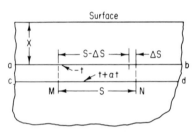

FIG. 16.12 Illustration of tensile stresses in a heating or cooling brick.

the start, the isothermal planes ab and cd, which are initially separated by the distance dx, are at the same temperature t_0. The surface of the brick is now cooled to a temperature t_2; and at a later time τ, the planes ab and cd possess temperatures t and $t + dt$, respectively, where the temperature is given by Eq. (16.1). Let us consider a gage length s between points M and N on plane cd. As a result of the

lower temperature on plane ab, the material on this plane tends to contract by an amount ΔS. This contraction is, however, prevented by the adjacent material, and a tensile strain is set up in plane ab of amount

$$\Sigma = \frac{\Delta S}{S} = \alpha \cdot dt \qquad (16.6)$$

where α = the coefficient of thermal expansion
Now, from Eq. (16.4),

$$\frac{dt}{dx} = \frac{k_1}{h}$$

and

$$\sigma = E \cdot \Sigma \qquad (16.7)$$

It follows from the preceding equations, then, that

$$\sigma = E \cdot \alpha \cdot dt = E \cdot \alpha \cdot \frac{k_1}{h} \cdot dx \qquad (16.8)$$

If σ_b is the breaking stress of the material or the so-called "tensile strength,"

$$\sigma_b = E \cdot \Sigma_b \tag{16.9}$$

where Σ_b = the tensile breaking strain

The tendency to spall then becomes

$$S_t = \frac{\sigma}{\sigma_b} = \frac{E \cdot \alpha \cdot k_1 \cdot dx}{Eh\Sigma_b} = k_1 \cdot dx \, \frac{\alpha}{h\Sigma_b} \tag{16.10}$$

It will be noted that Eq. (16.10) has the same units and is in the same form as Eq. (16.5) of the preceding section. Instead of the shear strain σ_b at fracture, we have the tensile strain Σ_b; and in place of the distance S from the neutral place of shear, we have the distance dx over which the temperature drop dt exists.

Let us now compare this equation with the Winkelmann-Schott equation, which is as follows:

$$\sigma = -\frac{A'A''}{A} \frac{t_0 \cdot E \cdot \alpha x}{2h \sqrt{\tau}} \tag{16.11}$$

where A', A'', A = constants

t_0 = the temperature of the outer layer

The tendency to spall is

$$S'_t = \frac{\sigma}{\sigma_b} = -\frac{A' \cdot A'' \cdot t_0 \cdot \alpha \cdot x}{2Ah \sqrt{\tau} \, (\sigma_b/E)} \tag{16.12}$$

or

$$S_t = k_3 \cdot x \cdot \left(\frac{\alpha}{h\Sigma_b}\right) \tag{16.13}$$

where

$$k_3 = -\frac{A'A''t_0}{2A \sqrt{\tau}} \tag{16.14}$$

It is thus seen that the Winkelmann-Schott equation can be put in the same form as given by Eq. (16.5) or (16.10). In the derivation given in the original article by Winkelmann and Schott, the size factor x was absorbed in another constant B, where B is essentially Kx. This, of course, can be done, since the purpose was to compare the spalling properties of various glasses for the same values of τ and x.

16.6 Experimental Confirmation of Theory. In order to determine the validity of the equation for tendency to spall, the physical characteristics of a number of different bricks shown in Table 16.1 were tested and the tendency to spall, as computed from the preceding formula, was compared with the actual spalling characteristics of the brick in Table 16.2.

TABLE 16.1　Characteristics of Brick Tested for Spalling

Material	Grog	Bond	Hardness	Approximate burning temperature °C	°F	Remarks
A. Fireclay.............	Very coarse	Soft	Soft	1200	2300	Commercial brick
B. Sillimanite...........	Medium	Little	Hard	1680	3050	Special brick
C. Fireclay.............	Very coarse	Soft	Medium	1430	2600	Same as A (hard burn)
D. SiC................	20-mesh	Molasses	Hard	1540	2800	Special brick
E. Fireclay.............	Coarse	Soft	Soft	1316	2400	Commercial brick
F. Bauxite.............	Fine	Little	Hard	1704	3100	Special brick
G. Fireclay.............	Medium	Soft	Soft	1316	2400	Commercial brick
H. Kaolin.............	Medium	Little, hard	Very hard	1650	3000	Commercial brick
I. Fireclay.............	Fine	Soft	Soft	1200	2300	Commercial brick
J. Fireclay.............	Medium	Hard	Hard	1200	2300	Commercial brick
K. Fireclay.............	Fine	Soft	Soft	1200	2300	Commercial brick
L. Spinel...............	Medium	Hard	Hard	1370	2500	Commercial brick
M. Silica..............	Fine	None	Hard	1480	2700	Commercial brick
N. Same as C preheated to 2900°F						

Flexibility in Shear. In order to measure the characteristics of the bricks as listed above, a number of 1 × 1 × 9-in. bars were sawed out of each specimen. These bars were then cemented into holders, and the angle and torque measured up to the breaking point at a temperature of 500°C (about 930°F).

Such a low temperature is justified because it has been noticed that spalling nearly always occurs at about red heat. In all cases where the specimen was carefully held, there was no permanent set after removing the torque until a point was reached just before failure; i.e., the specimens all acted as if perfectly elastic. The bricks tested are described in the table above and represent a wide range. Some are experimental bricks; others are of extensively used commercial brands.

The Coefficient of Expansion. The coefficient of thermal expansion of each type of brick was measured at 500°C (about 930°F). In general, the brick having the greatest coefficient of thermal expansion had the highest content of silica, but we are by no means justified in using the silica content of a brick as an index of its coefficient.

Diffusivity. The value of the diffusivity can, of course, be obtained by measurements of thermal conductivity and specific heat. These measurements, however, are laborious and slow; so a simple method of measuring the relative diffusivity was devised. A 2-in. cube of the material was cemented from four 1 × 1 × 2-in. blocks with a thermal junction at the

TABLE 16.2   Constants for the Bricks Tested

Brand	ϕ_b	α	h	$\alpha/\phi_b h$	Number of quenches
A	0.114	0.0000051	6.2	72×10^{-7}	26.5
B	0.052	0.0000039	6.5	115	8.0
B	0.052	0.0000039	6.5	115	24.0
C	0.050	0.0000050	6.3	157	23.0
D	0.072	0.0000031	9.9	44	*
E	0.057	0.0000053	6.5	143	15.6
F	0.032	0.0000037	6.2	110	2.0
F	0.032	0.0000037	6.2	110	15.0
G	0.090	0.0000104	6.2	187	12.2
H	0.050	0.0000063	6.4	197	10.6
I	0.107	0.0000133	6.3	197	8.4
J	0.050	0.0000071	6.0	237	5.2
K	0.089	0.0000126	6.2	229	2.7
L	0.032	0.0000116	6.4	567	2.4
M	0.094	0.0000405	6.7	644	0.3
N	0.060	0.0000072	6.1	197	

* No exactly comparable spalling value was available for this brick; but it is undoubtedly lower than should be expected, as an oxidizing atmosphere tends to convert the SiC into SiO_2 and thereby to raise the coefficient of thermal expansion.

center, was suspended by the junction wires in an electric furnace until it had reached a constant temperature of about 550°C (about 1020°F), and then was taken out and suspended in a room free from drafts. The temperature of the center was recorded, and the slope of the cooling curve at 500°C (about 930°F) taken as the value for the diffusivity. This method is rapid and accurate where only relative values are desired.

The values of the diffusivity given are entirely arbitrary, and the results are comparable only among themselves, but it will be noted that with the exception of the silicon carbide specimen, the values of h vary only a few percent and cannot be an important factor in spalling.

Comparison with Actual Spalling Tests. The results from laboratory air-spalling tests are also given. A comparison between this test and actual service results has been made, and the agreement is generally good. From five to ten bricks of each kind were spalled, and the average taken.

Attention should be called to the two values given for bricks *B* and *F*. These two bricks showed the common type of shear spalling at the corners for several quenches, then suddenly broke across the center with a tension failure. The first value in the table gives the quenches actually obtained, and the second is estimated from the amount lost by shear. The first values have been disregarded in drawing the curves.

If the value of $\alpha/h\phi_b$ is an index of the spalling tendency of a brick as indicated by the theory, then these values plotted against the number of quenches for each specimen should fall along a curve rather than be widely scattered. Such a plot has been made in Fig. 16.13, and it is quite clear that the points group themselves into a definite band. The deviations that occur can be explained by the fact that some of the bricks were altered from their original characteristics in heating.

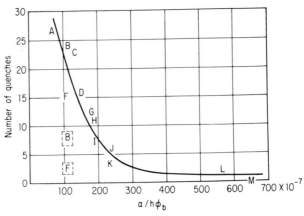

FIG. 16.13 A plot showing correlation between theoretical and experimental values of spalling. (*J. Am. Ceram. Soc.*)

Specimen K, which is considerably below the mean curve, was noticeably laminated in structure. The higher values for bricks B and F were only estimates.

The form of the curve is of interest, for it approaches a rectangular hyperbola asymptotic to the two axes. When the value of $\alpha/\phi_b h$ is high, the bricks show little difference in spalling resistance, as they are almost completely shattered at the first temperature change. On the other hand, the bricks having a low value of $\alpha/\phi_b h$ increase rapidly in spalling resistance as $\alpha/\phi_b h$ is decreased. This behavior is not unexpected, as it is simply an approach to a classification of bricks into two sets: those which spall on the first quench and those which do not spall under an infinite number of quenches.

The best spalling brick of those tested withstood 26 quenches and had a value for $\alpha/\phi_b h$ of 72×10^{-7}. Suppose that by slightly increasing the flexibility or by decreasing the coefficient of thermal expansion, a value of 50×10^{-7} were obtained; the brick should then withstand about 50 quenches. This illustrates the possibilities of making remarkable non-spalling brick.

On the whole, it would seem that the theoretical work agrees with practice as closely as the nature of the tests allows. Although this method cannot be recommended to replace the usual spalling tests without a more thorough trial, it is felt that more information is obtained by it as to the causes of spalling in a certain brick than by other methods.

It is interesting to note that Booze[2] and Phelps[3] independently came to the same conclusions after testing a number of experimental bricks for coefficient of thermal expansion, diffusivity, and flexibility. They found that the best spalling bricks had the greatest flexibility and that the diffusivity had no appreciable effect. Endell's work on magnesite brick checks this theory.[6]

16.7 Microstresses. The preceding discussion of stresses due to thermal gradients has been based on a homogeneous solid, such as glass. Nevertheless, the theory in most cases agrees in general with the observed cracking in normal refractories with a structure containing pores, glasses, and crystals.

In the last few years considerable information has been published which leads to the question of microstresses in the refractory structure. A thought-provoking paper by White[8] brings out the stresses due to differences in expansion coefficient of contiguous material particles, such as glass and crystal, which may cause microcracks even though the temperature is uniform. Jorgen[10] gives a simple method of calculating microstresses from the physical constants of the two phases. Charvat and Kingery[7] have shown that microcracks in sintered rutile are due to anisotropy of the crystallites. However, in general there seems to be no positive evidence that these microcracks are harmful. In fact, well-distributed microcracks may actually increase the spalling resistance because they enhance the structural flexibility, as proved by Kukolev et al.[13] There is need for more research in this field.

16.8 Laboratory Spalling Tests. The expense and the time required to determine the spalling characteristics of bricks in service are so great that a laboratory spalling test is highly desirable.

Many types of laboratory spalling test have been suggested and used; these may be divided into single-brick tests and panel tests. Silica brick have been tested by heating one end on a hot plate, as described by Hargreaves.[9] Recently, an interesting rotary-hearth furnace described by King and Walther[11] has shown good results for brick that are heated on more than one face, such as checkers. Testing of brick heated on one face as in most wall and arch construction is performed with panels, such as in ASTM C 38–58.

The ASTM panel spalling test equipment has been installed in a number of laboratories and, in spite of the considerable cost of operation and large floor space required, has been well worth while when reliable spalling results are required. This test is valuable to the producer of refractories for a control test of his product and for the development of improved refractories. It also is of aid to the consumer in assuring that the standard of quality of the refractories purchased is adhered to. Those who do not have enough testing to justify the cost of this equipment may have their product tested at a commercial testing laboratory.

The U.S. Navy simulative service furnace, as described in Chap. 18, can be used for a spalling test with results agreeing well with those obtained in service, but it has not been used extensively in the last few years.

Many specialized thermal-shock tests have been suggested. The test of Dugdale, Maskrey, and McVickers[12] uses intense gas discharges to cause failure.

16.9   Spalling Characteristics of Various Bricks. In Table 16.3 is given the spalling resistance of a number of fireclay brick as measured by the ASTM panel spalling test.

Silica brick are not at all resistant to spalling in the low-temperature range; but once they are above the inversion temperature, they are very resistant to wide changes in temperature because of the flatness of the expansion curve in this region. This is the reason that they give such good service in regenerative furnaces.

Chrome and magnesite brick, because of their rather high coefficient of thermal expansion, are not so resistant to spalling as fireclay brick, but

TABLE 16.3   Spalling Resistance of Various Classes of Refractories

Type of brick	*Resistance to spalling*
High-duty fireclay, regular	Fair
High-duty fireclay, spall-resistant	Good
Superduty, regular	Very good
Superduty, spall-resistant	Excellent
Semisilica	Fair to good
High-alumina	Good to excellent
Silica	Poor below 1200°F, excellent above
Chrome-magnesite, chemically bonded	Fair to good
Chrome-magnesite, fired	Fair to good
Magnesite-chrome, chemically bonded	Excellent
Magnesite-chrome, fired	Excellent
Forsterite	Fair
Silicon carbide	Excellent
Carbon and graphite	Excellent

NOTE: These ratings are average results experienced by users. There are many cases that deviate from these ratings.

great improvements have been made in the last few years in providing a more flexible structure[6] and thus increased resistance.

Silicon carbide brick have an excellent resistance to spalling.

16.10 Effect of the Brick Structure on Spalling. The structure of the brick and the method of manufacture have a considerable influence on spalling. Data given below illustrate the effect of various factors. These data are for one particular clay and will not be exactly the same for others, but the trend should be general.

Effect of Grog Size. A number of bricks were made up with various sizes of grog, other conditions being the same. The spalling resistance for each grog size is given in Table 16.4. The spalling resistance increases steadily with the grog size for a given burning temperature. These results were obtained on single bricks, and somewhat different values might be obtained by panel spalling methods.

Table 16.4   Effect of Grog Size on Spalling Resistance

Burned at	1649°C (3000°F)	1677°C (3051°F)	1704°C (3099°F)
2 mesh.........	20 cycles	12 cycles	5 cycles
3½ mesh......	11 cycles	6 cycles	3 cycles
4 mesh........	6 cycles	3 cycles	1 cycle
20 mesh........	1 cycle	1 cycle	1 cycle

Effect of Amount of Grog. In Fig. 16.14 is plotted a curve of spalling resistance against percentage of grog. It is interesting to note that the maximum spalling resistance occurs at 40 percent of bond, or at the condition where the pores are just filled up. Above or below this point, the spalling resistance falls off rapidly, probably owing to loss of flexibility.

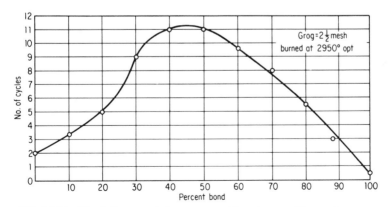

FIG. 16.14   The influence of the amount of bond on spalling resistance.

Effect of Burning Temperature.   Figure 16.15 shows a curve of spalling resistance plotted against burning temperature.   There is a minimum value at 1620°C (about 2950°F) and a maximum at 1650°C (about 3000°F).   This type of curve has been observed a number of times for this particular brick, but it may not apply to all others.   The minimum point is probably due to the decreasing flexibility of the structure with increased temperature.   Beyond 1600°C (about 2950°F), the free silica is dissolved and the coefficient of thermal expansion is lowered, which increases the spalling resistance up to 1635°C (about 3000°F).   After this, the structure gets more rigid and the spalling resistance falls off.

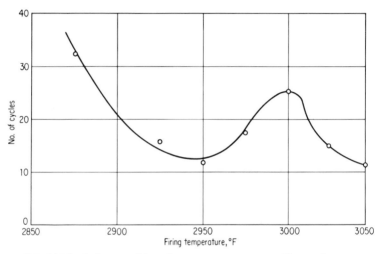

FIG. 16.15   Influence of burning temperature on spalling resistance.

Characteristics of a Good Spall-resisting Brick.   We usually have a specified material from which to make a brick.   With this material, it is impossible to alter the diffusivity appreciably, and only small changes can be made in the coefficient of thermal expansion.   Therefore, if we desire to make the best spall-resistant brick from a given material, we must do two things: (1) make a structure having the greatest possible flexibility and (2) eliminate all possible flaws and laminations.

Flexibility can best be obtained by using a large size of grog, a high degree of porosity, and a low degree of vitrification.   These characteristics cannot be obtained to a high degree without sacrificing others of considerable importance.   For example, bricks could easily be made to withstand 50 cycles in the air-spalling test; yet they would be mechanically weak and would probably have a high residual shrinkage.   It has often been noticed that when a series of bricks is tested for spalling, there will be a great variation between the best and the worst out of the same lot.   Where it is known that these bricks were made by identically the same process and that they all were burned at the same temperature, we can conclude only that the difference in spalling resistance is due to some lack of homogeneity in the structure of the brick. It is impossible to state definitely that any one method of making bricks gives the best or the worst spalling results, although it is generally found that re-pressed bricks, when properly tempered, give greater uniformity in the spalling test. We are greatly in need of a simple test to determine which bricks in a given lot will give a poor resistance to spalling. Some investigations have been made to test each brick by noting the sound emitted when struck with a hammer; but as yet, no very definite results have been obtained along this line.

16.11 Spalling Fractures in Service. The types of spalling fracture in service walls are quite varied.[14] Perhaps the most typical is that shown

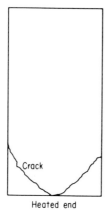

Heated end

FIG. 16.16 A type of spalling fracture often occurring in service. (*J. Am. Ceram. Soc.*)

Heated end

FIG. 16.17 Tension crack in end of a brick due to spalling. (*J. Am. Ceram. Soc.*)

in Fig. 16.16, which consists of two shear cracks coming together at the center of the brick. There is often a small tension crack in the end of the brick, as shown in Fig. 16.17. The fracture corresponds well with the determined stresses for this condition. This type of failure is not often serious, because the fragments are so large that they stay in place.

Another type of spalling failure is the separation of a thin layer of material from the face of the brick, as shown in Fig. 16.18. This usually occurs with bricks of low vitrifying points or with slagged surfaces, and the fracture runs between the vitrified material and the body of the brick. Often layer after layer will peel off, giving a considerable length of service before the wall becomes too thin.

Exposed corners in a furnace are quite subject to spalling and should be avoided in design where possible.   This type of failure by shear is shown in Fig. 16.19.

Spalling in a wall is greatly augmented by compression loads due to expansion.   Expansion joints should be left at least every 15 ft and should amount to ½ to 1 in.   Another method is to place a slip of thin cardboard between bricks in each vertical joint (or thicker pieces in alternate joints). These will burn out and allow each brick to expand freely, and yet there will be no general movement of the wall.   Thin pieces of steel, which are pulled out as the wall progresses, have sometimes been used instead of the cardboard.

FIG. 16.18 Separation of a thin layer due to vitrification. (*J. Am. Ceram. Soc.*)

FIG. 16.19 Spalling of exposed corners of a furnace.

In the case of arches, there is often a pinching off of the tips due to temperature strains.   Care, however, should be taken to see that there is clearance between the inner ends when temperature or load conditions are severe.

Another factor in spalling is the alteration in the structure of the hot face of the brick due to slag penetration, sintering, or gaseous reactions. This alteration changes the expansion coefficient at the hot face in relation to the original structure and thereby introduces shearing stresses which often cause fracture and shelling.   This may well be the reason why dense, high-burned brick often give better service than would be indicated by the laboratory spalling test.

Failure due to heating wet brickwork rapidly is sometimes considered spalling.   The remedy is obvious.   Mechanical breakage when slicing off slag deposits is also classed as spalling by some people. The brick manufacturer can do little to help this failure except to supply the hardest bricks that will stand the temperature changes.

There has perhaps been too great a tendency for manufacturers and users of refractories to specify brick with a good spall-resisting quality, as judged by the laboratory test. Many cases have come up where a hard, dense brick has stood up in service far better than one with a good spall-test record. This is probably due not to any fault in the test but to conditions of mechanical loads in service not encountered in the laboratory.

BIBLIOGRAPHY

1. Ingersoll, L. R., and O. J. Zobel: "The Mathematical Theory of Heat Conduction," Ginn and Company, Boston, 1913.
2. Booze, M. C.: A Study of the Spalling Test for Fireclay Brick, *Am. Soc. Testing Mater., Proc.*, **26**, 277, 1926.
3. Phelps, S. M.: A Study of Tests for Refractories with Special Reference to Spalling Tests, *Am. Refractories Inst., Tech. Paper* 1, 1926.
4. Preston, E. W.: The Spalling of Bricks, *J. Am. Ceram. Soc.*, **9**, 654, 1926.
5. Goodrich, H. R.: Spalling and Loss of Compressive Strength of Fire Brick, *J. Am. Ceram. Soc.*, **10**, 784, 1927.
6. Endell, K.: Gegen Temperaturänderungen unempfindliche Magnesitsteine, *Stahl u. Eisen*, **52**, 759, 1932.
7. Charvat, F. R., and W. D. Kingery: Thermal Conductivity, XIII, Effect of Microstructure on Conductivity of Single-phase Ceramics, *J. Am. Ceram. Soc.*, **40**, 306, 1957.
8. White, J.: Thermal Shock, *Trans. Brit. Ceram. Soc.*, **57**, 591, 1958.
9. Hargreaves, J.: The Spalling of Silica Bricks, *Trans. Brit. Ceram. Soc.*, **57**, 242, 1958.
10. Selsing, J.: Internal Stresses in Ceramics, *J. Am. Ceram. Soc.*, **44**, 419, 1961.
11. King, D. F., and F. H. Walther: Rotary Hearth Furnace for Testing Resistance of Refractories to Thermal Shock, *Bull. Am. Ceram. Soc.*, **40**, 456, 1961.
12. Dugdale, R. A., J. T. Maskrey, and R. C. McVickers: Some Effects of Thermal Shock Produced by Intense Gas Discharges, *Trans. Brit. Ceram. Soc.*, **60**, 427, 1961.
13. Kukolev, G. V., I. I. Nemets, and M. T. Khomyakov: A Method for Increasing the Life of Saggers, *Glass Ceram., USSR, English Transl.*, **19**, 486, 1963.
14. Lécrivain, L., B. Lambert, and M. Risse: A Study of the Mechanism of the Formation of Spalling Cracks in Refractory Ware, *Ind. Ceram.*, **548**, 15, 1963.

CHAPTER SEVENTEEN

Resistance to Slags and Glasses

17.1 **Introduction.** Slags and glasses are active in attacking refractories; and in fact, in many processes they are the most destructive influence. Unfortunately, it is difficult to make any satisfactory quantitative analysis of the reactions, although recently studies with pure materials under dynamic conditions are beginning to give some idea of the corrosion mechanism. The evaluation of refractories in the laboratory is very difficult, as it seems impossible to make a really satisfactory accelerated test.

17.2 **Chemical Effect of Slag Action.** In general, slag action can be divided into chemical effects and physical effects, the first of which will now be considered.

Equilibrium Conditions. It is very important to make a distinction between equilibrium conditions and the rate of reaction. For example, if an intimate mixture of magnesia and silica is maintained at a definite temperature, it will, after a certain length of time, be converted into magnesium silicate with a relatively low melting point. When the conversion is as complete as the proportion of the substances permits, equilibrium is attained and no further change will occur at that temperature. If the temperature is low, this reaction will progress with extreme slowness; but on the other hand, if the temperature is high, the reaction will take place rapidly; however, in each case, a state of equilibrium will be reached. It will first be advisable to investigate the equilibrium conditions for the combinations of various slags and refractories; for if the equilibrium conditions produce no material melting at a temperature lower than that at which the refractory is maintained, there can be no serious slag reaction. In such a case, the rate of reaction would not concern us. Usually, however, the slag or the material produced by the reaction of the slag with the refractory melts at a lower temperature than that maintained in the furnace.

Owing to the great complexity of the materials entering into slag and glass reactions, we do not have very complete information on the equilibrium diagrams concerned in these reactions. More data of this type are greatly needed. However, it can be stated in general that the slag reaction depends upon the formation of a low-melting compound or mixture.

Composition of Slags and Glasses. Table 17.1 shows typical analyses of commercial slags and glasses. In general, each one has so many elements present that it is difficult to tie them in with any simple system; and when the elements in the refractory are added, the complexity is still further increased.

Equilibrium Diagrams. It will be seen from Table 17.1 that the systems particularly applicable to slag-refractory reactions are

$$SiO_2—Al_2O_3—CaO$$
$$SiO_2—Fe_2O_3—CaO$$
$$SiO_2—CaO—Na_2O$$

It should not be forgotten that small amounts of some other materials such as PbO or F may greatly influence the slag action.

Rate of Reactions. The second factor in slag reactions is the rate at which reactions occur. Suppose that a refractory is washed by a slag capable of combining with the refractory to form a low-melting compound. If the reaction occurs rapidly, these low-melting compounds will be formed in considerable quantities and will run off the face of the wall, thereby exposing fresh surfaces for the attack of the slag; therefore, the slag resistance of the refractory will be poor. On the other hand, let us suppose that the same low-melting-point compounds are formed but that the rate of reaction is slow. Then the compounds will be formed in small quantities, and the resistance of the refractory to the slag may be good. This is the reason why the prevalent idea that basic refractories must be used with basic slags and acid refractories with acid slags is not necessarily true. Although the acidic and basic qualities determine the formation products, they do not necessarily determine the rates of reaction. For example, magnesite will stand up well under the action of certain highly siliceous slags, and silica bricks are known to give good service in lime kilns under highly basic conditions. Also the viscosity of the slag and reaction products has an important bearing on the reaction rate.

The rate of reaction is greatly influenced by the temperature. In

TABLE 17.1 Analyses of Typical Slags and Glasses

	Portland cement clinker	Lead blast furnace	Copper reverberatory	Coal ash, Ind.	Coal ash, Ill.	Acid, open hearth	Basic, open hearth	Blast furnace	Heating furnace	Window glass	Opal glass	Boro-silicate glass	Fuel oil	Basic oxygen furnace
SiO_2	21.9	32.6	47.4	42	43	50.1	23.3	36.8	17.0	72.10	65.8	80.6	44.0	26.6
Al_2O_3	5.6	3.1	29.0	15	17	3.0	0.3	11.7	3.9	0.80	6.6	2.0	1.3	2.7
Fe_2O_3	4.6	30.4*	34.5*	33	27	30.3*	26.6*	3.4*	75.3	0.08	.....	0.2	.....	10.5
TiO_2	.....	.....	.....	.....	1	1.5	1.0	0.5	0.7	.....	.....	.....	27.7	.....
CaO	65.8	19.1	9.4	3	7	2.3	38.0	43.4	0.8	9.10	10.1	0.2	22.5	49.6
MgO	1.4	1.8	1.4	Trace	1	0.9	7.0	2.5	0.1	1.30	.....	0.3	.....	3.7
MnO	.....	0.9	0.4	.....	.....	11.7	4.2	0.9	1.9	.....	.....	.....	.....	4.7
Na_2O	.....	.....	.....	} 7	} 4	} Trace	} 0.1	} 0.6	} 0.1	16.20	3.8	3.8	} 3.3	.....
K_2O	.....	.....	.....							.....	9.6	0.6	1.0	.....
V_2O_5	.....	.....	.....	.....	.....	.....	.....	.....	.....	.....	.....	.....	.....	.....
ZnO	.....	8.3	1.5	.....	.....	.....	.....	.....	.....	.....	.....	.....	.....	.....
CuO	.....	0.2	0.5	.....	.....	.....	.....	.....	.....	.....	.....	.....	.....	.....
PbO	.....	1.2	0.3	.....	.....	.....	.....	.....	.....	.....	.....	.....	.....	.....
SO_2	0.2	.....	.....	.....	.....	.....	.....	.....	.....	0.5	.....	0.7	.....	.....
As_2O_5	.....	.....	.....	.....	.....	.....	.....	.....	.....	Trace	.....	.....	.....	.....
F	.....	.....	.....	.....	.....	.....	.....	.....	.....	.....	5.3	.....	.....	.....
P_2O_5	.....	.....	.....	.....	.....	.....	.....	.....	.....	.....	.....	.....	.....	1.2

* Chiefly FeO.

many instances, an increase of 10°C (18°F) will double the rate of reaction. This is the principal reason why certain refractories will resist slag very well at the usual working temperatures but will be rapidly eaten away if the furnace temperature is increased by 50°C (about 90°F). Unfortunately, practically no information is available, owing mainly to the experimental difficulties, on the rate of reaction of the materials in which we are interested at high temperatures. However, the reaction rate may be approximately expressed by the following equation of Arrhenius:

$$\log \frac{K_{t_2}}{K_{t_1}} = A \left(\frac{1}{T_1} - \frac{1}{T_2} \right)$$

where K_{t_1} = the reaction velocity at temperature T_1, °abs
$\qquad K_{t_2}$ = the reaction velocity at temperature T_2, °abs
$\qquad A$ = a constant of the reaction
The value of A varies but is in the neighborhood of 10,000 for clays and fluxes.

17.3 Physical Effects of Slag Action. *Velocity of Flow.* As the rate of reaction is dependent upon the concentration of the slagging material on the surface of the refractory, it is evident that the more rapid the movement of the liquid slag over the refractory surface the more rapidly the fresh slag will be presented to the surface. It is well known that a liquid may flow in two ways. The first is a streamline flow, where the direction of the flow is parallel to the surface over which it is flowing; and the second a turbulent flow, where the particles of the liquid are undergoing motion in various directions. The transition from streamline flow to turbulent flow depends upon the size, velocity, and viscosity of the melt. It is evident that turbulent flow is much more effective in bringing fresh slag to the surface than is streamline flow; so it is desirable to keep the velocity of flow low and the viscosity high. As the viscosity is dependent upon the temperature, we have another reason why high temperatures promote more rapid slag reaction. It may be said that in general, all slag flows are of the streamline type where the slag is flowing in thin layers over the refractory but, in the cases where the slag is in large quantities, as in glass tanks, the flow may be of the turbulent type for the more fluid glasses.

Diffusion through the Melt. It has been shown by many experimenters such as Reed and Barrett[17] that a reaction zone is built up between the solid refractory and the fluid slag. This zone is a few millimeters in thickness (more or less) but becomes thinner as the flow velocity of the slag increases. This reaction zone may be thought of as a barrier through which the ions diffuse and is therefore the rate-determining factor. With higher relative velocities between the slag and refractory the reaction zone becomes thinner.

Figure 17.1 is a schematic diagram showing the increase of corrosion with time. During time t_0, the rate decreases as the boundary layer is built up, but during the remainder of the time the diffusion-controlled rate is constant. The corrosion rate can be decreased by

1. Decreasing the rate of slag flow and thus increasing the boundary-layer thickness
2. Decreasing the temperature and therefore increasing the viscosity, which decreases the diffusion rate
3. Providing larger atom groups in the melt to decrease the diffusion rate
4. Reducing the active surface of the refractory by eliminating roughness

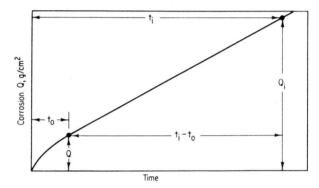

FIG. 17.1   Mechanics of slagging.

In the last few years many ceramists have studied solution of pure refractories, such as alumina, in slags or glasses of controlled composition. Samaddar et al.[18] tested a number of refractories by revolving a cylinder in glass, finding alumina the most resistant. Busby and Eccles[19] studied the corrosion of single crystals of sapphire in several types of glass. Reed and Barrett[17] worked with single and multicrystalline specimens in varying compositions of glass. Vago and Griffith[9] worked with static and revolving specimens of refractory in glass, while Förster and Knacke[12] immersed refractory rods in molten slag contained in a graphite crucible, inducing slag motion by forcing argon through holes on the axes of the specimens. Vago and Smith[14] tested zircon specimens which revolved in calcium alumino-borosilicate glass, and found that the corrosion rate depended upon surface speed at a single temperature.

Slag Penetration. In the previous discussion, we have considered the reaction occurring between the surface of the refractory and the molten slag. If the refractory is a dense material, this surface will be smooth; but on the other hand, if the refractory is porous, the surface may be considerably increased owing to penetration of the slag into the pores. A number of refractories are made from a hard grog and a softer bond, in which case the slag will react with the bond more rapidly than with the grog particles and will allow the latter to project from the surface. It often happens that these grog particles are carried away in the melt before they have become dissolved. This has been noticed in glass melts and, of course, is harmful to the glass. Other types of refractories, even though composed of homogeneous material, will allow the rapid penetration of the slag into the interior.

Busby and Eccles[10] made up a number of slip-cast sillimanite blocks with porosities of 12 to 40 percent by varying the firing temperature and by the addition of naphthalene. They found, as would be expected, greater corrosion rates in the specimens with higher porosities. This work should be repeated where the pore size is systematically varied.

Wetting of the Surface by Slag. There is another side to slag reactions which has not, it is believed, been generally discussed, i.e., the wetting of the refractory surface by the molten slag. It has been noticed that when certain slags are applied to some refractories, they will melt on the surface and form in drops without wetting the surface. A refractory, under such conditions, is practically unaffected by the slag. This condition is just the opposite of many other cases, where the slag is drawn by capillary attraction deep into the pores. It seems quite reasonable to believe that the wetting by the molten slag in contact with the refractory may have a very important influence on the slag resistance of that refractory, and data on the surface tension under these conditions would be of considerable value.

17.4 Laboratory Slag Tests. Of all laboratory tests for refractories, the least satisfactory is the slag test, for it is difficult in a reasonably short time to evaluate the resistance of a refractory. To accelerate the action by higher temperatures or by a greater flow of slag may materially alter the conditions of reaction.

Early Methods of Test. Some of the proposed methods of slag testing consisted in measuring the penetration of slag into the refractory, the fusion point of powdered slag and refractory when mixed together, and the petrographic study of mixtures of slag and refractory after definite heat treatments. These methods measure some factors of slag reaction but not by any means all of them.

Refractory in Slag Baths. A number of experimenters have measured the resistance of a refractory to slags by suspending the specific refractory in a bath of molten slag, with motion to bring new slag to the surface, as referred to in the preceding section.

Running Slag Test.   Another method suggested by Prof. G. B. Wilkes consists of heating a single brick or a column of bricks to a uniform temperature and regularly dropping powdered slag on top, allowing it to run down the face in a groove, as shown in Fig. 17.2.   The depth to which this groove has been cut at the end of the test gives a very good indication of the slag resistance of the brick, because the test is carried on under conditions closely approximating those of service.   Although this test gives excellent relative values, it is not a quantitative test and does not maintain a proper temperature gradient from the slag surface to the other face. Brunner[15] recently used a similar method with a small specimen in an electric furnace.

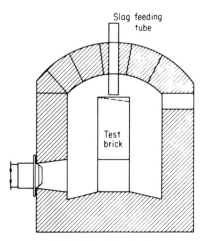

FIG. 17.2   Slag test.

Spray Slag Tests.   Perhaps one of the most satisfactory tests consists in spraying powdered slag against the heated refractory.   A number of experimenters have used this method, but the most practical furnace is that used by Hursh[1] as shown in Fig. 17.3.   Here the powdered slag is fed through a revolving burner impinging on the test bricks set on the inside of the cylindrical furnace.   Thus, all the bricks get the same treatment and the temperature gradient through the bricks is similar to that found in service.   If the test is run for several days, truly quantitative values can be obtained.

Tanks for Testing Glass Refractories.   The only really satisfactory way to test glass-tank blocks is to use them in a small tank that is fed and drawn at regular intervals.   A satisfactory tank for this purpose, shown in Fig. 17.4, is about the minimum size that can be used successfully.   A 2-weeks' run will give a fair indication of the resistance to the glass, but a longer run is sometimes desirable. McMullen[4] describes a small test tank for glass tank blocks that is heated by globars.

17.5 Slag Resistance of Refractories. There are so few really quantitative data on the resistance of refractories to slags that it would seem impossible at the present time to give a table of values that would

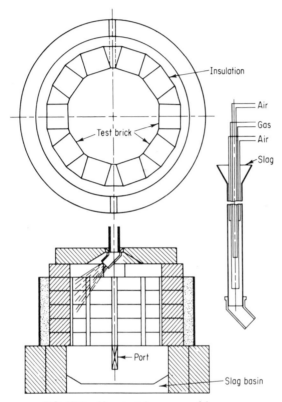

FIG. 17.3 Slag-test furnace and burner.

mean very much. However, from our experience in the use of refractories, certain conclusions can be drawn, which will be discussed in Part 5.

17.6 Products Formed in Slag Reactions. The study of reaction zones between the refractory and slag by petrographic methods is not entirely satisfactory as the specimens cannot be quenched and therefore do not represent the conditions at the use temperature. Nevertheless, they do show penetration and some other factors.

17.7 Gases and Refractories. Many investigators have studied the disintegrating effect of CO on various types of refractory. Berry,

Ames, and Snow[5] show that in the temperature range of 400 to 700°C (about 750 to 1300°F) ferric oxide in localized spots is converted to the carbide which acts as a catalyst. Between temperatures of 400 and 565°C (about 750 and 1050°F) $Fe_{20}C_9$ is the catalyst, while between 565 and 700°C (about 1050 and 1300°F) Fe_3C is the catalyst. The catalyst changes CO to C. The formation of the catalyst may be suppressed if small amounts of sulfur or ammonia are in the gases, whereas water vapor or hydrogen increases their formation rate.

Davis, Slawson, and Rigby[6] have shown that the carbon is formed in threads of rolled-up ribbon 50 to 150 Å in diameter, much like the organic

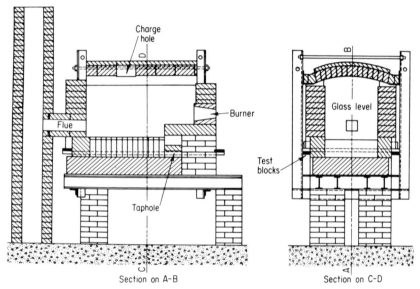

Section on A-B   Section on C-D

FIG. 17.4 Small tank with a 3 × 5 ft melting chamber. (*Babcock & Wilcox Company.*)

chain molecules. When constrained, they are in helical form and produce bursting forces. When free-growing, they are straight.

Hubble[20] has studied the cracking of basic brick in the open-hearth roof. He shows that the best results are obtained with a low-iron chrome, a direct-bonded brick, or a special unburned brick with agglomerated chrome ores.

Grant and Williamson[11] have studied the embrittlement of silica brick in H and CO, but came to no conclusion as to the cause.

Steinhoff[16] shows that magnesite brick reacts strongly with boron containing vapor at temperatures between 1400 and 1450°C (about 2550 and 2640°F) causing SiO_2, CaO, and Al_2O_3 to be carried away. This effect might be a useful method of purifying MgO. Lécrivain and

TABLE 17.2 Maximum Temperature, °F, at Which Two Refractories Can Be in Contact without Reaction*

	Magnesite	Chrome-magnesite	Chrome	90% alumina	70% alumina	Super-duty fireclay	Clay-bonded silicon carbide	Silica	Zircon
Magnesite........	...	+3100	+3000	+3100	2900	2400	2700	2700	2700
Chrome-magnesite....	...		+3100	2900	2700	2900	2700	2900	
Chrome........	...			2900	2700	2700	2700	2900	
90% alumina........	...						2900	2900	
70% alumina........	...						2900	2700	
Superduty fireclay....	...						2900		
Clay-bonded silicon carbide........	...								
Silica........	...							2900	3000
Zircon........	...								

* Abstracted from "Modern Refractory Practice," Harbison-Walker Refractories Company.

Maretheu[21] believe that $FeCl_2$ is released from the burden and condenses in the pores of the cooler refractories, where it is reduced to Fe by zinc vapors. The Fe then catalyzes the CO, causing disintegration. Eusner and Bachman[7] review the hydration tests for magnesite brick and suggest an 80-psi autoclave test for 1 to 10 hr, and a burn-in test for ramming mixes. Coleman and Ford[22] indicate that hydration is proportional to the area of micropores of the burned magnesite. Hubble and Lackey[13] propose hydration tests for dead-burned dolomite.

Jorgensen et al.[8] gives data on the oxidation of silicon carbide in dry oxygen and derives an expression showing the reaction is diffusion-controlled. The oxidized product is first amorphous silica, then converts to cristobalite above 1200°C (about 2190°F). Water vapor in the gas greatly influences the rate of oxidation.

Oxygen. Silicon carbide refractories have a tendency to oxidize slowly. Many efforts have been made to prevent this by glazing the surface as described by Walton[3] or by removing catalytic agents as described by Hartmann.[2]

17.8 Reaction between Refractories in Contact. It is often desirable to know the maximum safe temperature at which one refractory can be heated when set on another. Our available data are not very extensive, and many factors, such as purity and atmosphere, have not been under complete control. However, Table 17.2 gives some data on commercial refractories.

BIBLIOGRAPHY

1. Hursh, R. K., and C. E. Grigsby: A Laboratory Furnace for Testing Resistance of Firebrick to Slag Erosion, *Univ. Illinois Bull. Circ.* 17, 1928.
2. Hartmann, M. L.: Making Silicon Carbide Refractory Articles, U.S. Patent 1,790,474, 1931.
3. Walton, S. F. (Exolon Co.): Method of Applying Protective Refractory Glaze to Refractory Bodies Containing Silicon Carbide, U.S. Patent 1,868,451, 1932.
4. McMullen, J. C.: A Miniature Glass Test Tank, *Bull. Am. Ceram. Soc.,* **26,** 365, 1947.
5. Berry, T. F., R. N. Ames, and R. B. Snow: Deposition of Carbon from Carbon Monoxide, *J. Am. Ceram. Soc.,* **39,** 308, 1956.
6. Davis, W. R., R. J. Slawson, and G. R. Rigby: CO Attack on Firebricks: The Physical State of the Deposited Carbon, *Trans. Brit. Ceram. Soc.,* **56,** 67, 1957.
7. Eusner, G. R., and J. R. Bachman: Hydration of Basic Refractories, *Bull. Am. Ceram. Soc.,* **37,** 213, 1958.
8. Jorgensen, P. J., et al.: Oxidation of Silicon Carbide, *J. Am. Ceram. Soc.,* **42,** 613, 1959.
9. Vago, E., and C. F. Griffith: Corrosion of Refractories by Molten Glass, III, A Comparison between the Results Obtained from Stirring and Static Finger Tests, *Glass Technol.,* **2,** 238, 1961.

10. Busby, T. S., and J. Eccles: The Relationship between Porosity, Firing Temperature, and Corrosion Resistance of Slip-cast Sillimanite, *Glass Technol.*, **2**, 159, 1961.
11. Grant, K., and W. O. Williamson: The Embrittlement of Silica Bricks Re-heated in Hydrogen or Carbon Monoxide, *Trans. Brit. Ceram. Soc.*, **60**, 647, 1961.
12. Förster, E., and O. Knacke: The Solution of Refractories in Slags, *Arch. Eisenhuettenw.*, **33**, 141, 1962.
13. Hubble, D. H., and W. J. Lackey: Hydration Test for Dead-burned Refractory Dolomite, *Bull. Am. Ceram. Soc.*, **41**, 442, 1962.
14. Vago, E., and C. E. Smith: The Corrosion of Zircon Refractories by Molten Glass, III, The Effect of Varying the Speed and Duration of Stirring Tests in a Calcium Alumino-borosilicate Glass, *Glass Technol.*, **4**, 129, 1963.
15. Brunner, M.: Method for the Study of Slag Attack on Refractories, *Trans. Brit. Ceram. Soc.*, **62**, 813, 1963.
16. Steinhoff, E.: The Behaviour of Basic Regenerator-packing towards Boron-containing Gases of a Glass Furnace, III, The Attack of Boric Oxide, Borax, and Rasorite Vapour on Low-iron Magnesite Bricks, *Glastech. Ber.*, **36**, 16, 1963.
17. Reed, L., and L. R. Barrett: The Slagging of Refractories, II, The Kinetics of Corrosion, *Trans. Brit. Ceram. Soc.*, **63**, 509, 1964.
18. Samaddar, B. N., W. D. Kingery, and A. R. Cooper: Dissolution in Ceramic Systems, II, Dissolution of Alumina, Mullite, Anorthite, and Silica in a Calcium-Aluminum-Silicate Slag, *J. Am. Ceram. Soc.*, **47**, 249, 1964.
19. Busby, T. S., and J. Eccles: A Study of the Solution of Single Crystals of Corundum in Molten Glass, *Glass Technol.*, **5**, 115, 1964.
20. Hubble, D. H.: Resistance of Basic Brick to Deterioration from Changes in Atmosphere, *Bull. Am. Ceram. Soc.*, **43**, 506, 1964.
21. Lécrivain, L., and A. Maretheu: Contribution to the Study of the Destruction of Blast-furnace Refractories, *Trans. 9th Intern. Ceram. Congr.*, p. 377, 1964.
22. Coleman, D. S., and W. F. Ford: The Effect of Crystallite Size and Micro-porosity on the Hydration of Magnesia, *Trans. Brit. Ceram. Soc.*, **63**, 365, 1964.

CHAPTER EIGHTEEN

Expansion and Shrinkage

18.1 Introduction. There are a number of reasons why the volume changes of refractories on heating are of interest. In the first place, a knowledge of the reversible expansion and contraction is needed to design proper expansion joints and also to predict the spalling resistance. Secondly, the irreversible change gives a good idea of the maximum temperature at which the refractory may be used. The expansion curve may also be used to detect reversible crystal inversions, as are found, for example, in silica.

18.2 Method of Measurement. The measurements of expansion are usually made on a linear basis using the length change in a bar-shaped specimen heated with a uniform temperature distribution. The length of the specimen may be measured by bringing the movement out of the furnace by means of sapphire rods to a dial indicator or electrical trans-ducer, a method satisfactory up to 1500°C (about 2730°F). A setup of this type is shown in Fig. 18.1, as used by Ruh and Wallace.[12]

A preferable method for higher temperatures is to use a pair of tele-scopes to sight on the ends of the specimen. With a good optical system and a gage length of 6 in., a precision of 6 parts per million is possible. This system is not subject to errors in bringing the reading out of the furnace but does require somewhat more skill in making the readings than the dial method. A furnace for using this method is shown in Fig. 18.2 for temperatures up to 1600°C (about 2910°F), as used by Whittemore and Ault.[8] For temperatures up to 2200°C (about 3990°F) the furnace shown in Fig. 18.3 was used by Nielsen and Leipold.[11]

18.3 Expansion of Typical Refractories. The curves in Figs. 18.4 to 18.11 from a paper by Ruh and Wallace[12] give the expansion properties of typical refractories up to a temperature of 1320°C (about 2400°F). The description of these refractories is given in Table 18.1. It is unfortunate that these tests were not carried to higher temperatures to show when rapid shrinkage or bloating started. For example, expansion

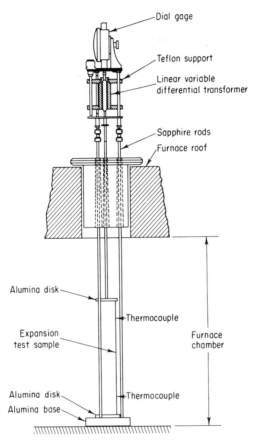

Dial gage

Teflon support

Linear variable
differential transformer

Sapphire rods
Furnace roof

Alumina disk

Thermocouple

Expansion
test sample

Furnace
chamber

Alumina disk
Alumina base

Thermocouple

FIG. 18.1 Schematic of dilatometer assembly. (*Harbison-Walker Refractories Co.*)

curves of two high-heat-duty fireclay brick are shown in Fig. 18.12 when carried up to 1400°C (about 2550°F). The brick made of plastic fireclay shows shrinkage, while the flint-clay brick expands.

18.4 Expansion of Silica Minerals. The thermal-expansion characteristics of the various forms of silica are of interest, because the inversion points are clearly shown. An expansion curve of a silica refractory will indicate roughly the constituents present. In Fig. 18.13 are shown expan-

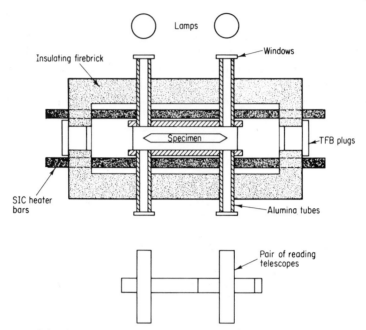

FIG. 18.2 Horizontal section of thermal-expansion furnace.

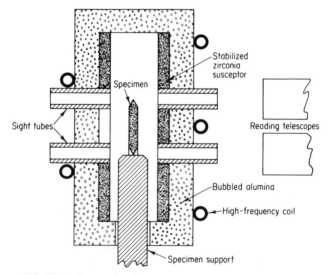

FIG. 18.3 Expansion furnace for use up to 2000°C in air.

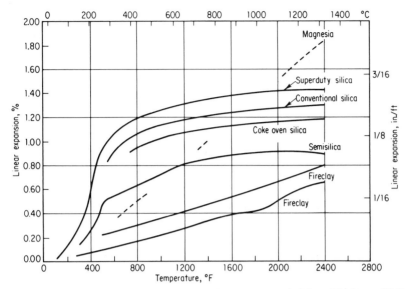

FIG. 18.4 Linear thermal expansion of silica-alumina brick. (*Harbison-Walker Refractories Co.*)

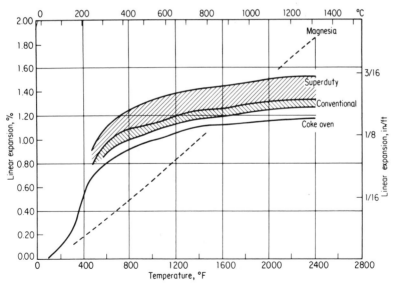

FIG. 18.5 Linear thermal expansion of silica brick. (*Harbison-Walker Refractories Co.*)

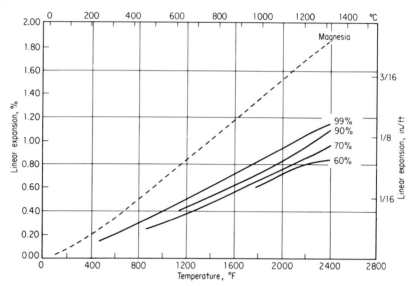

FIG. 18.6 Linear thermal expansion of high-alumina brick. (*Harbison-Walker Refractories Co.*)

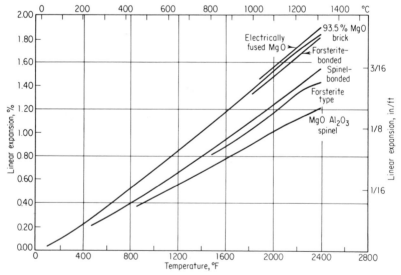

FIG. 18.7 Linear thermal expansion of burned magnesia brick. (*Harbison-Walker Refractories Co.*)

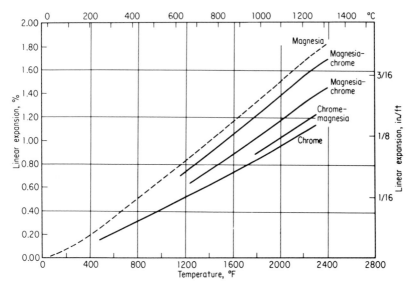

FIG. 18.8 Linear thermal expansion of burned magnesia-chrome brick. (*Harbison-Walker Refractories Co.*)

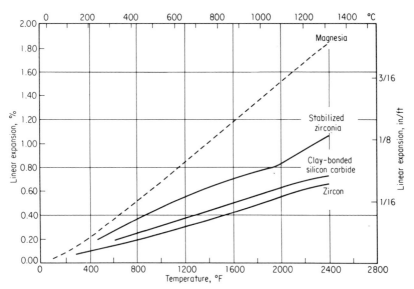

FIG. 18.9 Linear thermal expansion of zircon, stabilized zirconia, and silicon carbide. (*Harbison-Walker Refractories Co.*)

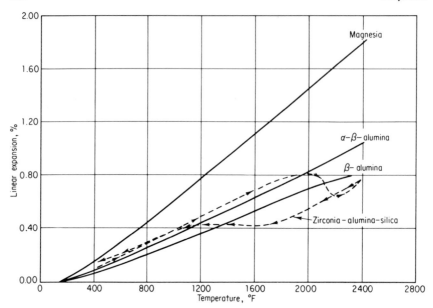

FIG. 18.10   Linear thermal expansion of fusion cast brick.   (*Harbison-Walker Refractories Co.*)

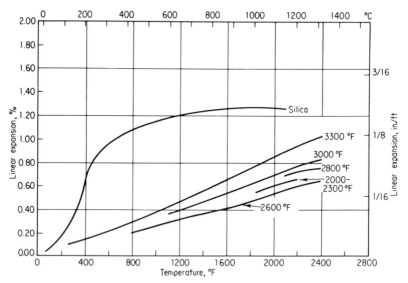

FIG. 18.11   Linear thermal expansion of insulating brick.   (*Harbison-Walker Refractories Co.*)

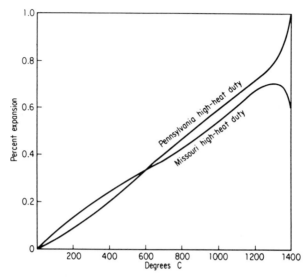

FIG. 18.12 Expansion and contraction of fireclay brick.

sion curves of the four forms of silica. The low expansion of the quartz glass and the sudden changes at the inversion points are noteworthy. All the high-temperature forms of silica have a low coefficient of expansion.

The volume of all forms of silica is comparatively constant above 600°C (about 1100°F); therefore, no spalling troubles will be encountered if the furnace is never allowed to fall below this temperature. Open-hearth steel furnaces, for example, which are constructed to a considerable extent of silica brick, experience large and sudden temperature changes. The temperature, however, is never allowed to fall below 600°C (about 1100°F) or 800°C (about 1470°F) even for repairs.

Unfortunately, quartz glass is not stable above about 1200°C (about 2190°F). If it were, a refractory extremely resistant to temperature changes could be made of it.

18.5 Expansion of Special Refractory Materials. *Graphite.* The coefficient of expansion of graphite is shown in Fig. 18.14 from the work of Allen.[10] It is seen that the coefficient increases with temperature, and in extruded pieces is influenced by the direction due to preferred orientation of the platelets.

Pure Oxides. The expansion of pure oxides in air has been measured by Nielsen and Leipold.[11] These oxides are described in Table 18.2, and the expansion coefficients in Table 18.3.

Negative Expansion. There has been much interest in materials with a zero or negative expansion. One class studied by Smoke[6] is lithium

TABLE 18.1   Physical Properties of Refractory Brick

Description or type	Approximate analysis, %	Bulk density, pcf	Modulus of rupture, psi	Cold crushing strength, psi	Apparent porosity, %
Electrically fused MgO....	99.1 MgO	206	No data	No data	Nil
Magnesia................	93.6 MgO, 4.1 SiO₂	174	3,050	9,110	19.7
Forsterite-bonded magnesia..............	88.4 MgO, 9.6 SiO₂	176	2,370	....	19.1
Forsterite type..........	50.3 MgO, 29.5 SiO₂	164	1,230	4,950	22.1
Spinel-bonded magnesia...	88.5 MgO, 8.3 Al₂O₃	173	970	....	21.8
MgO·Al₂O₃ spinel........	25.0 MgO, 70.0 Al₂O₃	174	3,860	....	19.3
Magnesia-chrome........	79.6 MgO, 6.4 Cr₂O₃	182	530	....	17.4
Magnesia-chrome........	62.5 MgO, 13.3 Cr₂O₃	182	580	....	19.6
Chrome-magnesia........	34.2 MgO, 25.9 Cr₂O₃	188	1,120	4,030	22.4
Chrome.................	34.1 Al₂O₃, 29.7 Cr₂O₃	194	2,680	14,580	15.8
99 % alumina...........	99.4 Al₂O₃, 0.3 SiO₂	180	1,700	5,300	23.4
90 % alumina...........	91.2 Al₂O₃, 7.3 SiO₂	174	1,600	10,370	21.3
70 % alumina...........	69.2 Al₂O₃, 24.8 SiO₂	151	1,750	9,150	22.9
60 % alumina...........	59.8 Al₂O₃, 33.2 SiO₂	144	1,910	7,460	19.5
Superduty silica.........	96.3 SiO₂, 0.2 Al₂O₃	117	1,040	4,990	19.6
Conventional silica.......	95.8 SiO₂, 0.5 Al₂O₃	113	1,130	4,140	22.5
Coke-oven silica.........	94.9 SiO₂, 1.0 Al₂O₃	107	1,180	4,240	25.6
Semisilica..............	77.0 SiO₂, 20.4 Al₂O₃	122	450	2,300	25.5
Superduty fireclay.......	52.9 SiO₂, 42.0 Al₂O₃	149	2,340	11,940	9.8
Superduty fireclay.......	52.2 SiO₂, 43.2 Al₂O₃	146	2,800	9,000	11.0
Stabilized zirconia.......	93.8 ZrO₂ + HfO₂, 4.0 CaO	268	940	6,000	23.2
Clay-bonded silicon carbide..............	89.2 SiC, 7.3 SiO₂	166	3,770	9,430	9.0
Zircon.................	65.8 ZrO₂ + HfO₂, 31.9 SiO₂	235	1,370	11,510	17.4
3300°F low-iron IFB*.....	79.3 Al₂O₃, 17.4 SiO₂	78	230	700	
3000°F low-iron IFB......	65.9 Al₂O₃, 30.9 SiO₂	57	45	700	
2800°F low-iron IFB......	61.7 Al₂O₃, 34.4 SiO₂	53	45	800	
2600°F low-iron IFB......	38.4 Al₂O₃, 57.1 SiO₂	50	170	140	
2300°F low-iron IFB......	38.3 Al₂O₃, 44.8 SiO₂	31	150	140	
2000°F low-iron IFB......	38.4 Al₂O₃, 44.2 SiO₂	29	120	100	
Silica insulating firebrick..	0.9 Al₂O₃, 92.5 SiO₂	59	140	170	

* Insulating firebrick.

aluminum silicate. Some composition areas show very low or negative expansions, but they are not refractory, as 800°C (about 1470°F) is approximately the upper use limit. A more promising material is aluminum titanate. Buessem et al.[7] describe its very low or negative expansion even up to temperatures of 1350°C (about 2460°F). The crystal of aluminum titanate has one of the highest thermal-expansion anisotropies known, and it is felt that this causes microcracks in cooling and thus produces an overall size stability. This microcrack structure also causes low mechanical strength at room temperature; so no use as a refractory seems to have developed from this interesting material.

18.6   Reheat Shrinkage of Refractories.   To determine the fitness of a particular brick for service, it is often tested for shrinkage under tempera-

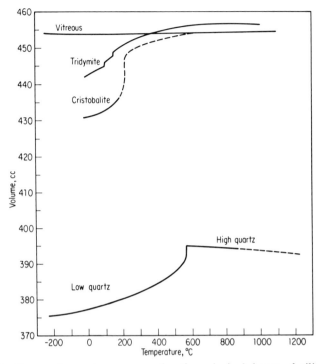

FIG. 18.13 Temperature-volume curves for the principal forms of silica. (*R. B.· Sosman, "The Properties of Silica," Reinhold Publishing Corporation, New York, 1927.*)

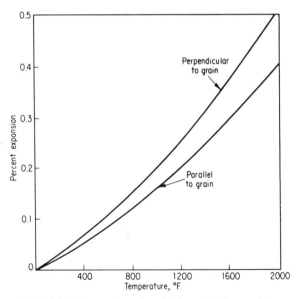

FIG. 18.14 Expansion of molded synthetic graphite.

TABLE 18.2   Properties of Oxides

Treatment or property	MgO	CaO	Al$_2$O$_3$*	MgO·Al$_2$O$_3$
% theoretical density......	93.3	96.3	92.5	98.5
Grain size...............	95	111	66	3
Forming method.........	Isostatic, 30,000 psi	Slip-cast	Slip-cast	Hot-pressed
Fired at.................	1750°C, 24 hr	1650°C, 10 hr	1650°C, 8 hr	
Atmosphere..............	Air	Air	Air	Argon

* Plus 2% Y$_2$O$_3$.

Table 18.3   Mean Coefficient of Linear Thermal Expansion from 25°C to Indicated Temperature ($\times$ 10^{-6})

Material	400°C	800°C	1200°C	1600°C	2000°C	2300°C
MgO....................	13	14	15	16	17	19
CaO....................	13	14	15	16	16	
MgO·Al$_2$O$_3$..............	8	9	9	10		
Al$_2$O$_3$.................	8	8	9	10		

ture conditions equivalent to those which it would receive in use.   This aftershrinkage may be determined by the bar method, as described earlier in this chapter, which has the advantage of giving a continuous record of the shrinkage throughout the range of temperatures, but often it is desirable to make a test on a whole brick.

This is done by first determining the length or volume of the brick by measurement and then subjecting it to a prolonged heating at the desired temperature.   After the brick has cooled, it is again measured and the length, volume, and shrinkage determined.

The standard ASTM test for reheat shrinkage is C 113–36.

Another test that determines the aftershrinkage of a brick is the simulative service test used by the U.S. Navy.   According to the description by Rogers,[1] it consists in building up two walls of a furnace, one of the brick to be tested and the other of a standard brick.   The furnace is then run for 24 hr at a temperature of 1590°C (about 2900°F) and again at 1650°C (about 3000°F), with an examination of the walls between the tests.   An examination of the wall after it is cooled will indicate by the width of the open joints the approximate shrinkage of the bricks.   It is believed that a wall which shrinks a small amount at each joint is better than a wall which concentrates its shrinkage into a few large cracks.   It should also be realized that a wall may have a considerable amount of open joints when cool and yet be substantially tight when at a high temperature.

As previously mentioned, bricks may show either an expansion or a shrinkage when heated to the temperatures of service. In general, bricks are not burned so high in the process of manufacture as the temperature of service; consequently, with the usual type of clay, the brick will decrease in volume at service temperatures. This is undesirable, especially in crowns and arches, because the open joints resulting give corners for spalling and slag erosion to start. The open cracks also concentrate the load in the arch on a relatively small area of the block, which greatly decreases the stability of the structure. A slight expansion is not particularly harmful if the wall or arch is properly designed.

Theory of Secondary Expansion. There are three phenomena that may produce an increase in volume of ware subjected to the reheat test. They are

1. Overfiring, i.e., development of a vesicular structure, or bloating
2. Opening of the laminations of the clay
3. Transformations or reactions in the crystalline phases during the firing, yielding products of lower true density than the reactants

In general, these phenomena take place in certain definite temperature ranges. When the temperature at which a brick is first fired is below this range, the brick may expand if subjected to a reheat temperature in the expansion range. When the temperature of the initial firing is in the expansion range, the ware may shrink, expand, or stay constant in volume in the reheat test, depending upon the time and the atmosphere of firing.

The reactions that take place to cause overfiring have been discussed in Chap. 12. Overfiring occurs in certain clays, especially plastic clays, and it is probable that the decomposition of some materials, perhaps sulfides and sulfates, is inhibited so gases are not given off until a temperature near the fusion point is reached. These gases form bubbles in the glassy portion of the clay, and an expansion may result.

True secondary expansion may be caused by either 2 or 3 above. The former was investigated by Everhart,[2] who concluded that the increase in bulk volume was due to the opening of natural laminations or bedding planes in the clay grains. Fine grinding, blending of expanding and shrinking clays, slow firing, oxidizing atmosphere, and high forming pressure are suggested as means to decrease or eliminate the tendency to expand. This secondary expansion is characteristic of many clay deposits. The secondary expansion of fireclay bricks is undoubtedly due to the opening of the laminations in the clay grains. Reference should also be made to the paper on this subject by Heindl and Mong.[3]

The discussion of the changes in kyanite during heating and the firing properties of clay-kyanite bodies in Chap. 12 is an excellent example of secondary expansion due to 3 above. Referring to curve D in Fig. 12.9, it is clear that a brick of that composition fired first at 2400°F (about 1320°C) would show considerable expansion if reheated at 2700°F (about 1480°C). However, if the reheat temperature were above 2950°F (about 1620°C), shrinkage would result.

The expansion of refractories having high-alumina grog (corundum and mullite) bonded with fireclay or kaolin is due to transfer of silica into the grog grains by diffusion, which converts some of the corundum to the more bulky mullite, as clearly shown by Hall.[4] On the other hand, McGee and Dodd[9] have determined the expansion of high-alumina refractories made with bauxite due to secondary mullitization. The crystal growth seemed to expand the structure, giving more porosity.

The role of sulfur in controlling the volume changes in reheat tests is discussed by Lesar[5] and his associates. They show that soluble sulfur compounds added to the clay can be made to compensate for the shrinkage of the normal clay or can be made to produce considerable expansion.

BIBLIOGRAPHY

1. Rogers, G. L.: Use of the Simulative Test Furnace as a Means of Making Comparative Tests of Fire Brick, *J. Am. Ceram. Soc.*, **11**, 323, 1928.
2. Everhart, J. O.: Secondary Expansion in Refractory Clays, *Ohio State Univ. Expt. Sta. Bull.* 98, p. 23, 1938.
3. Heindl, R. A., and L. E. Mong: Length Changes and Endothermic and Exothermic Effects during Heating of Flint and Aluminous Clays, *J. Res. Natl. Bur. Stds.*, **23**, 427, 1939.
4. Hall, J. L.: Secondary Expansion of High-alumina Refractories, *J. Am. Ceram. Soc.*, **24**, 349, 1941.
5. Lesar, A. R., C. A. Krinbill, Jr., W. D. Keller, and R. S. Bradley: Effect of Compounds of Sulphur on Reheat Volume Change of Fireclay and High Alumina Refractories, *J. Am. Ceram. Soc.*, **29**, 70, 1946.
6. Smoke, E. J.: Ceramic Compositions Having Negative Linear Thermal Expansion, *J. Am. Ceram. Soc.*, **31**, 87, 1951.
7. Buessem, W. R., et al.: Thermal Expansion of Aluminum Titanate, *Ceramic Age*, **60**, 38, 1952.
8. Whittemore, O. J., Jr., and N. N. Ault: Thermal Expansion of Various Ceramic Materials, *J. Am. Ceram. Soc.*, **39**, 443, 1956.
9. McGee, T. D., and C. M. Dodd: Mechanism of Secondary Expansion of High-alumina Refractories Containing Calcined Bauxite, *J. Am. Ceram. Soc.*, **44**, 277, 1961.
10. Allen, R. D.: Thermal Expansion of Synthetic Graphites between 80° and 2000°F, *Bull. Am. Ceram. Soc.*, **41**, 460, 1962.
11. Nielsen, T. H., and M. H. Leipold: Thermal Expansion in Air of Ceramic Oxides to 2000°C, *J. Am. Ceram. Soc.*, **46**, 381, 1963.
12. Ruh, E., and R. W. Wallace: Thermal Expansion of Refractory Brick, *Bull. Am. Ceram. Soc.*, **42**, 52, 1963.

CHAPTER NINETEEN

Heat Transmission by Refractories

19.1 Introduction. In certain parts of a furnace, such as the walls or roof, a minimum transmission of heat to the outside is desired; hence, a refractory with a low thermal conductivity would be necessary for these conditions. On the other hand, such parts as muffles or underfired hearths require a refractory with a high conductivity value. It will therefore be seen that the furnace designer must have accurate data on the thermal conductivity of his refractories before he can intelligently select his materials or predict the performace of his furnace.

19.2 Laws of Heat Conduction. *Definitions and Units.* The following symbols have been generally adopted in reference to heat flow:

q = the quantity of heat conducted in Btu per hr
t_0 = the initial temperature in variable flow
t_1 = the temperature of the hotter surface, °F
t_n = the temperature of the cooler surface, °F
L = the thickness of the wall, in. (or length of heat path)
k = the conductivity of the material, Btu/(hr)(ft²)(in.)(°F)
A = the area of the section normal to the flow, sq ft
x = the distance from the hotter surface, in.
h^2 = the diffusivity = k/pc
ρ = bulk density, pcf
c = specific heat
τ = time, hr

Table 19.1 gives factors to convert the common units of heat flow into others. Throughout the chapter, engineering units will be used, except for the values of single crystals, glasses, and other pure materials.

TABLE 19.1 Conversion Factors*

Units of conductance of energy: thermal and electrical

	g-cal/(sec)(cm²)(cm)(°C)	kg-cal/(hr)(cm²)(cm)(°C)	watt/(cm²)(cm)(°C)	watt/(in.²)(in.)(°C)	kw/(ft²)(in.)(°C)	kw/(ft²)(in.)(°F)	Btu/(hr)(in.²)(in.)(°F)	Btu/(hr)(ft²)(in.)(°F)	Btu/(hr)(ft²)(ft)(°F)	Btu/(hr)(ft²)(in.)(°C)
1 g-cal/(sec)(cm²)(cm)(°C)	1	3.600	4.186	10.63	1.531	0.8506	20.16	2903	241.9	5225.8
1 kg-cal(hr)(cm²)(cm)(°C)	0.2778	1	1.163	2.953	0.4253	0.2363	5.600	806.4	67.20	1452
1 watt/(cm²)(cm)(°C)	0.2389	0.8600	1	2.540	0.3658	0.2032	4.816	693.5	57.79	1248
1 watt/(in.²)(in.)(°C)	0.09405	0.3386	0.3037	1	0.1440	0.08000	1.896	273.0	22.75	491.5
1 kw/(ft²)(in.)(°C)	0.6531	2.351	2.734	6.944	1	0.5556	13.17	1896	158.0	3413
1 kw/(ft²)(in.)(°F)	1.176	4.232	4.921	12.50	1.800	1	23.70	3413	284.4	6143
1 Btu/(hr)(in.²)(in.)(°F)	0.04960	0.1786	0.2076	0.5274	0.07595	0.04219	1	144.0	12.00	259.2
1 Btu/(hr)(ft²)(in.)(°F)	3.445×10^{-4}	0.001240	0.001442	0.003663	5.274×10^{-4}	2.930×10^{-4}	0.006944	1	0.08333	1.800
1 Btu/(hr)(ft²)(ft)(°F)	0.004134	0.01488	0.01730	0.04395	0.006329	0.003516	0.08333	12.00	1	21.60
1 Btu/(hr)(ft²)(in.)(°C)	1.914×10^{-4}	6.889×10^{-4}	8.011×10^{-4}	0.002035	2.930×10^{-4}	1.628×10^{-4}	0.003858	0.5556	0.04630	1

* Based on International Critical Tables; mean values used for kg-cal and Btu; from White, American Ceramic Society.

Steady Flow. For conditions of steady flow, the amount of heat passing through a wall is expressed by

$$q = \frac{kA(t_1 - t_n)}{L}$$

or if we consider a unit area of 1 sq ft, then

$$q = \frac{t_1 - t_n}{L/k} \tag{19.1}$$

for a simple wall.

In a composite wall made up of several layers, the value of q is obviously the same for each layer under steady conditions; therefore,

$$q = \frac{t_1 - t_n}{(L_1/k_1) + (L_2/k_2) + \cdots + (L_n/k_n)} \tag{19.2}$$

where $L_1, L_2, \ldots, L_n$ and $k_1, k_2, \ldots, k_n$ refer to the properties of each layer, respectively.

Since the temperature drop across each layer is equal to the heat flow times the actual resistance of the respective layers, we may say

$$t_1 - t_2 = q\frac{L_1}{k_1} \tag{19.3}$$

$$t_2 - t_3 = q\frac{L_2}{k_2} \cdots \tag{19.4}$$

From these temperature drops, the interface temperatures can be computed. It should also be remembered that the value of q entering the hot face and leaving the cool face is equal to the q of each layer.

Variable Flow. In practice, many cases of unsteady heat flow occur, as in batch-type furnaces. In a simple case, suppose that a wall is suddenly heated from t_0 to t_1 on one face. The temperature at a point x distant from the face is given by Fourier's equation

$$t = \frac{2t_0}{\sqrt{\pi}} \int^{\infty} \frac{x}{2h\sqrt{\tau}} e^{-\beta^2} \, d\beta \tag{19.5}$$

where $\beta = $ a variable of integration.

It will be seen that the penetration of heat will depend on time and the value of diffusivity. An infinite time is required theoretically to reach equilibrium, and practical equilibrium requires many hours in thick walls. In Fig. 19.1 are shown curves of temperature rise of the cool face in refractory walls of various thicknesses that have suddenly been heated on one face. When the hot face is not brought up suddenly but at a definite schedule, the calculations are difficult. They may be approximated, however, by adding an imaginary layer to the hot face of the wall, as

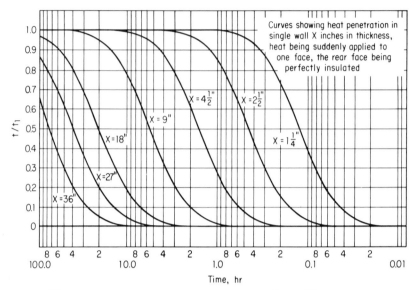

FIG. 19.1 Heat-penetration curves; t_1 = temperature of heated face, t = temperature of rear face, x = wall thickness.

suggested by Ingersoll and Zobel,[1] or the Schmidt method may be used as illustrated by Trinks,[2] or a computer may be used.

19.3 Theory of Heat Transfer through a Porous Body. Since most refractories are more or less porous, it is of interest to see how this porosity influences the transfer rate.

Conduction by the Solid. Some of the heat is conducted through the body by virtue of the close contact of the particles making up the structure. This conduction will be decreased as the porosity increases.

Conduction by the Gas. In highly porous structures the conductance of heat by the gas in the pores may be a sizable proportion of the total conductance, as discussed later in Sec. 19.5.

Convection. If the pores are filled with air or combustion gases, as would usually be the case, some heat is transferred by convection currents in the pore itself. This transfer is so small as to be negligible for the usual size of pores.

Radiation. A considerable portion of the heat flow at high temperatures may be due to direct radiation across the pores. For a given amount of porosity, the large pores will transfer more heat by radiation.

The radiant heat transferred is also proportional to the emissivity of the refractory surface. These total emission coefficients are now accurately known at high temperatures. During 1948, J. M. Brownlow, working under Professor Wilkes in the Heat Measurements Laboratory of the

Massachusetts Institute of Technology, developed an ingenious method of measuring total emission coefficients. This consisted of revolving a cylinder of refractory in a globar-heated furnace, with a narrow portion of the cylinder exposed to the open. The true temperature of the cylinder, kept uniform by rapid turning, was measured by a platinum thermocouple on the axis, and the apparent surface temperature at the exposed point, with a radiation pyrometer. From these two readings the coefficient could be computed readily.

The results on a few materials are shown in Fig. 19.2. The fused magnesia, alumina and zirconia specimens were made by the Norton Company of their regular material. The kaolin insulating brick was a Babcock and Wilcox K-28, and the dense kaolin was this same specimen coated with a kaolin cement. Most of the specimens have values quite close together except for the silicon carbide, which averages 0.9. The influence of surface roughness is very evident in the kaolin materials, as the porous

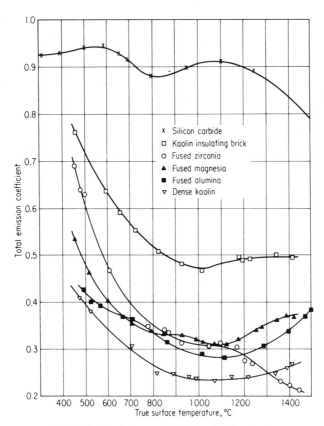

FIG. 19.2 Emission coefficients of refractories.

structure has twice the value of the dense one.   Further work should be carried out in this field.

The theory set up by Loeb[8] now allows the calculation of the k value of porous refractories if the k value is known for the solid material, along with the geometry of the pore structure and the emission coefficient.   It is often more exact to calculate a k value than to measure it.

19.4   Measurement of Thermal Conductivity.   *The High-temperature Calorimeter Method.*   Most of the high-temperature tests on refractories for thermal conductivity have been made by setting up a uniform parallel flow of heat normal to the faces of a slab.   The quantity of heat flowing is measured by a water calorimeter on the cold face of the specimen, and the temperature gradient by the thickness and hot- and cold-face temperatures.

The establishment of a uniform flow is by no means easy, and only by careful guarding can this be accomplished.   Also the temperatures on the isothermal planes must be read with great care.   At the same time, an equilibrium flow must be established by a constant hot-face temperature for a long time interval.   Perhaps no other physical measurement is beset with so many pitfalls.   The papers by Norton[3] and Patton and Norton[4] give an excellent discussion of these errors and their correction.

This method can best be explained by an example of a high-temperature plate tester originally suggested by Wilkes but later constructed in many laboratories with minor changes.   The following description applies to an apparatus recently constructed in the refractories laboratory of the Babcock and Wilcox Company and later made ASTM Standard Method C 202–47.

A drawing of this apparatus is shown in Fig. 19.3.   In Fig. 19.4 is shown the arrangement of water passages in the calorimeter and guard ring.

Special features are the constant-temperature water source, the temperature control of the furnace to maintain the temperature for the 24-hr run within $\pm 2°F$ (about 1°C), and the flow passages in the calorimeter.

Other methods have been used for measuring the thermal conductivity. In the radial method, heat may be generated inside a hollow cylinder and the temperature drop in the wall measured.   However, this method requires end corrections.   The same method may be applied to a hollow sphere where there are no corrections needed, but the temperature readings are difficult.   A shape having unique advantages is the hollow ellipsoid shown in Fig. 19.5.   The thermocouples lie in the isothermal planes where the wall is thickest.   Loeb[6] has shown that such a shape has equal amounts of heat flowing through the walls per unit area at any point, and has given equations for simple calculation of k values.

The Northrup method consists in passing a uniform, linear flow of heat through two layers of refractory in contact.   If the temperature drop is

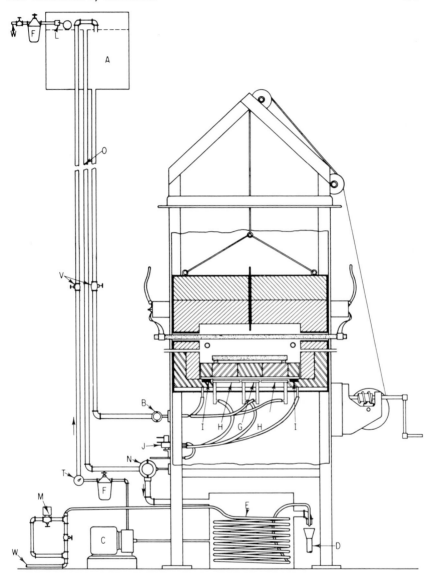

FIG. 19.3 Diagram showing essential parts of thermal-conductivity apparatus. *A*, constant-head water supply. *B*, inlet manifold and thermometer. *C*, circulating pump. *D*, to drain. *E*, cooling coil. *F*, water filter. *G*, center calorimeter. *H*, inner guard calorimeters. *I*, outer guard calorimeters. *J*, microregulating valves. *L*, water-level valve. *M*, magnetic control valve. *N*, outlet manifold. *O*, overflow pipe. *T*, thermostat (controls *M*). *V*, valves. *W*, water inlet.

FIG. 19.4   Water passages in the calorimeter and guard ring.

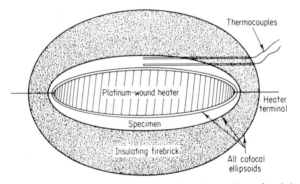

FIG. 19.5   Absolute method for measuring thermal conductivity.

known through each layer and the value of k is known for one material, the conductivity of the other layer can be readily determined, as the value of q must be the same for both. The following relation holds:

$$\frac{\text{Temperature drop in specimen } A}{\text{Temperature drop in specimen } B} = \frac{k_B}{k_A}$$

Harmonic-flow Method.   A somewhat different method has been used to determine the thermal conductivity by making use of Fourier's equation for nonuniform heat flow. If we consider Eq. (19.5), it will be noticed that the value of h^2, the diffusivity, can be determined by studying the temperature changes in a given material. From the value of h, the thermal conductivity can be computed if the density and specific heat are known, properties that are not difficult to measure. This method has been carried out practically by applying to one surface of the test material a temperature varying harmonically. The temperature is then recorded inside the solid at several distances from the surface; and from these values, h can be computed. It will be noticed that this method does not necessitate a measurement of the quantity of heat flowing. On the other hand, the same difficulties are encountered in ensuring a linear flow of heat as in the other methods, and the problem of providing a sinusoidal heat supply at high temperatures is a serious one. Another method of measurement of this type consists in computing the diffusivity by measuring the temperature change on the axis of a cylinder, the outside temperature of which is suddenly changed by quenching. This may be found quite convenient and precise for small specimens such as pure oxides.

19.5 Thermal-conductivity Values for Refractories. A great deal of work has been carried out on the measurement of thermal conductivity in the last 10 or 15 years. The long-continued research in this field in the Ceramics Group at the Massachusetts Institute of Technology is noteworthy. More recently, government bureaus have been fostering work at extremely high temperatures for aerospace materials. For the heavy refractories, many companies have published results from the ASTM apparatus.

Pure Oxides. Table 19.2 gives the k values of a number of pure sintered oxides taken from various M.I.T. papers. It will be noticed

TABLE 19.2 Thermal Conductivity for Pure Oxides
Corrected to zero porosity, cgs Units

Oxide	Mean temperature, °C							
	200	400	600	800	1000	1200	1400	1600
Al_2O_3.........	0.054	0.031	0.022	0.017	0.015	0.013	0.013	0.015
BeO..........	0.417	0.222	0.112	0.065	0.049	0.041	0.039	0.036
CaO..........	0.027	0.022	0.020	0.019	0.019			
MgO..........	0.068	0.039	0.028	0.020	0.017	0.015	0.014	0.014
NiO..........	0.024	0.017	0.014	0.011	0.011			
TiO_2..........	0.012	0.009	0.009	0.008	0.008	0.008		
ThO_2.........	0.204	0.143	0.010	0.008	0.007	0.006		
ZnO..........	0.041	0.027	0.017	0.016	0.013			
ZrO_2..........	0.005	0.005	0.005	0.005	0.005	0.006	0.006	
UO_2..........	0.019	0.014	0.011	0.009	0.008			
$Al_6Si_{12}O_{13}$.......	0.013	0.011	0.010	0.010	0.009	0.009		
$MgAl_2O_4$.......	0.031	0.024	0.019	0.016	0.014	0.013		
$ZrSiO_4$.........	0.014	0.012	0.011	0.010	0.010	0.009		

that the thermal conductivity of the oxides is high for those containing light elements and low for those with heavy elements.

The conductivity of crystals, glass, and a combination of both is discussed by Lee and Kingery.[10] Grain boundaries serve as scattering centers, so that multicrystalline structures are generally poorer conductors than single crystals of the same material.

Heavy Refractories. Ruh and McDowell[11] have tested many types of refractory brick by ASTM C 201–47 apparatus, which gives us the best overall picture of their properties. These values are given in Table 19.3. It will be seen that the fireclay and silica brick are in a narrow range of 7 to 15 units, while alumina, magnesia, and silicon carbide are in much higher ranges.

Washburn and Bart[15] give *k* values for silicon carbide brick with clay

TABLE 19.3 Mean Thermal Conductivity of Heavy Refractories*
Expressed in Btu/(hr)(ft²)(°F)(in.)

Type	Apparent porosity	Mean temperature, °F			
		500	1000	1500	2000
99% Al₂O₃	23	30	21	18	18
90% Al₂O₃ (dense)	16	23	20	18	18
90% Al₂O₃	21	14	13	13	13
70% Al₂O₃	22	11	10	10	10
60% Al₂O₃	19	10	10	10	10
Superduty	10	10	11	11	12
Superduty, blast-furnace	8	11	12	12	13
High-duty	14	8	8	9	9
High-duty	20	9	9	10	10
Semisilica	24	6	7	8	8
Silica, regular	23	8	9	10	12
Silica, superduty	20	9	10	11	12
Silica, coke-oven	26	7	8	9	10
Silica, insulating	60	2.4	3.2	4.1	5.2
Magnesia	20	60	40	32	28
Chrome	16	14	14	14	14
MgO·Cr₂O₃	23	12	12	12	12
Forsterite	22	11	10	10	10
MgO·Cr₂O₃ (unburned)	10	16	14	13	12
Silicon carbide (clay-bonded)	9	120	110	100	90
Zircon	17	23	18	17	17
Tar-bonded magnesia	...	39	30	26	25
Tar-bonded dolomite	...	26	23	18	16
Impregnated magnesia	...	62	46	38	34

* Abstracted from Ref. 11.

bond, Si_3N_4 bond, and Si_2ON_2 bond with good agreement with Ruh and McDowell.

Clements and Vyse[9] have published the k values of a number of refractories, including carbon, with a good discussion of the literature in general.

Insulating Materials. These insulating materials are porosified by various means in order to break up the path by which the heat travels through the solid. It is convenient to divide them into two classes: first, hot-face insulation such as insulating firebrick or castables, and second, backup insulation.

The insulating refractories have been gradually raised in use limit until now 3300°F has been reached, but of course, the higher the use temperature, the higher will be the k value as shown in Table 19.4.

TABLE 19.4   Mean Thermal Conductivity of Insulating Firebrick*
Expressed in Btu/(hr)(ft²)(°F)(in.)

	Weight, pcf	Mean temperature, °F			
Type		500	1000	1500	2000
2300.....................	30	0.9	1.2	2.0	2.3
2600.....................	49	2.0	2.4	2.7	3.2
2800 fireclay.............	57	2.1	2.5	2.8	3.3
2800 high-alumina........	56	2.0	2.4	2.7	3.2
3000 high-alumina........	64	3.0	3.2	3.3	
3300 bubble alumina......	87	6.5	6.2	6.1	6.1

* Abstracted from Ref. 11.

Castables vary in density[12] over a wide range, the k value generally increasing with the density as shown by Hansen and Livovich,[7] who tested a number of refractory concretes and plotted k values against density as shown in Fig. 19.6.

Backup insulation is of many types, all with relatively low densities and k values. Some of the commonly used materials are listed in Table 19.5.

It has long been known that the thermal conductivity of a porous solid varies with the gas composition in the pores. Young et al.[13] measured the thermal conductance of insulators in various gases and developed a relation to predict the conductance of a brick in any gas. It happens that the conductivity of combustion gas is very close to that of air but in furnaces with helium or hydrogen atmosphere there is a marked increase in conductance of a porous solid.

The mechanism of heat transfer through a porous solid has been discussed by many writers such as Loeb[8] and Barrett.[5] The total con-

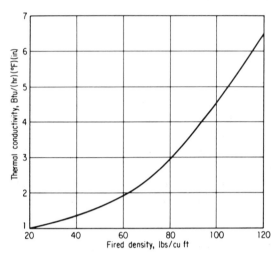

FIG. 19.6   Thermal conductivity of insulating concretes.

TABLE 19.5   Thermal Conductivity of Insulators
Expressed in Btu/(hr)(ft²)(°F)(in.)

Type of insulation	Weight, pcf	Mean temperature, °F		
		200	500	1000
Insulating blanket, mineral wool..............	8–12	0.4	0.6	
Insulating blanket, glass wool.................	3	0.3	0.5	
Insulating blanket, kaolin wool................	6	0.3	0.4	0.9
Insulating cement, diatomite base.............	40	0.7	0.8	0.9
Insulating cement, 85% magnesia.............	15	0.5	0.6	
Insulating cement, vermiculite base............	15	0.8	0.9	1.2
Block insulation, diatomite...................	23	0.6	0.7	0.8
Block insulation, vermiculite base.............	18	0.6	0.7	0.9

ductance may be broken down into four channels: conduction through the solid, conduction through the gas, convection in the gas, and radiation across the pores.   Their relative value varies with the porosity, size of pores, emission coefficient of the solid, conductivity of the gas and solid, and of course, the temperature.   For an insulating refractory of 70 percent porosity and the mean temperature of 1000°F, values have been given[13,16] in Table 19.6.

While this is not a strict comparison between Young and Barrett's work, as the exact characteristics of the refractories are not known, it does show that there is a great difference in the share taken by the solid in the two cases.   More work is needed in this field.

Clements[16] gives an excellent discussion of heat transfer in insulating

TABLE 19.6   Heat Transfer in a Porous Solid

Type of transfer	Barrett	Young
Total conductance.....................	100	100
Solid conductance....................	74	22
Gas conductance.....................	11	39
Heat transfer by convection...........	0	3
Heat transfer by radiation.............	15	36

firebricks.   He feels that conduction through the solid is important and that finer pores would be desirable at higher temperatures.

19.6   Flow of Heat through Walls under Steady Conditions. *Computation of Heat Flow in a Simple Wall.*   The flow of heat through simple walls is dependent on the thermal conductivity at the mean wall temperature and the temperature difference between the hot and cold face.   Also it should be remembered that the quantity of heat passing through the wall must be equal to the heat leaving the wall.   The latter is also dependent on the cold-face temperature and the velocity of the air outside the wall.

The calculation of the heat flow will be made clearer by a specific example.   Assume a 9-in. refractory wall heated on the hot side to 2200°F.   The outside air is 80°F and not moving.   The values of k for this particular refractory are given by curve A in Fig. 19.7.   The quantity

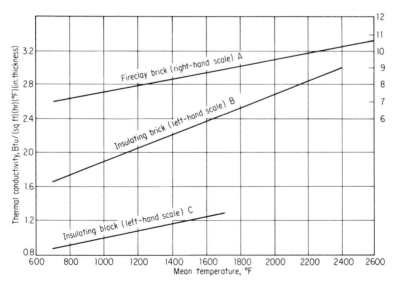

FIG. 19.7   Average curves of thermal conductivity used in calculations.   (*From R. H. Heilman, ASTM Committee C 8 on Refractories.*)

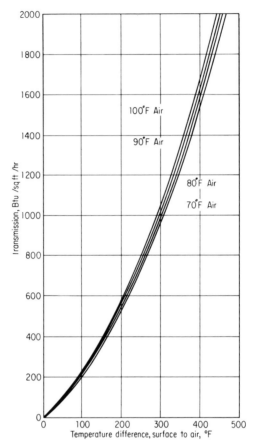

FIG. 19.8   Heat transmission from a vertical surface.   (*From R. H. Heilman, ASTM Committee C 8 on Refractories.*)

of heat lost from the outside of the wall q_{rc} is given by Heilman's curve in Fig. 19.8.   We may write from Eq. (19.1)

$$q = \frac{t_1 - t_n}{L/k} = \frac{2200 - t_n}{9/k} = q_{rc}$$

but as k varies with $(t_1 - t_n)/2$ and q_{rc} varies with t_n, the equation cannot be solved directly.   Therefore, we must assume a mean temperature and an outside temperature.   If q and q_{rc} are not equal, another and closer assumption of values is made until the required precision is reached. Assume $t_n = 300°F$ and $k = 8.1$, then

$$q = \frac{2200 - 300}{9/8.1} = 1710 \qquad q_{rc} = 620 \text{ from curve}$$

The cold face has been taken too low; so 500 will be taken for the next trial.

$$q = \frac{2200 - 500}{9/8.1} = 1720 \qquad q_{rc} = 1530 \text{ from curve}$$

This gives a mean temperature of $1350°F$; hence, a more precise value of k from the curve is 8.2. Choosing a lower cold-face temperature of $480°F$ gives

$$q = \frac{2200 - 480}{9/8.2} = 1570 \qquad q_{rc} = 1590$$

which is sufficiently close.

Computation of Heat Flow in a Compound Wall. In this case, the value of q is the same for each layer of the wall of unit area and equal to the heat leaving the cold face. As another example, a 9-in. firebrick wall with a $2400°F$ hot face is backed up with 4.5 in. of insulating brick and 2 in. of insulating block. The air conditions are the same as before. From Eq. (19.2)

$$q = \frac{2400 - t_n}{(9/k_1) + (4.5/k_2) + 2/k_3} = q_{rc}$$

As a first trial, assume $t_n = 280°F$ and k_1, k_2, and k_3 as 9.5, 2.0, and 0.9, respectively, from Fig. 19.7. Then

$$q = \frac{2400 - 280}{(9/9.5) + (4.5/2.0) + (2/0.9)} = \frac{2120}{0.95 + 2.25 + 2.23} = 390$$
$$q_{sc} = 545 \text{ from curve}$$

Next assume $t_n = 245$, then

$$q = 398 \qquad q_{rc} = 417$$

which is sufficiently close for a first approximation.

Now from Eq. (19.3)

$$
\begin{aligned}
t_1 - t_2 &= 398 \times 0.95 = && 375 \\
t_2 - t_3 &= 398 \times 2.25 = && 895 \\
t_3 - t_n &= 398 \times 2.23 = && \underline{890} \\
t_1 - t_n &= && 2160
\end{aligned}
$$

The interface temperatures will be 2025 and $1130°F$; the mean temperatures of each layer will be 2212, 1577, and $685°F$. Now go back to the conductivity curves and select more precise values of k, which will be, respectively, 9.90, 2.35, and 0.85. The heat flow is now worked out again, with $t_n = 245$.

$$q = \frac{2400 - 245}{(9/9.9) + (4.5/2.35) + (2/0.85)} = \frac{2155}{0.91 + 1.91 + 2.35}$$
$$= \frac{2155}{5.17} = 417$$

which is the same as q_{rc}. Now the temperature drops are recalculated:

$$t_1 - t_2 = 417 \times 0.91 = \quad 379$$
$$t_2 - t_3 = 417 \times 1.91 = \quad 759$$
$$t_3 - t_n = 417 \times 2.35 = \quad \underline{981}$$
$$t_1 - t_n = \qquad\qquad\qquad 2155$$

The interface temperatures will then be 2021 and 1226°F.

The mean temperatures will be 2210, 1623, and 612°F. Taking new values for k and t_n of 242°F,

$$q = \frac{2400 - 242}{(9/9.9) + (4.5/2.39) + (2/0.81)} = \frac{2158}{0.91 + 1.88 + 2.47}$$
$$= \frac{2158}{5.26} = 410$$

and $q_{rc} = 410$ as closely as can be read.

The values of temperature drops are

$$t_1 - t_2 = 410 \times 0.91 = \quad 373$$
$$t_2 - t_3 = 410 \times 1.88 = \quad 772$$
$$t_3 - t_n = 410 \times 2.47 = \underline{1013}$$
$$t_1 - t_n = \qquad\qquad\qquad 2158$$

Therefore, the interface temperatures are 2027 and 1256°F, and the mean temperatures are 2213, 1641, and 627°F. By further trial, any degree of precision can be obtained, but the fourth trial here is sufficiently exact for ordinary purposes, as we do not know the conductivity values very accurately. Computing walls is a tedious process; but with experience, the initial selection of temperature will be fairly close to the final values. The heat flow and heat storage in a number of standard wall constructions are shown in Table 19.7.

There are now available simple analog computers that make possible rapid calculation of compound walls.

19.7 Computation of Heat Flow in Cylinders. *Through Circular Structures.* Although the same general method is used for determining heat flow through circular structures, the equation for heat flow through a plane wall cannot be applied to flow through a cylinder if an accurate result is desired. The reason for this may readily be understood when it is considered that the inside circumference of a cylinder is smaller than the outside circumference. Consequently, heat spreads out as it travels from the inside to the outside, and conversely, it converges when the direction of flow is from the outside toward the inside.

Heat flow outward through a cylinder may be calculated by the following formulas:

1. For a wall of one material:

$$q = \frac{2\pi(t - t_n)}{(12/k_n) \ln (D_n/D)} \text{ Btu/(lin ft)(hr)}$$

or

$$q = \frac{t - t_n}{(12D_n/2k_n) \ln (D_n/D)} \text{ Btu/(sq ft)(hr)}$$

2. For a composite wall:

$$q = \frac{2\pi(t - t_n)}{\dfrac{12}{k_1} \ln \dfrac{D_1}{D} + \dfrac{12}{k_2} \ln \dfrac{D_2}{D_1} + \cdots + \dfrac{12}{k_n} \ln \dfrac{D_n}{D_{(n-1)}}} \text{ Btu/(lin ft)(hr)}$$

or

$$q = \frac{t - t_n}{\dfrac{12D_n}{2k_1} \ln \dfrac{D_1}{D} + \dfrac{12D_n}{2k_2} \ln \dfrac{D_2}{D_1} + \cdots + \dfrac{12D_n}{2k_n} \ln \dfrac{D_{(n-1)}}{D_n}}$$

$$\text{Btu/(sq ft)(hr)}$$

where q = the quantity of heat, Btu/hr

t_1 = the hot-face temperature

t_n = the cold-face or casing temperature

D = the inside diameter of cylinder, in.

D_1 = the outside diameter of inside lining material, in.

D_2 = the outside diameter of secondary lining material, in.

$D_{(n-1)}$ = the outside diameter of next to last lining material, in.

D_n = the outside diameter of outside lining material, in.

k_1 = the thermal conductivity of inside lining material

k_2 = the thermal conductivity of secondary lining material

k_n = the thermal conductivity of outside lining material

k_1, k_2, and k_n are expressed in Btu/(hr)(sq ft)(in.)(°F)

Interface temperatures between various layers are expressed as

t_1 = the temperature at D_1, °F

t_2 = the temperature at D_2, °F

$T_{(n-1)}$ = the temperature at $D_{(n-1)}$, °F

For a case in which the flow of heat is from the outside of the cylinder toward the center, D would become the outside or largest diameter and D_n the inside or smallest diameter.

19.8 Flow of Heat through Furnace Walls under Variable Conditions.

The temperature distribution in a simple or composite wall can be computed approximately for conditions of unsteady heat flow mathematically or graphically by the method of Schmidt. Such calculations require values of diffusivity for each material over the whole temperature range and, at best, are only approximate. For a discussion of this method,

TABLE 19.7 Heat Losses and Heat Storage Capacities of Wall Structures under Equilibrium Conditions* †

Hot-face temperature, °F

Thick-ness of wall, in.	Thickness of B & W insulating and fireclay brick (FB), in.	600 HL	600 HS	800 HL	800 HS	1000 HL	1000 HS	1200 HL	1200 HS	1400 HL	1400 HS	1600 HL	1600 HS	1800 HL	1800 HS	2000 HL	2000 HS
4½	4½K-16	67	398	100	564	138	742	180	928	226	1,130	278	1,320				
	4½K-20	89	538	128	760	175	1,000	230	1,250	290	1,505	350	1,775	420	2,040	468	2,320
	4½K-23	81	574	117	820	160	1,085	211	1,350	262	1,645	320	1,930	381	2,230	447	2,525
	4½K-26	137	867	202	1,230	273	1,620	350	2,010	434	2,430	528	2,850	626	3,280	736	3,720
	4½K-28	136	905	198	1,280	266	1,680	341	2,100	420	2,530	508	2,960	599	3,410	701	3,878
	4½K-30	176	1,170	261	1,660	356	2,170	460	2,700	574	3,270	697	3,830	835	4,420	985	5,000
	4½FB	547	3,970	809	5,610	1,095	7,260	1,400	9,000	1,727	10,720	2,082	12,450	2,442	14,240	2,820	16,080
7	7K-16	44	603	66	858	90	1,130	118	1,410	148	1,700	182	2,000				
	4½K-20 + 2½K-16	51	894	75	1,278	105	1,695	144	2,128	172	2,587	208	3,045	245	3,515	286	4,000
	4½K-23 + 2½K-16	47	945	67	1,355	94	1,790	126	2,235	163	2,710	195	3,195	235	3,669	280	4,165
	4½K-26 + 2½K-16	64	1,410	95	2,030	129	2,680	166	3,350	207	4,070	251	4,780	299	5,510	350	6,280
	4½K-28 + 2½K-16	65	1,470	96	2,120	130	2,800	169	3,510	212	4,260	260	5,040	313	5,860	372	6,690
	4½K-30 + 2½K-16	73	1,900	108	2,730	148	3,620	192	4,540	241	5,500	295	6,510	355	7,540	421	8,600
	4½K-30 + 2½K-23	80	1,920	125	2,750	167	3,620	209	4,580	267	5,580	322	6,629	385	7,650	437	8,750
	4½FB + 2½K-16	100	5,820	150	8,360	204	11,030	264	13,790	328	16,590	399	19,490	475	22,430	555	25,350
	4½FB + 2½K-23	114	6,062	170	8,500	230	11,100	293	13,890	366	16,625	442	19,550	522	22,485	608	25,500
	4½FB + 2½K-26	179	5,730	261	8,210	350	10,850	445	13,570	545	16,390	652	19,150	768	22,070	888	24,930
9	4½K-20 + 4½K-16	39	1,085	56	1,560	78	2,075	102	2,637	127	3,210	157	3,805	190	4,415	229	5,050
	4½K-23 + 4½K-16	37	1,140	54	1,625	73	2,155	99	2,735	122	3,320	150	3,935	178	4,565	210	5,210
	4½K-26 + 4½K-16	46	1,590	68	2,370	93	3,160	121	3,970	151	4,810	186	5,690	223	6,590	266	7,690
	4½K-28 + 4½K-16	46	1,730	67	2,470	92	3,230	119	4,080	148	4,930	180	5,810	215	6,730	253	7,650
	4½K-28 + 4½K-23	51	1,780	75	2,515	103	3,360	132	4,228	165	5,140	201	6,200	236	6,990	273	7,990
	4½K-30 + 4½K-16	50	2,160	74	3,120	102	4,130	132	5,190	166	6,270	203	7,400	244	8,570	290	9,790
	4½K-30 + 4½K-23	57	2,245	86	3,810	117	4,260	148	5,400	183	6,570	224	7,780	267	9,000	313	10,230
	4½FB + 4½K-16	61	6,100	91	8,900	126	11,610	163	14,540	204	17,530	248	20,580	294	23,680	344	26,860
	4½FB + 4½K-23	72	6,218	107	9,010	144	12,000	190	15,030	235	18,100	282	21,200	330	24,220	385	27,400
	4½FB + 4½K-26	118	6,280	172		231	11,930	294	14,920	362	18,000	433	21,170	510	24,350	592	27,560
9	9K-16	34	768	51	1,090	70	1,440	92	1,790	116	2,170	142	2,550				
	9K-20	46	1,050	67	1,505	86	1,995	117	2,470	147	2,955	177	3,450	210	3,985	248	4,510
	9K-23	41	1,085	58	1,550	80	2,080	108	2,640	135	3,170	164	3,365	192	4,280	225	4,880
	9K-26	72	1,650	106	2,340	142	3,060	182	3,810	225	4,600	273	5,400	324	6,230	380	7,080
	9K-28	71	1,710	104	2,440	140	3,160	178	3,970	220	4,800	264	5,630	312	6,480	362	7,360
	9K-30	94	2,190	138	3,110	188	4,090	242	5,070	302	6,140	366	7,230	436	8,330	513	9,450
	9FB	325	7,180	473	10,120	634	13,260	801	16,450	977	19,690	1,160	22,980	1,345	26,250	1,536	29,600

11½	11½K-16	27	977	40	1,390	55	1,830	72	2,280	91	2,740	111	3,230				
	9K-20 + 2½K-16	33	1,417	50	2,020	67	2,660	86	3,331	107	4,000	129	4,718	152	5,420	180	6,137
	9K-23 + 2½K-16	31	1,527	46	2,190	62	2,880	81	3,646	100	4,410	122	5,213	146	6,090	172	6,960
	9K-26 + 2½K-16	45	2,340	67	3,410	91	4,500	117	5,660	145	6,860	177	8,115	213	9,410	252	10,780
	9K-28 + 2½K-16	45	2,450	66	3,520	89	4,680	114	5,840	141	7,070	171	8,340	203	9,640	237	11,010
	9K-30 + 2½K-16	53	3,220	79	4,660	108	6,200	140	7,770	174	9,430	213	11,130	256	12,940	303	14,770
	9FB + 2½K-16	89	10,810	131	15,460	177	20,420	227	25,510	282	30,790	340	36,080	402	41,460	468	46,900
	9FB + 2½K-23	102	10,702	146	15,400	198	20,350	256	25,350	316	30,400	376	35,600	441	41,140	510	46,450
13½	9K-20 + 4½K-16	27	1,700	40	2,410	54	3,210	71	4,060	90	4,860	109	5,830	130	6,840	154	7,710
	9K-23 + 4½K-16	26	1,764	38	2,520	51	3,350	68	4,235	85	5,330	103	6,067	122	7,045	143	8,040
	9K-26 + 4½K-16	35	2,820	52	3,910	70	5,170	91	6,500	114	7,890	140	9,320	168	10,840	200	12,360
	9K-28 + 4½K-16	35	2,920	51	4,050	69	5,360	89	6,710	111	8,110	134	9,550	160	11,050	188	12,590
	9K-30 + 4½K-16	40	3,650	59	5,290	81	7,000	105	8,790	131	10,640	160	12,600	192	14,630	228	16,640
	9FB + 4½K-16	57	11,430	84	16,460	115	21,720	149	27,090	185	32,650	224	38,390	265	44,370	309	49,930
	9FB + 4½K-23	67	11,692	100	16,800	132	22,250	168	27,770	209	33,100	251	38,450	296	43,930	341	49,350
	9FB + 4½K-26	103	11,100	150	15,990	199	21,010	252	26,300	309	31,660	368	37,200	431	42,810	496	48,410
13½	9K-16 + 4½FB	33	1,040	49	1,570	68	2,080	89	2,650	112	3,260	137	3,760				
	9K-20 + 4½FB	43	1,466	64	2,050	88	2,725	113	3,422	140	4,200	168	5,070	198	6,020	231	7,010
	9K-23 + 4½FB	40	1,770	59	2,320	80	2,960	102	3,650	127	4,440	153	5,270	182	6,130	213	7,045
	9K-26 + 4½FB	67	2,350	98	3,410	132	4,480	169	5,610	209	6,790	253	8,000	300	9,240	353	10,570
	9K-28 + 4½FB	66	2,420	96	3,450	129	4,530	165	5,750	204	6,870	243	8,070	287	9,410	334	10,720
	9K-30 + 4½FB	85	3,160	125	4,500	169	5,930	218	7,420	271	9,030	330	10,750	393	12,360	462	14,270
	13½FB	233	10,230	338	14,520	448	18,950	563	23,510	684	28,180	809	33,000	934	37,800	1,067	42,700
16	13½FB + 2½K-16	80	15,390	118	22,160	159	29,200	202	36,410	250	43,850	300	51,420	352	58,990	407	66,770
18	9K-16 + 9FB	32	1,580	48	2,350	65	3,130	85	4,100	108	5,000	132	5,920				
	9K-20 + 9FB	42	2,283	63	3,215	84	4,270	106	5,325	131	6,500	158	7,800	190	8,860	227	10,050
	9K-23 + 9FB	38	2,527	50	3,420	73	4,403	99	5,505	124	6,470	149	7,800	177	9,000	205	10,250
	9K-26 + 9FB	62	3,390	90	4,850	122	6,410	156	8,100	193	9,900	235	11,760	280	13,710	329	15,780
	9K-28 + 9FB	61	3,470	89	4,940	120	6,540	153	8,210	188	9,960	226	11,750	266	13,720	310	15,690
	9K-30 + 9FB	77	4,480	114	6,420	154	8,550	198	10,700	246	13,120	300	15,600	356	18,180	419	20,940
	13½FB + 4½K-16	53	16,470	79	23,600	106	31,250	136	38,980	169	46,930	204	55,150	241	63,380	280	71,630
	13½FB + 4½K-23	62	16,432	91	23,200	123	31,000	156	39,222	192	47,200	230	55,500	271	64,150	315	72,700
20½	18FB	182	13,160	263	18,700	350	24,500	439	30,380	532	36,600	628	42,700	726	49,100	828	55,500
	18FB + 2½K-16	73	19,800	106	28,310	143	37,380	182	46,540	224	55,900	268	65,770	314	75,540	362	85,400
22½	18FB + 4½K-16	50	21,200	73	30,530	99	40,260	127	50,140	156	60,430	188	70,960	222	81,500	257	92,040
	18FB + 4½K-23	57	20,908	85	29,450	114	39,000	142	48,895	175	58,750	208	69,250	244	79,935	280	90,800
22½	22½FB	150	16,180	216	22,850	286	29,800	358	37,100	434	44,700	512	52,400	591	60,200	674	68,100

* HL = heat loss in Btu per sq ft per hr. Based on still air temperature of 80°F. HS = heat storage capacity in Btu per sq ft.
† From *Bull.* R-2-H, Babcock and Wilcox Co.

TABLE 19.7 (Continued)*,†

Hot-face temperature, °F

Thickness of wall, in.	Thickness of B & W insulating and fireclay brick (FB), in.	2200 HL	2200 HS	2300 HL	2300 HS	2400 HL	2400 HS	2500 HL	2500 HS	2600 HL	2600 HS	2700 HL	2700 HS	2800 HL	2800 HS	2900 HL	2900 HS
4½	4½K-16																
	4½K-20																
	4½K-23	510	2,830	545	2,985												
	4½K-26	855	4,170	921	4,400	987	4,630	1,062	4,870	1,135	5,180						
	4½K-28	806	4,350	864	4,570	921	4,770	980	5,050	1,042	5,290	1,105	5,540	1,173	5,780		
	4½K-30	1,146	5,610	1,235	5,920	1,325	6,220	1,424	6,530	1,520	6,840	1,628	7,170	1,735	7,470	1,857	7,790
	4½FB	3,210	17,770	3,410	18,750	3,580	19,600	3,780	20,500	3,960	21,400	4,160	22,280	4,350	23,150	4,550	24,100
7	7K-16																
	4½K-20 + 2½K-16																
	4½K-23 + 2½K-16	330	4,670	357	4,940												
	4½K-26 + 2½K-16	406	7,050	436	7,450	468	7,840	502	8,240	533	8,640						
	4½K-28 + 2½K-16	437	7,540	473	7,980	511	8,420	553	8,890	596	9,360	642	9,840	694	10,370		
	4½K-30 + 2½K-16	495	9,690	535	10,250	576	10,790	620	11,360	663	11,940	712	12,520				
	4½K-30 + 2½K-23	507	9,860	552	10,430	604	10,985	652	11,600	709	12,200	769	12,830	832	13,590	897	14,240
	4½FB + 2½K-16	639	28,260	682	29,790	797	31,670	845	32,200	895	34,700						
	4½FB + 2½K-23	701	28,550	747	29,280												
	4½FB + 2½K-26	1,020	27,870	1,084	29,280	1,155	30,740	1,223	32,220	1,295	33,700	1,372	35,200	1,448	36,560	1,525	38,050
9	4½K-20 + 4½K-16																
	4½K-23 + 4½K-16	249	5,865	274	6,195												
	4½K-26 + 4½K-16	313	8,510	339	8,990	366	9,500	396	10,020	427	10,550						
	4½K-28 + 4½K-16	294	8,600	316	9,050	339	9,550	364	10,050	422	10,960						
	4½K-28 + 4½K-23	316	8,960	340	9,450	368	9,960	395	10,480			450	11,480	478	11,930		
9	4½K-30 + 4½K-16	340	11,010	367	11,640	396	12,280	429	12,930								
	4½K-30 + 4½K-23	362	11,560	388	12,200	418	13,000	450	13,490	482	14,200	518	14,880	558	15,580		
	4½FB + 4½K-16	399	29,940	467	32,200	496	33,820	523	35,750								
	4½FB + 4½K-23	440	30,450	721	32,400	766	33,970	812	35,600	860	37,170	907	38,750	958	40,380	1,010	42,020
	4½FB + 4½K-26	678	30,780														
9	9K-16																
	9K-20																
	9K-23	358	5,475	275	5,770												
	9K-26	442	7,930	474	8,370												
	9K-28	416	8,230	444	8,700	474	9,160	504	9,620	535	10,080	567	10,540	601	11,020		
	9K-30	596	10,580	640	11,180	686	11,750	734	12,400	785	12,980	840	13,600	889	14,180	946	14,800
	9FB	1,740	33,050	1,835	34,800	1,935	36,450	2,042	38,200	2,155	39,950	2,260	41,700	2,345	43,400	2,450	45,200

11½	11½K-16 9K-20 + 2½K-16 9K-23 + 2½K-16 9K-26 + 2½K-16	200 297	7,850 12,170	214 321	8,310 12,870	347	13,590	375	14,370	404	15,110						
	9K-28 + 2½K-16 9K-30 + 2½K-16 9FB + 2½K-16 9FB + 2½K-23	275 354 538 577	12,380 16,650 52,430 52,100	294 382 574 616	13,070 17,610 55,070 55,000	314 412 611 658	13,790 18,590 57,910 57,710	335 443 649 676	14,490 19,570 60,740 60,800	358 476 691 712	15,220 20,570 63,790 64,900	385 512 746	15,980 21,630 66,800	408 548 784	16,700 22,640 70,000	589 822	**23,680** 73,100
13½	9K-20 + 4½K-16 9K-23 + 4½K-16 9K-26 + 4½K-16 9K-28 + 4½K-16 9K-30 + 4½K-16	168 235 218 268	9,000 13,980 14,160 18,790	184 255 234 290	958 14,830 14,970 19,890	275 251 312	15,670 15,780 21,000	297 269 337	16,490 16,580 22,140	219 286 363	17,360 17,420 23,260	306 390	18,260 24,400	327 420	19,120 25,610	451	26,810
	9FB + 4½K-16 9FB + 4½K-23 9FB + 4½K-26	357 390 564	55,900 55,800 54,120	382 418 600	58,890 58,900 56,970	447 634	61,920 59,810	480 670	65,200 62,560	503 708	68,500 65,420	532 746	71,750 68,360	784	71,110	825	73,850
13½	9K-16 + 4½FB 9K-20 + 4½FB 9K-23 + 4½FB 9K-26 + 4½FB	252 410	7,960 11,960	275 440	8,440 12,770	474	13,440	508	14,150	543	14,940						
	9K-28 + 4½FB 9K-30 + 4½FB 13½FB	384 540 1,213	12,080 16,100 47,700	411 580 1,276	12,790 17,010 50,200	439 622 1,349	13,520 18,060 52,700	466 667 1,425	14,220 19,080 55,200	495 714 1,500	14,930 20,040 57,800	526 762 1,580	15,610 21,040 60,400	557 812 1,650	16,400 22,170 63,000	862 1,720	23,150 65,600
16	13½FB + 2½K-16	463	74,340	491	78,170	522	81,900	552	85,940	584	89,680	615	93,710	645	[illegible]7,370	678	101,380
18	9K-16 + 9FB 9K-20 + 9FB 9K-23 + 9FB 9K-26 + 9FB	236 383	11,450 18,040	255 412	12,130 19,220	443	20,340	475	21,580	507	22,900						
	9K-28 + 9FB 9K-30 + 9FB 13½FB + 4½K-16 13½FB + 4½K-23	356 488 320 360	17,650 23,870 79,870 81,600	380 524 342 382	18,820 25,320 83,950 86,300	406 562 364 406	19,900 26,830 88,220 90,783	432 602 385 441	21,010 28,390 92,300 95,550	459 643 460	22,160 30,000 100,300	487 686 485	23,300 31,700 110,200	515 731 513	24,430 33,420 120,100	782 541	**35,120** 130,100
20½	18FB 18FB + 2½K-16	930 412	62,100 95,270	982 438	65,300 100,400	1,036 464	68,600 105,340	1,090 490	72,000 110,270	1,145 518	75,500 115,400	1,197 546	78,700 120,340	1,255 573	82,000 125,370	1,313 600	85,400 130,200
22½	18FB + 4½K-16 18FB + 4½K-23	294 317	102,880 102,250	313 341	108,250 108,200	333 362	113,710 114,165	352 384	119,290 120,300	372 407	124,160 126,500	431	132,700	456	139,000	482	**145,350**
22½	22½FB	757	76,200	802	80,200	844	84,200	886	88,400	930	92,400	971	96,500	1,015	100,700	1,060	105,000

* HL = heat loss in Btu per sq ft per hr. Based on still air temperature of 80°F. HS = heat storage capacity in Btu per sq ft.
† From *Bull.* R-2-H, Babcock and Wilcox Co.

which requires too much space to treat here, reference should be made to Trinks,[2] Holman,[14] or Schack.[17]   Composite walls require a very complicated set of calculations.

Practically, the problem can be solved quite readily by constructing a small section of the wall as the side or door of a furnace.   Couples on the faces and between the layers will give the temperature distribution accurately.   The value of q passing out from the wall can be estimated fairly closely from the outside temperature.   Recently electrical analogs of heat-flow systems have been made up for solving unsteady-state problems.

BIBLIOGRAPHY

1. Ingersoll, L. R., and O. J. Zobel: "An Introduction to the Mathematical Theory of Heat Conduction," Ginn and Company, Boston, 1913.
2. Trinks, W.: "Industrial Furnaces," John Wiley & Sons, Inc., New York, vol. I, 1923; vol. II, 1925.
3. Norton, C. L., Jr.: Apparatus for Measuring Thermal Conductivity of Refractories, *J. Am. Ceram. Soc.*, **25**, 451, 1942.
4. Patton, T. C., and C. L. Norton, Jr.: Measurement of Thermal Conductivity of Fireclay Refractories, *J. Am. Ceram. Soc.*, **26**, 350, 1943.
5. Barrett, L. R.: Heat Transfer in Refractory Insulators, I, *Trans. Brit. Ceram. Soc.*, **48**, 235, 1949.
6. Loeb, A. L.: A Theory of the Envelope Type of Thermal Conductivity Test, *J. Appl. Phys.*, **22**, 282, 1951.
7. Hansen, W. C., and A. F. Livovich: Thermal Conductivity of Refractory Insulating Concrete, *J. Am. Ceram. Soc.*, **36**, 359, 1953.
8. Loeb, A. L.: A Theory of Thermal Conductivity of Porous Materials, *J. Am. Ceram. Soc.*, **37**, 96, 1954.
9. Clements, J. F., and J. Vyse: Thermal Conductivity of Some Refractory Materials, *Trans. Brit. Ceram. Soc.*, **56**, 296, 1957.
10. Lee, D. W., and W. D. Kingery: Radiation Transfer and Thermal Conductivity of Ceramic Oxides, *J. Am. Ceram. Soc.*, **43**, 594, 1960.
11. Ruh, E., and J. Spotts McDowell: Thermal Conductivity of Refractory Brick, *J. Am. Ceram. Soc.*, **45**, 189, 1962.
12. Ruh, E., and A. L. Renkey: Thermal Conductivity of Refractory Castables, *J. Am. Ceram. Soc.*, **46**, 89, 1963.
13. Young, R. C., F. J. Hartwig, and C. L. Norton, Jr.: Effect of Various Atmospheres on Thermal Conductance of Refractories, *J. Am. Ceram. Soc.*, **47**, 205, 1964.
14. Holman, J. P.: "Heat Transfer," McGraw-Hill Book Company, New York, 1964.
15. Washburn, M. E., and R. K. Bart: Thermal Conductivity of Silicon Carbide Refractories, *Bull. Am. Ceram. Soc.*, **44**, 555, 1965.
16. Clements, J. F.: Characteristics of Refractory Insulating Materials, *Trans. Brit. Ceram. Soc.*, **65**, 479, 1966.
17. Schack, A., "Industrial Heat Transfer," John Wiley & Sons, Inc., New York, 1965.

CHAPTER TWENTY

Miscellaneous Properties
of Refractories

20.1 Introduction. There are many properties and measurements that are occasionally needed in the production or use of refractories. These include particle size, permeability, specific surface, electrical conductivity, and various other properties. These will be discussed briefly in this chapter, together with references for those who wish more detailed information.

20.2 Measurement of Particle Size. *Screen Analysis.* The determination of the particle-size distribution in a granular material is of considerable importance in ceramic work. For materials coarser than 150 to 300 mesh, screens are used to separate the material into any desired number of grades; but for the finer materials, air- or water-separation methods are most satisfactory.

The United States standard series of sieves is based upon a 1-mm opening, the larger and smaller screens varying as $\sqrt{2}$, or 1.414. The characteristics of this series of screens are given in Table 20.1.

The Tyler screens are more commonly used. They have as a basis a 200-mesh screen with an opening of 0.0029 in., and the sizes of openings also vary as the $\sqrt{2}$. The intermediate sizes are available to give closer spacing. In Table 20.2 are given the characteristics of this series.

A sample to be screened should be completely dry, and the lumps broken up. It can then be passed through any required number of screens in series. Care must be taken, however, to shake the screens long enough to pass all undersized material through. A convenient piece of apparatus for

TABLE 20.1   United States Standard Sieve Series

Meshes per lineal inch	Sieve number	Sieve opening, in.	Sieve opening, mm	Wire diameter, in.	Wire diameter, mm
2.58	2½	0.315	8.00	0.073	1.85
3.03	3	0.265	6.73	0.065	1.65
3.57	3½	0.223	5.66	0.057	1.45
4.22	4	0.187	4.76	0.050	1.27
4.98	5	0.157	4.00	0.044	1.12
5.81	6	0.132	3.36	0.040	1.02
6.80	7	0.111	2.83	0.036	0.92
7.89	8	0.0937	2.38	0.0331	0.84
9.21	10	0.0787	2.00	0.0299	0.76
10.72	12	0.0661	1.68	0.0272	0.69
12.58	14	0.0555	1.41	0.0240	0.61
14.66	16	0.0469	1.19	0.0213	0.54
17.15	18	0.0394	1.00	0.0189	0.48
20.16	20	0.0331	0.84	0.0165	0.42
23.47	25	0.0280	0.71	0.0146	0.37
27.62	30	0.0232	0.59	0.0130	0.33
32.15	35	0.0197	0.50	0.0114	0.29
38.02	40	0.0165	0.42	0.0098	0.25
44.44	45	0.0138	0.35	0.0087	0.22
52.36	50	0.0117	0.297	0.0074	0.188
61.93	60	0.0098	0.250	0.0064	0.162
72.46	70	0.0083	0.210	0.0055	0.140
85.47	80	0.0070	0.177	0.0047	0.119
101.01	100	0.0059	0.149	0.0040	0.102
120.48	120	0.0049	0.125	0.0034	0.086
142.86	140	0.0041	0.105	0.0029	0.074
166.67	170	0.0035	0.088	0.0025	0.063
200.00	200	0.0029	0.074	0.0021	0.053
238.10	230	0.0024	0.062	0.0018	0.046
270.26	270	0.0021	0.053	0.0016	0.041
323.00	325	0.0017	0.044	0.0014	0.036

making screen analyses is shown in Fig. 20.1.   In some cases, wet screening is used, for example, to separate clay from grog.

A convenient method to make an analysis is to weigh out a sample of, say, 100 g.   In obtaining this sample, the greatest care is necessary to ensure that it is representative.   The sample is then passed through the

TABLE 20.2   Tyler Screen Series

Openings, in.	Openings, mm	Mesh per lineal inch	Diameter of wire, in.
1.050	26.67	. . . . .	0.148
0.883	22.43	. . . . .	0.135
0.742	18.85	. . . . .	0.135
0.624	15.85	. . . . .	0.120
0.525	13.33	. . . . .	0.105
0.441	11.20	. . . . .	0.105
0.371	9.423	. . . . .	0.092
0.312	7.925	2½	0.088
0.263	6.680	3	0.070
0.221	5.613	3½	0.065
0.185	4.699	4	0.065
0.156	3.962	5	0.044
0.131	3.327	6	0.036
0.110	2.794	7	0.0328
0.093	2.362	8	0.0320
0.078	1.981	9	0.0330
0.065	1.651	10	0.0350
0.055	1.397	12	0.0280
0.046	1.168	14	0.0250
0.0390	0.991	16	0.0235
0.0328	0.833	20	0.0172
0.0276	0.701	24	0.0141
0.0232	0.589	28	0.0125
0.0195	0.495	32	0.0118
0.0164	0.417	35	0.0122
0.0138	0.351	42	0.0100
0.0116	0.295	48	0.0092
0.0097	0.246	60	0.0070
0.0082	0.208	65	0.0072
0.0069	0.175	80	0.0056
0.0058	0.147	100	0.0042
0.0049	0.124	115	0.0038
0.0041	0.104	150	0.0026
0.0035	0.088	170	0.0024
0.0029	0.074	200	0.0021
0.0024	0.061	250	0.0016
0.0021	0.053	270	0.0016
0.0017	0.043	325	0.0014
0.0015	0.038	400	0.0010

FIG. 20.1 Ro-Tap sieve shaker. (*The W. S. Tyler Company.*)

TABLE 20.3 Example of Screen Data

Screen	Weight, g	%	Cumulative, %
On 4...............	0.5	0.5	0.5
On 6...............	1.5	1.5	2.0
On 8...............	3.0	3.0	5.0
On 10..............	6.5	6.5	11.5
On 14..............	10.0	10.0	21.5
On 20..............	15.5	15.5	37.0
On 28..............	20.0	20.0	57.0
On 35..............	21.5	21.5	78.5
On 48..............	14.0	14.0	92.5
On 65..............	5.5	5.5	98.0
On 100.............	1.0	1.0	99.0
Through 100........	1.5	1.5	100.5

desired screens, and the amount remaining on each screen is weighed. This weight in grams will give the percentage directly. A typical analysis is shown in Table 20.3.

There are a number of methods of plotting the results of a screen analysis. The fraction of each particular size of particle may be plotted on a frequency curve, as in Fig. 20.2, or the cumulative percentage may be plotted on logarithmic paper as is often done when using the Tyler screens

(Fig. 20.3). Many attempts have been made to express the size distribution by means of a single factor. The work of Hatch and Choate[2] gives, perhaps, the best representation. They found that if logarithms of the sizes were plotted against the percentages of each size, a very close approach to a probability curve resulted in all cases. This curve can be expressed by two parameters, M_g and σ_g. M_g and σ_g can determined graphically by plotting experimental data on log-probability paper. The references at the end of the chapter will give many other viewpoints of this problem.

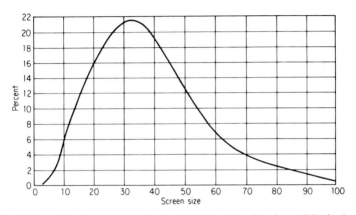

FIG. 20.2 The result of a screen analysis. The fraction of each particle size is plotted as a frequency curve.

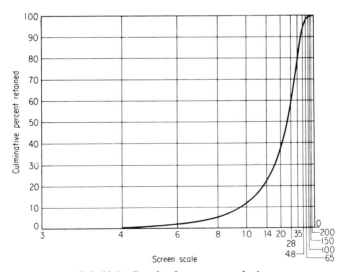

FIG. 20.3 Result of a screen analysis.

Settling Methods. In order to determine the particle-size distribution in a finely divided material such as a clay, the settling rate of the particles in water is found to be their most useful property. Using Stokes' law, which states that a sphere of diameter D will fall at a velocity V in a liquid of viscosity η when the density of the particle is S_1 and of the liquid S_2, the diameter is found by

$$D = \sqrt{\frac{18\eta V}{(S_1 - S_2)g}}$$

in cgs units.

It might be thought that the irregular or platelike particles would settle at a different rate from equivalent spheres; but with the exception of extreme plates such as are found in mica, Stokes' law is found to hold down to 1 μ diameter and probably lower. The work of Wadell[3] and Norton and Speil[4] confirms this.

One of the most important factors in this measurement is the complete dispersion of the particles in water. If they are flocculated to any extent, the results will be greatly in error. In the case of clays, considerable agitation is needed to break up clumps in addition to a deflocculating agent such as silicate of soda, sodium pyrophosphate, sodium carbonate, or Daxad. The state of dispersion should be checked with the microscope before the settling test.

When the particles are below 1 μ in diameter, the settling rate becomes so slow that it must be accelerated by centrifugal force. A number of centrifuges are now available for this purpose.

There are a number of methods of measuring the rate of settling. In the Andreassen pipette, small samples of the suspension are drawn off at a constant distance below the surface at various time intervals. These samples are then dried down and the solids weighed. The Casagrande method uses a hydrometer to obtain the specific gravity of the suspension at various times, using a nomographic chart to obtain the size distribution.

The diver method of measuring particle size as developed by Berg[5] is most ingenious and consists of adding to the suspension tiny glass floats of known density which will always maintain a level consistent with the density of the suspension at the point. Therefore, by plotting the settling rate of these floats, the density changes in the suspension can be followed with no disturbance. The details of these methods are given in the references.

Gaudin and Hukki suggest plotting the logarithm of the cumulative percent finer against the logarithm of the size. This has the great advantage of producing a straight line in most cases.

Other Methods. Elutriation with either water or air as a medium can be used for the measurement of particle size, but this method is more useful for producing definite-size fractions.

The microscope allows direct measurement of the particle sizes with a calibrated eyepiece. The lower limit, however, is about 1 μ, but certain problems can well be handled by this method.

For finer sizes, the electron microscope gives us the only direct means of particle-size measurement down to less than 0.1 μ.

The x-ray diffraction pattern allows the calculation of particle sizes from 0.05 μ and less by the line broadening that occurs with small particles. This method may also be used to measure the size of crystals in a lump of refractory.

Interpretation of Particle Size. The mean diameter of a single particle is clear when considering a sphere or cube where the diameter or edge respectively is a definite measure. In the case of a parallelepiped with sides a, b, and c, we can express four mean diameters as follows:

$$dm = \frac{a + b + c}{3} \qquad \text{(Statistical)}$$

$$dm = \sqrt{\frac{2ab + 2ac + 2bc}{6}} \qquad \text{(Side of a cube of equal area)}$$

$$dm = \sqrt[3]{abc} \qquad \text{(Side of a cube of equal volume)}$$

$$dm = \frac{3abc}{ab + ac + bc} \qquad \text{(Harmonic mean relating surface to volume)}$$

For platelike particles the various dm's vary a great deal; so the type of mean diameter should always be specified. The spherical mean diameter is usually taken as the diameter of a sphere settling at the same rate as the given particle.

When considering an evenly graded series of particles, the following mean diameters for the series may be written:

$$d_{ms} = \frac{\Sigma nd}{\Sigma n} = \frac{1}{2}\frac{d_1^2 - d_2^2}{d_1 - d_2} = \frac{1}{2}(d_1 + d_2) \qquad \text{(Statistical mean)}$$

$$d_{me} = \frac{\Sigma nd^2}{\Sigma nd} = \frac{2}{3}\frac{d_1^3 - d_2^3}{d_1^2 - d_2^2} \qquad \text{(Length mean)}$$

$$d_{ma} = \frac{\Sigma nd^3}{\Sigma nd^2} = \frac{3}{4}\frac{d_1^4 - d_2^4}{d_1^3 - d_2^3} \qquad \text{(Area mean)}$$

$$d_{mv} = \frac{\Sigma nd^4}{\Sigma nd^3} = \frac{4}{5}\frac{d_1^5 - d_2^5}{d_1^4 - d_2^4} \qquad \text{(Volume mean)}$$

where n = the total number of particles
d_1 = the smallest diameter in the series
d_2 = the largest diameter in the series

The so-called "Mellor mean" is often used:

$$dm = \sqrt[3]{\frac{(d_1 + d_2)(d_1{}^2 + d_2{}^2)}{4}}$$

but does not seem to be based on any experimental evidence.

To illustrate the use of the various mean diameters, an example based on an actual microscopic count on an evenly graded series of particles is given in Table 20.4.

TABLE 20.4

	Fraction diameter, μ							
	60	50	40	30	20	10	6	2
No. of particles.............	87	100	156	660	1,750	6,200	25,000	155,000
Percentage of total based on:								
Number $\dfrac{n}{\Sigma n}$.............	0.05	0.05	0.1	0.3	0.9	3.3	13.5	81.8
Length $\dfrac{nd}{\Sigma nd}$.............	0.9	0.9	1.1	3.5	6.1	10.9	22.4	54.3
Area $\dfrac{nd^2}{\Sigma nd^2}$.............	7.8	6.3	6.3	14.9	17.6	15.5	16.1	15.6
Volume $\dfrac{nd^3}{\Sigma nd^3}$.............	22.5	14.9	11.9	21.3	16.7	7.4	3.8	1.5

Calculating the mean diameter for the whole series gives

$$d_{ms} = \sum \left(\frac{n}{\Sigma n} d\right) = \frac{\Sigma nd}{\Sigma n} = 3.0 \ \mu$$

$$d_{me} = \sum \left(\frac{nd}{\Sigma nd} d\right) = \frac{\Sigma nd^2}{\Sigma nd} = 7.0 \ \mu$$

$$d_{ma} = \sum \left(\frac{nd^2}{\Sigma nd^2} d\right) = \frac{\Sigma nd^3}{\Sigma nd^2} = 21.0 \ \mu$$

$$d_{mv} = \sum \left(\frac{nd^3}{\Sigma nd^3} d\right) = \frac{\Sigma nd^4}{\Sigma nd^3} = 36.4 \ \mu$$

The physical significance of these values can be visualized by assuming that all the particles in the sample were laid out in a line, each particle touching the next and each being smaller than the one preceding it. If the particle in this line was selected that had equal numbers of particles above and below it, its diameter would be 3.0 μ. On the other hand, if a particle were selected exactly in the center of the line, it would have a diameter of 7.0 μ. A particle selected so that the total area or volume of all the particles above or below it were equal would have a diameter respectively of 21.0 and 36.4 μ. This brings out the fact that a mean particle size has no significance unless the type of mean is specified.

20.3 Structure. The macrostructure of refractories is of importance because a homogeneous material is seldom encountered. The size and shape of the pores, the shape of the grog particles, and the extent of the bonding are of interest. By a study of the structure, the manufacturing methods may be improved to produce a better product.

A cross section of a brick can be ground to a flat surface on a cast-iron wheel with coarse carborundum and water. The brick is then dried and heated enough to flow red sealing wax over the surface and into the pores. This surface can now be polished with a fine abrasive. The contrast between the light refractory and the dark sealing wax should bring out the structure very clearly.

20.4 Permeability. *Importance in Refractories.* The flow of gases through refractory walls is important in many furnaces, especially where large pressure differences exist, as in recirculating types. Also the permeability of such parts as thermocouple tubes, muffles, and recuperators is of interest in efficient design.

Permeability is also of value as a check on the manufacture, as laminations, voids, and soft spots are shown up in the permeability test. In the case of heat insulators, this property is useful in indicating the extent of. interconnecting pores.

Measurement of Permeability. Absolute permeability is the property of the refractory and does not in any way depend on the fluid. It may be expressed as

$$\lambda = \eta \frac{Q}{A} \frac{l}{(P_1 - P_2)} \frac{1}{t}$$

where λ = the absolute permeability
   η = the viscosity of the fluid, poises
   Q = the volume of gas flowing at mean pressure $(P_1 + P_2)/2$, cc
   A = the area, sq cm
   l = the length of path, cm
   P_1 = the exit pressure, dynes/sq cm
   P_2 = the entrance pressure, dynes/sq cm
   t = the time, sec

The dimensions of this equation are L^2 = area. However, our usual measurements of permeability are with a particular fluid such as air at room temperature, so that the measured permeability k is expressed by

$$\frac{\text{Volume of gas} \times \text{length of path}}{\text{Area} \times \text{time} \times \text{pressure difference}}$$

which has the dimension

$$\frac{L^2}{ML^{-1}T^{-1}} = \frac{\text{area}}{\text{viscosity}}$$

k is usually expressed as cc/(cm)(sq cm)(sec)(cm of water) or as cu ft/(in.) (sq ft)(min)(in. of water pressure).

One of the serious problems in measuring the permeability of a specimen is to prevent air from escaping along the sides. Most of the modern test equipment contains the specimen[9] (a cube, cylinder, or brick) in an inflatable rubber sleeve. Sufficient pressure is used to seal the edges completely. The volume of gas flow may be measured by water displacement or by a standard gas meter, while the pressure difference is measured by a simple manometer. For low permeabilities the testing time may be of considerable duration to obtain good precision.

In Table 20.5 are given the permeability values for a number of refractory brick tested by the U.S. Bureau of Standards.[8] The values

TABLE 20.5 Edgewise Permeability of Some Refractory Brick

Type	Forming method	Average porosity	Average permeability, f^*
High-alumina, 60%.......	DP	28.0	0.117
Superduty..............	DP	19.6	0.183
Superduty..............	DP	11.2	0.037
Superduty..............	DP	14.8	0.085
High-duty..............	DP	14.3	0.033
High-duty..............	DP	19.1	0.101
High-duty..............	DP	30.3	0.100
High-duty..............	DP	16.5	0.032
High-duty..............	DP	17.0	0.041
High-duty..............	HM	25.8	0.262
Mullite.................	Rammed	21.5	0.015
Mullite.................	SC	26.3	0.016
Mullite.................	DP	20.8	0.032
Chrome-magnesite........	Unburned	15.0	0.005
Magnesite-chrome........	Unburned	13.5	0.001
Magnesite..............	DP	24.2	0.143
Silica, standard..........	....	30.3	0.461
Silica, standard..........	HM	29.2	0.364
IFB 20.................	....	85.8	2.14
IFB 26.................	....	75.6	3.22
IFB 30.................	....	70.5	14.6
IFB bubble A1..........	....	60.5	1.98

$$^*f = \frac{(\text{cm}^3)(\text{cm})}{(\text{sec})(\text{cm}^2)(\text{gm}/\text{cm}^2)}$$

were measured for flow along the three axes of the brick but are reported here only for the edgewise flow.   It should be remembered that laminations or other structural discontinuities may greatly affect the values.

From Table 20.5 it may be concluded that all dry-pressed fireclay brick have substantially the same permeability; silica brick are higher; and IFB much higher.   As these values were determined on brick made in the 1930s and 1940s, it would be expected that brick made today with more careful grog sizing, higher molding pressures, and higher firing temperatures would have somewhat lower permeabilities.   It is clear that furnaces made of IFB linings must be cased with a steel shell or a hard cement coating to prevent extensive flow of gas through them.

20.5 Specific Surface.   This property can now be measured for powders or permeable specimens by nitrogen-adsorption methods.[10]

FIG. 20.4   Transverse testing of a refractory.   (*Harbison-Walker Refractories Co.*)

20.6 Strength of Refractories.   The cold strength of fired refractories is not generally of importance in itself; however, it often serves as a guide to other characteristics such as vitrification, burning temperature, or purity of material.   In the case of heat insulators, the strength is of value in determining the ability to withstand handling and shipping.   The ASTM Standard method for cold crushing strength is given in C 133–39, a method that should be followed to obtain consistent results.   For insulating brick, C 93–46 should be used.   Cold-strength values really have little significance but may usually be found in manufacturers' catalogs.   Figure 20.4 shows modern testing equipment for this purpose.

20.7 True Density. The following method has been found satisfactory for the determination of true density: The material is powdered through at least 100 mesh, dried at 105°C until thoroughly dry, and placed over $CaCl_2$ until at room temperature. About 1 g is weighed into the previously calibrated and weighed pycnometer bottle. The bottle is filled about one-half full of freshly boiled distilled water and is kept at 60°C for about ½ hr, and the contents are agitated to wet the particles thoroughly and remove entrapped air. The bottle is next placed under a vacuum to make certain that all air has escaped. The reduced pressure should not be strong enough to boil the water violently, since loss is likely to occur. About 20 min of treatment using an ordinary aspirator pump is sufficient.

The bottle is then filled to the mark with distilled water and placed in a constant-temperature bath for 1 hr, whereupon the bottle is carefully capped; the capillary and the outside of the bottle wiped clean; and the whole weighed.

The specific gravity is calculated from

$$\text{Specific gravity} = \frac{W - P}{(W_1 - P) - (W_2 - W)}$$

where P = the weight of pycnometer and stopper
   W = the weight of pycnometer and stopper and sample
   W_1 = the weight of pycnometer and stopper full of water
   W_2 = the weight of pycnometer and stopper and sample and water

Tetrahychonaphthalene or tetrachlorethane, because of their low viscosity and high wetting properties, are useful in the case of nonplastics in place of water, as the boiling and deairing steps are eliminated. Values on the same sample should check to ± 0.005.

20.8 Abrasion Resistance. Refractories used in the hearths of furnaces or other places where objects are slid along them must have a good resistance to abrasion. Furthermore, refractories are often worn away by particles in a rapidly moving gas. Abrasion tests at room temperature, such as are regularly used for paving brick and tiles, mean little for refractories because the conditions are far different at high temperatures. In the first place, the refractory must have a strong, well-bonded structure, and second, it must not become plastic at the working temperature. In general, it has been found that a brick which shows good resistance to load will also resist abrasion.

Abrasion has been measured at room temperature by the Bauschinger disk grinder, the sandblast of Bradshaw and Emery, and the rattler test of the National Paving Brick Manufacturers' Association. None of these tests, however, is used at the working temperature of the refractory. Perhaps the most logical tests are those proposed by Hancock and King.[1] Here samples of the refractory are rubbed together inside a furnace, and the loss in weight is determined for a certain time of test. To show the effect of temperature, a specimen was tested as follows:

Temperature of test	Loss in weight, g
20°C (68°F)	0.115
1050°C (1922°F)	0.255

This shows that the abrasion resistance decreases with the temperature. It was also concluded that a fine-grained structure and a high-burning temperature increase the abrasion resistance.

The abrasion resistance of an insulating firebrick is naturally less than for a heavy refractory, although hard coatings are helpful. Abrasion tests have been made on this material by circulating fine particles by means of a high-velocity air stream through a duct lined with a particular brick.

20.9 Electrical Resistivity. In electric-furnace work, the electrical resistance of refractories is of considerable importance, and there are many other cases where a high-temperature electrical insulator is desirable.

The method of measuring the resistivity is theoretically simple but at high temperatures is beset by numerous practical troubles; however, at temperatures up to 1000°C (1832°F), the apparatus is comparatively simple. The greatest difficulty is to reduce the contact resistance, and a

TABLE 20.6   Electrical Resistivity of Commercial Refractories

Type	Porosity	Resistivity, ohm-cm		
		800°C	1200°C	1400°C
Fireclay, superduty.............	18	19,000	1,550	720
Fireclay, high-heat-duty.........	20	20,500	2,000	970
Fireclay, intermediate-heat-duty.	24	10,000	920	445
High-alumina (70%) 37–38 PCE.	23	11,800	900	460
Alumina, fused cast.............	3.1	3,800	740	290
Mullite, 38+ PCE.............	26	210,000	16,000	7,200
Silica, superduty, 97% SiO_2.....	26	360,000	10,500	3,300
Silica, standard, 90% SiO_2.......	22	150,000	43,000	1,650
Mullite, fusion-cast.............	1.5	25,000	17,000	760
Basic, 90–95% MgO............	17	15,000,000	210,000	11,000
Basic, chrome-magnesite........	14	2,100,000	130,000	2,400
Forsterite.....................	21	1,450,000	11,500	680
Basic, magnesite-chrome........	19	370,000	3,900	400
Silicon carbide, clay-bonded.....	12	37,000	4,600	1,700
Zircon, 65% ZrO_2..............	30	1,250,000	21,000	3,600
Zirconia......................	....	....	1,250	300
Carbon.......................	....	0.003	0.003	
Graphite......................	....	0.0007	0.0008	0.0009

number of methods have been used to prevent this error. Hensler and Henry[6] show an excellent method.

In Table 20.6 are given some values of electrical resistivity obtained by different experimenters. It will be noticed that the values are rather discordant in some instances. This may be due in part to the variation in specimens and in part to errors in experimentation.

At low temperatures, refractories conduct very slightly. As the temperature is raised, the conductivity increases rapidly until the melting point is reached, where electrolytic conduction occurs. With a pure material, the conduction would be low up to the melting point; but with the average refractory, liquid phases are formed by the impurities at relatively low temperatures, so that we get an early falling off of the resistance. In general, a pure, high-melting-point material will give the best high-temperature electrical resistivity, as shown by Chiochetti and Henry.[7]

BIBLIOGRAPHY

1. Hancock, W. C., and W. E. King: Note on the Abrasion of Fireclay Materials, *Trans. Brit. Ceram. Soc.*, **22**, 317, 1922–1923.
2. Hatch, T., and S. P. Choate: Statistical Description of the Size Properties of Non-uniform Particulate Substances, *J. Franklin Inst.*, **207**, 369, 1929.
3. Wadell, H.: The Coefficient of Resistance as a Function of Reynolds Number for Solids of Various Shapes, *J. Franklin Inst.*, **217**, 459, 1934.
4. Norton, F. H., and S. Speil: The Measurement of Particle Sizes in Clays, *J. Am. Ceram. Soc.*, **21**, 89, 1938.
5. Berg, S.: Studies on Particle-size Distribution, *Pub.* 2, *Lab. Mortar, Glass and Ceramics, Royal Tech. Univ.*, Copenhagen.
6. Hensler, J. R., and E. C. Henry: Electrical Resistance of Some Refractory Oxides and Their Mixtures in the Temperature Range 600° to 1500°C, *J. Am. Ceram. Soc.*, **36**, 76, 1953.
7. Chiochetti, V. E. J., and E. C. Henry: Electrical Conductivity of Some Commercial Refractories in the Temperature Range 600° to 1500°C, *J. Am. Ceram. Soc.*, **36**, 180, 1953.
8. Massengale, G. B., L. E. Mong, and R. A. Heindl: Permeability and Some Other Properties of a Variety of Refractory Materials, I, *J. Am. Ceram. Soc.*, **36**, 222, 1953; II, *J. Am. Ceram. Soc.*, **36**, 273, 1953.
9. Astbury, N. F., J. F. Clements, and L. Sabiston: The Definition and Measurement of Permeability of Consolidated Materials, *Trans. Brit. Ceram. Soc.*, **60**, 658, 1961.
10. Haynes, J. M.: The Specific Surface of Clays, *Trans. Brit. Ceram. Soc.*, **60**, 691, 1961.

PART FIVE

Use

CHAPTER TWENTY-ONE

Refractories in the Iron and Steel Industry

21.1 Introduction. Since the Third Edition of "Refractories" was published in 1949, there has been a great revolution in the production of iron and steel. Contrary to the condition in the United States and Great Britain, where the greater part of the steel tonnage has been produced in the basic open-hearth furnace, the continental steel plants have depended on the basic bottom-blown converter for the bulk of their tonnage. When oxygen became available at less than $100 a ton, efforts were made to speed up the reactions in the open hearth with an oxygen lance, but it was soon realized that to take full advantage of this process, a more refractory roof was needed; and so the not-too-promising development of basic roofs was speeded up, resulting in the direct-bonded chrome-magnesite brick of greatly increased hot strength. This, together with better roof construction, gave an open hearth that could be run several hundred degrees hotter than before. Also, the blast for the bottom-blown converter was enriched with oxygen to speed up this process.

The next step was the development of the top-blown converter using an oxygen lance. This process, in several forms, proved very attractive, both in first cost and in refractory expense. Thus the major steel plants have recently installed large units, which means that the bottom-blown converter and the open-hearth furnace will soon be obsolete except for special purposes.

This upheaval has had a profound effect on the refractories industry.

331

The demand for silica brick has fallen off, and the costly chrome-magnesite brick has been largely replaced by the low-cost tar-bonded dolomite brick for converter linings, where the consumption of refractories has dropped to 6 to 15 lb of refractory per ton of steel, compared with over a hundred pounds for the open hearth.   Another important development is the increasing use of the fusion-cast basic refractory in areas of the furnace where conditions are very severe.

A significant change in the production of steel billets, which is proceeding with increased activity, is continuous casting of steel, which again will influence the use of refractories, replacing large amounts of low-cost brick by a small amount of special high-cost refractory.

The vacuum casting and vacuum treatment of steels is coming into greater use, and the transfer of iron and steel through tubes is increasing. This demands a refractory of very high quality.

This all means that the manufacturer of refractories must always be alert to the changes that are coming about in the field of refractory consumers.

In the United States the development of refractories takes place largely in the laboratories of the refractory producers, with the close cooperation of the steel mills.   In Europe, however, the steel and refractories companies are often small; thus cooperative research is more realistic there. An example is the Institut de Recherches de la Sidérurgie (The Iron and Steel Research Institute of France), located in fine buildings, completely equipped, and well staffed with able scientists and engineers. In Paris are found the fine research laboratories of Société Française de Céramique, cooperatively supported by the ceramic industry.

21.2 Blast Furnace. *Principles of Operation.*   The blast furnace is the primary unit for reducing iron ore to iron.   The charge, which is a mixture of iron ore, limestone, and coke, is blasted with preheated air in the correct amount to generate heat and form carbon monoxide, which reduces the ore to melted iron and slag.   The composition of the iron depends a great deal on the type of ore used, but control of the final composition is possible in the blast furnace itself, except for manganese and phosphorus, which cannot be greatly varied.

A Typical Furnace.   In Fig. 21.1 is shown the cross section of a modern blast furnace with the various parts labeled.   It takes the form of a vertical shaft from 50 to 100 ft high, with a hearth diameter of 18 to 27 ft. The refractories are completely surrounded by a steel shell.   The charge is fed into the top of the furnace from a skip hoist through the double bell at the top and gradually passes down through the furnace as the reactions take place forming the iron and slag which collect in the crucible at the bottom.   Periodically, the iron and slag are tapped off through separate notches.

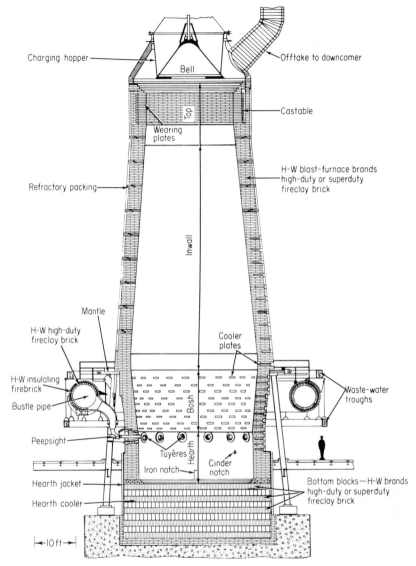

FIG. 21.1   Cross section of a typical blast furnace.   (*Harbison-Walker Refractories Co.*)

It will be noted that the refractories in hotter parts of the furnace and sometimes in the lower portion of the inwall are heavily water-cooled by hollow metal castings extending well into the wall around the bosh, and a water jacket extends down around the bottom, which is found necessary to give the long life needed for a satisfactory campaign, often lasting 2 to 5 years with a production of 1 to 2 million tons of iron.   Water cooling has made feasible the use of a thinner wall than previously thought possible.

The gas from the top of the blast furnace, which is mainly carbon monoxide and nitrogen, is cleaned, and part is burned in the stoves, which consist of vertical, brick-lined steel cylinders partially filled with checker brick. These checkers are first heated up by burning the blast-furnace gas at the bottom, and then air is drawn through them in the opposite direction to be fed to the tuyères, producing an average preheat temperature of 500 to 1000°C (about 930 to 1840°F). As soon as the temperature of the stove reaches a certain minimum value, the flow is switched over to another stove that has previously been heated so that the cycle can be continued. Usually three stoves are used with one modern blast furnace. Modern design calls for taller stoves with insulation inside the shell.

The blast furnace and its stoves are large users of refractories, as approximately one million bricks would be employed in the construction of a modern unit.

Types of Refractories Used. Practically all the refractories used in the blast furnace and stoves are fireclay brick. In the hearth and bosh are used high-flint clay bricks, which are generally vacuum-extruded, dry-pressed or air-rammed, and high-fired, to give a dense structure and low reheat shrinkage that will be resistant to slag and load under the severe conditions encountered in this part of the furnace. Accurate sizing of the blocks is desirable. The hearth bottoms are 9 to 12 ft thick and are usually constructed of blocks larger than the standard 9-in. brick in order to reduce the number of joints. In spite of this construction, however, the bottom bricks are gradually eaten away or floated out during a long campaign and are replaced by iron, which solidifies and forms the salamander.

The use of carbon blocks in the hearth of blast furnaces up to the center-line of the tuyères has now become common, especially in Europe. Although this is by no means a new development, it has been strangely slow in coming into general use in this country. Large, carefully sized blocks are used to give stability. Air cooling under the bottoms has prolonged the life of the carbon blocks and permitted thinner bottoms. Some carbon linings have been used up to 20 ft into the lower stack. With proper cooling such constructions may be found economical.

De La Rosa[9] finds carbon tap holes satisfactory. Collison[10] summarizes experience with carbon hearth and bosh linings. Bonnot[11] gives some results in Europe, while Kerr[12] gives data on carbon linings in four furnaces of the Kaiser Steel Corporation.

The bricks in the shaft of the furnace need not be so refractory as those in the lower part because the temperature decreases to approximately 400°

at the exit. However, the bricks should be hard and resistant to the abrasion of the moving charge; and most important of all, they should be stable under the disintegrating effect of carbon monoxide, which was discussed more fully in Chap. 17. It is generally believed that if the iron in the bricks has been converted to iron silicate by high firing, little disintegration will occur. The shaft bricks also are subject to disintegration in the upper portions from metallic zinc, which is sometimes reduced from the ore and, volatilizing, settles in the cooler portions of the lining, causing swelling and disintegration. Alkalies that are vaporized in the lower sections of the furnace also condense in the cooler parts and decrease the life of the lining.

Fused-cast α-alumina blocks[23] have been tried in the blast-furnace lining, but the results are not yet available. Remarkable results have been obtained by building up the inside of the worn lining, even up to a thickness of a foot or more, by gunning. This repair is far cheaper than relining and has shown excellent life. Aldred and Hinchliffe[6] describe the aim of very dense, hard lining brick and comment on the spiral method of setting in the stack to save cutting the hard brick for the key in each ring.

Life of Refractories. The nature of the blast furnace is such that little or no repairing can be done on the structure during a campaign, and therefore the life of the weakest part of the structure determines the total life. Usually the furnace must be shut down because of wearing away of the wall somewhere in the shaft due to slag erosion and abrasion. When the wall becomes so thin that the temperature on the outside casing cannot be held down even by extra cooling, it is impossible to continue the furnace in operation. The modern furnace with thorough water cooling gives excellent life, and campaigns of 3 to 5 years are commonly reported. In general, it would seem that the refractories available for the blast furnace are satisfactory for their purpose, but perhaps higher densities for the brick in the lower part of the furnace can still be achieved by more careful sizing of the raw materials, vacuum treatment, and higher pressures. Higher firing of the brick in the shaft will make them more stable in regard to disintegrating influences, but many problems still need study as far as these factors are concerned.

In Fig. 21.2 are shown the wear lines in a blast furnace at the end of a campaign. The first zone of wear is at the top under the stock-line wear plates and is due mainly to abrasion. The second zone is above the mantle and is caused by disintegrating influences of carbon monoxide, alkalies, and zinc, as well as slag erosion.

The Stoves. A typical stove is shown in Fig. 21.3. The scrubbed gases from the top of the blast furnace enter the bottom of the combustion chamber, impinge on the dome, and then pass down through the checker area. The lower two-thirds of the structure, including the checkers, is

made of dense high-heat-duty brick with about 42 percent Al_2O_3 content.
For the usual blast temperature of 750°C (about 1380°F), the dome and
upper checkers must run about 1200°C (about 2190°F).   Dense brick of
the superduty type are often used in the United States and Europe.
However, silica brick are reported to give good service in France.   Blast

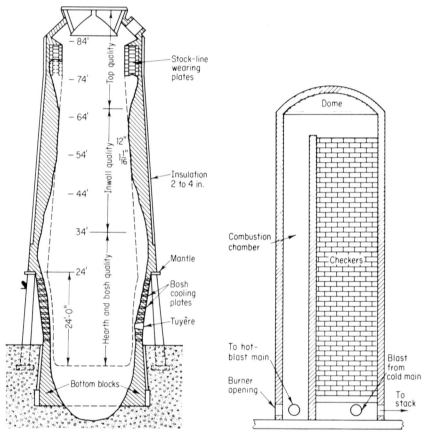

FIG. 21.2   Burn-out lines of blast fur-   FIG. 21.3   Blast-furnace stove.
nace.   (*Courtesy of W. R. McLain, Bull.*
Am. Ceram. Soc.)

temperatures of 1000°C (about 1830°F) are used for ferromanganese production and in the future will be used in most operations, which means
dome temperatures up to 1400°C (about 2550°F).   Under these conditions high-alumina refractories are required.

   Storey[7] discusses the type of brick needed for high blast temperatures
and shows laboratory tests.   Richardson[19] gives a good picture of the

conditions encountered in blast-furnace stoves. Petit[20] suggests siliceous refractories for the dome. In the United States semisilica checkers have shown excellent results at high blast temperatures.

Two developments in blast-furnace operation may have an important influence on the life of the lining. One of these is the use of oxygen-enriched air, and the other is high pressure in the shaft. Also the increased use of sintered ore in the burden seems to give a more severe attack on the lining. However, pelletized ore, now coming into extensive use, gives less refractory attack than bulk ore.

21.3 Cupola. *Principle of Operation.* The cupola is strictly a melting furnace using as a charge cast iron in the form of pigs, stove plate, and other scrap cast iron and in some cases scrap steel. The charge is alternately metal, coke, and some limestone as a flux. In a well-operated cupola, roughly 8 to 12 tons of iron can be melted to 1 ton of coke. The cupola itself is quite similar to the blast furnace in shape, consisting of a hollow shaft with a charge entering through a door near the top and the molten iron being tapped out periodically. Air for combustion enters the lower part of the shaft through tuyères in the same way as for the blast furnace, and preheated air is now often used in cupola melting. Unlike the blast furnace, the cupola is not used on continuous operation but is dumped each evening, repaired by patching, and started up the next day.

Typical Furnace. In Fig. 21.4 is shown a cross section of a typical cupola for melting away gray iron for foundry use. The shaft is completely lined with fireclay blocks molded to fit the curvature of the cupola. The door for charging is near the top of the shaft, and no attempt is made to collect the hot gases from the top. In the bottom of the shaft is a crucible for collecting molten iron, as clearly shown, together with the tap hole for draining off the melted iron. The bottom of the cupola can be dropped down at the end of a run for quicker cooling and easier repairs.

Types of Refractory Used. The refractories used in the cupola are high-duty fireclay blocks made with a dense structure by hard pressing or by deairing. Considerable improvement has been made in the cupola refractory in the last few years in the way of a more uniform and denser structure. Such a structure is more capable of withstanding the abrasion and slagging; but on the other hand, because of the sudden temperature changes, the block must also have considerable resistance to spalling. Although the complete lining of the cupola lasts for a considerable length of time, perhaps for a year or more, it is necessary to make rather extensive repairs to the portion of the shaft around the tuyères every day by the addition of new blocks or simply by patching with a plastic material, which is usually made of crushed ganister and plastic fireclay.

Attempts have been made to line cupolas with a castable refractory consisting of grog, ganister, and high-alumina cement. Although this

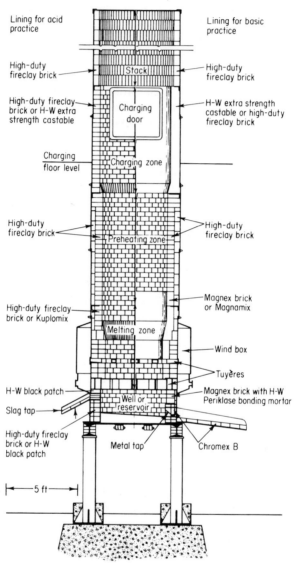

FIG. 21.4   A typical cast-iron cupola.   (*Harbison-Walker Refractories Co.*)

mixture has been used for patching in this country, the use of complete
linings of this material has been confined in the past to European practice.
Rammed linings are also used abroad because of their low cost.   Gunned
linings are described by Barlow and Humont.[2]   Carbon linings have been
used in the lower parts of the shaft with considerable success in completely
water-cooled shells.

21.4 Open-hearth Furnace. *Principles of Operation.*

The open-hearth furnace is a reverberatory type with a flame playing over a relatively shallow layer of metal. The charge may be pig iron, either solid or in the molten state directly from the blast furnace; most generally pig iron and scrap; or scrap and carbon. The reactions occurring in this furnace are mainly an oxidation of the carbon, elimination of the phosphorus in the basic process, and a decrease in the silicon, manganese, and sulfur. In the acid process (siliceous slag), it is impossible to eliminate the phosphorus; accordingly, only low-phosphorus materials can be used, which in this country limits the production of acid open-hearth steel to about 5 percent of the total. For this reason, only the basic process will be considered here.

The output of the open-hearth furnace is mostly used for casting steel ingots, which are later rolled, hammered, or pressed into the desired shapes.

The fuel used in the open hearth is most commonly producer gas, but furnaces are fired by natural gas, fuel oil, tar, coke-oven gas, and even pulverized coal. The type of fuel governs the design of ports in the furnace and has a considerable influence on the life of the refractories. For example, producer gas gives a long flame that has a tendency to heat the end walls and downtakes, whereas the coke-oven gas has a hot flame that tends to heat the roof near the burner end of the furnace. All open-hearth furnaces are of the regenerative type, in which the outgoing gases heat checker chambers. After a period of time, the flow is reversed and the incoming air and gas are preheated by passing through these hot chambers and then combined at the ports to form the combustion flame. Neither coke-oven nor natural gas can be preheated because of the tendency to crack.

Typical Furnace. Most of the furnaces are of the stationary type, but a few tilt on a horizontal axis for pouring and to make up the back wall. The tilting makes the operation of the furnace somewhat simpler, but the initial cost is much greater, and there is some difficulty in making the air and gas connections tight. If anyone is interested in more detailed construction of the open-hearth furnace, he should refer to the excellent book by Buell.[1]

The hearth of the furnace is built up on steel plates, with a layer of insulating brick, magnesite or chrome bricks, or a rammed mixture and then a fritted magnesite bottom, which is constantly being repaired during the campaign to maintain its normal thickness. The front and back walls of the furnace are usually made of silica brick with burned magnesite brick in the lower courses; but recently, chemically bonded brick of magnesite and metal-cased magnesite brick have been successfully used. A layer of chrome brick is usually introduced between the silica and magnesite brick.

The end walls and bulkheads are built up of silica, magnesite, or chrome brick; but here again, the metal-case brick have proved quite satisfactory in many installations. The ports are usually constructed of silica brick, but chrome and magnesite brick are used to withstand the slag and cutting action of the high-velocity dust-laden gases. Chromite ramming mixtures are sometimes used over the silica brick.

The doors of the furnace usually have water-cooled frames, which protect the brickwork from the mechanical abuse of the charging machine and give sufficient cooling to prevent deterioration of the front wall—a difficult portion of the furnace to repair.

The roof of the open-hearth furnace is one of the most important portions of the structure, because the life of the furnace as a whole is more or less determined by the life of the roof. Generally, the refractories used in the sprung arch of the roof are silica brick of $13\frac{1}{2}$ or 18 in. thickness. As the temperature required to maintain a low-carbon steel for pouring is 1585°C (about 2880°F) and the temperature at which the silica brick in contact with iron oxide begins to soften is 1650°C (about 3000°F), it will be seen that only a small operating interval is available and that for this reason the temperature control must be carefully maintained. In the modern plants, the roof temperature is read with optical pyrometers; and in some cases, temperature controllers operating on a radiation-pyrometer principle are permanently installed to cut off the fuel whenever the temperature exceeds the upper limit.

It might be thought that the silica brick, because of its tendency to spall, would be unsatisfactory in a regenerative furnace where a periodic change in temperature occurs at each reversal. However, if one would examine the expansion curve of a silica brick as shown in Fig. 18.7, it will be noticed that above the red-heat range, the volume of the brick is substantially constant and changes in temperature in the higher zones will cause no appreciable volume change. In other words, the silica brick is unexcelled for these particular conditions, and this constant volume at high temperatures explains why such good results are obtained with them.

The open-hearth roof, originally of silica brick, has been improved— first by using superduty silica brick (with low alumina, titania, and alkalies) and second by using basic brick. The latter, when of the direct-bonded type together with contour-controlling roof construction, gives excellent life at operating temperatures considerably above those possible with the silica roof.

The downtakes to the checker chambers are usually lined with silica brick, as is the upper part of the checker chamber itself. The checkers usually consist of hard-burned fireclay blocks such as $9 \times 6 \times 3$ or $10\frac{1}{2} \times 4\frac{1}{2} \times 4\frac{1}{2}$ in. in size. Often, however, the upper courses are made of silica because of the higher temperature conditions. A great many different types of checker block have been suggested and tried, and much theoretical work has been given over to the consideration of the checker efficiency.   The fact remains, however, that most of the furnaces now built use the simple, rectangular checker block.   The life of the furnace, in some cases, depends on the checkers, because as the campaign continues, the checker passages decrease as a result of building up of slag on the checkers, collection of dust in the passages, and spalling off of fragments of the checker bricks themselves, which fall down and tend to plug the passages.   It is usual to design the checkers to give a life equal to or considerably longer than that of the furnace roof without too great loss in regenerative efficiency.   Anyone interested in more detailed construction of the checker design should again refer to Buell.

While many open-hearth furnaces are now operating in the United States and Great Britain, it is doubtful if any new ones will be built.

21.5   Pouring-pit Refractories.   *Hot-metal Transfer Ladles and Hot-metal Mixers.*   The iron from the blast furnace is often transferred by large ladles or transfer cars, sometimes for long distances.   These vessels carry up to 300 tons and must have a rigid shell and means for pouring. The refractory lining in the United States is usually 9 or $13\frac{1}{2}$ in. of super-duty fireclay brick with a $4\frac{1}{2}$-in. safety lining against the shell.

It is common practice to store molten iron in large vessels, called metal mixers, holding up to 1,500 tons.   The metal temperature may run from 1180°C (about 2150°F) to 1260°C (about 2300°F), so that high-duty or superduty brick may be used as a lining.   In a few cases, where the temperatures are above this range, the lining is made from a fired basic brick. In Europe mixer linings are nearly always magnesite brick, as described by Latour[5] and Fitchett and Richardson.[8]   The latter comment on the better life with Austrian magnesite blocks than those made in England with seawater material.   On the Continent, mixers have been known to have a through-put of a million tons of iron on one lining.

Ladles.   The steel ladle is used to receive the molten steel from the open-hearth furnace.   In the ladle, the steel may receive some treatment such as the addition of deoxidizers or alloying elements, after which it is teemed into the ingot molds through a nozzle and stopper in the bottom of the ladle.   The refractories used to line the ladle present a real problem because they must fulfill the following requirements:

1. Good resistance to temperature shock due to the sudden heating when the molten steel is poured into the ladle
2. Good resistance to the slag layer on top of the steel
3. Resistance to abrasion of the molten steel
4. Freedom from particles of the refractory mixing in the molten steel and forming inclusions

5. Low thermal conductivity to prevent the formation of sculls

6. Tight joints to prevent adhesion of the sculls to the bottom and sides of the ladle

In some open-hearth steel plants, one-third of the yearly requirements of refractories goes into ladle linings, so that it seems strange that more effort has not gone into producing a better lining. In the United States linings are usually of "ladle brick," composed of a low-softening high-siliceous clay, which bloats in use. Also, some linings are made with high-heat-duty or superduty brick. On the other hand, many ladles on

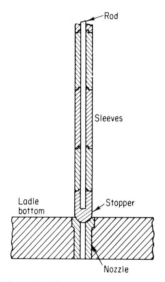

FIG. 21.5   Nozzle and stopper assembly.

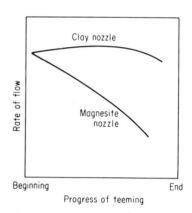

FIG. 21.6   Flow of steel through nozzles.

the Continent have basic linings which can be used if the ladle is well preheated. Smaller ladles are made of a rammed mixture of graded ganister and clay bond. Workman[3] believes a low porosity is more important than composition. Savioli[13] states rammed linings are more costly than brick, but it would seem possible to develop an automatic ramming machine similar to that in Fig. 10.30 to give dense and uniform wall structure. Bloch[14] has a patent (British Patent 907,007, Sept. 26, 1962) for a centrifugal thrower for building up linings.

Nozzles, Stoppers, and Sleeves.   The nozzle used in teeming steel from the ladle is a very important part of the ladle. A cross section of a typical nozzle and stopper assembly is shown in Fig. 21.5. A good stopper and nozzle should give a clean flow of steel without spattering at a substantially constant rate throughout the pouring operation, should be able to shut off the stream between ingot molds cleanly without dribbling, and it should be possible to open the nozzle on the first ingot without sticking.

In order to get a uniform rate of flow from the ladle, a material that will slowly erode and increase the diameter during the pouring operation must be used in the nozzle. It has been found that fireclay nozzles will have this property of giving a comparatively uniform flow rate from the beginning to the end of the pouring. On the other hand, a nozzle of magnesite is so resistant to erosion that it maintains substantially a constant diameter and therefore the rate of flow decreases, as shown in Fig. 21.6.

In order to get a tight fit between the nozzle and stopper, it is usually desirable to have one of them slightly plastic, but not both, or sticking is certain to result. For this reason, either the nozzle or the stopper will generally be made of a more refractory clay than the other; or in many cases, the stopper would be made of graphite and clay, which are more rigid at the pouring temperature than the clay alone.

The steel rod attached to the stopper must pass down through the molten steel in the ladle and therefore must be protected with refractory sleeves. They are made of a fireclay, and it is particularly important that they resist cracking under the sudden heat shock. It is also important that the joints between the sleeves be carefully fitted so that the molten steel cannot force them apart.

In recent years, a great deal of interest has been shown in steel inclusions. It is beginning to be realized that many of these are due to particles of refractory from the ladle lining or from the nozzle and that the ability to produce really clean steel would seem to depend largely on finding a refractory that would be more stable than the fireclay refractory now used.

Tap Holes. The tap hole in the open-hearth furnace must consist of a material strong enough to hold the pressure of the molten steel without breaking out and, at the same time, sufficiently soft to be knocked out with the tapping bar. It is general practice to make up a mixture of raw and burned dolomite for this purpose, although in some cases chrome-base cements have been used.

Hot Tops and Plugs. The plugs are used at the bottom of certain ingot molds to receive the direct impact of the stream of molten steel from the ladle. They must therefore be resistant to spalling and at the same time sufficiently dense to prevent erosion. They are usually made of a fireclay of comparatively low refractoriness and generally give little trouble.

Underpoured ingots require special shapes for pouring funnel, gate brick, spider brick, and runner brick. They are usually made of high-duty firebrick mixes.

The hot tops are placed on top of the ingot mold to prevent too rapid solidification of the top of the ingot and consequently a deep pipe. The hot tops must have a sufficiently close structure so that the steel will not penetrate them, and yet it is desirable to have them of light weight for low specific heat and low thermal conductivity with, at the same time, good spalling resistance. As they are generally used only once, the cost must be low. Usually, a comparatively low-burned fireclay mix is used, but some success has been had with lightweight refractories coated with a material to prevent penetration of the steel into the pores.

21.6 Air Furnace. *Principles of Operation.* The production of malleable iron is generally carried out in the air furnace. The malleable iron is produced from mixtures of pig iron, steel, and malleable foundry scrap, which is refined in the furnace to produce the proper percentages of carbon, silicon, sulfur, manganese, and phosphorus. When the malleable iron is first cast, it has practically all the carbon in the form of carbide of iron, which gives a comparatively brittle structure. The cast objects are then subjected to a heat-treating process while packed in iron boxes, perhaps seven or eight deep, for a considerable length of time, after which the carbon is precipitated and gathers into regular nodules known as "temper carbon." Sometimes the castings are simply annealed without packing. In the final condition, the metal is ductile and malleable and shows considerable elongation.

Typical Air Furnace. Figure 21.7 shows a section of a typical air furnace, which consists of a shallow hearth made of silica sand or firebrick and a firebrick wall and crown to cover it. An average-size hearth is 6 by 20 ft with a capacity of 15 tons, but furnaces up to 50-ton capacity have been made. Unlike most reverberatory furnaces, the crown is not fixed but is composed of separate bungs in which a ring of bricks comprising a section of the arch is clamped in an iron framework. In this way, one or more bungs can be lifted from the furnace at any time with a crane and the furnace charged through the opening thus produced. The fuel used for heating this type of furnace may be coal, oil, or gas, but powdered fuel is, perhaps, most used. This produces rapid heating and comparatively high temperatures; consequently the crown of the furnace receives considerable abuse. A furnace is brought up to heat in $1\frac{1}{2}$ hr and held 5 to 8 hr. The bungs have a comparatively short life of 20 to 30 heats as a result of the rapid heating and cooling, but the side and end walls give generally good service.

21.7 Bessemer and Thomas Converters. *Construction and Operation.* The Bessemer converter is a pear-shaped, refractory-lined vessel on trun-

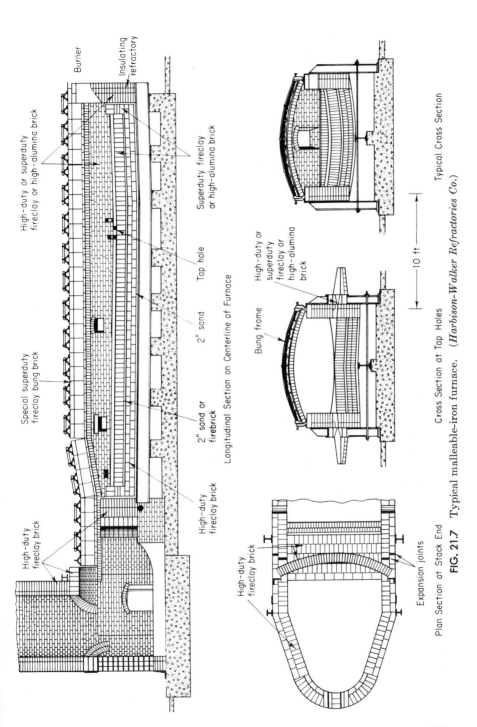

Burner

Insulating refractory

High-duty or superduty fireclay or high-alumina brick

Superduty fireclay or high-alumina brick

Tap hole

2" sand

2" sand or firebrick

High-duty fireclay brick

Longitudinal Section on Centerline of Furnace

Special superduty fireclay bung brick

High-duty fireclay brick

Bung frame

High-duty or superduty fireclay or high-alumina brick

Typical Cross Section

← 10 ft →

Cross Section at Tap Holes

High-duty fireclay brick

Expansion joints

Plan Section at Stack End

FIG. 21.7 Typical malleable-iron furnace. (*Harbison-Walker Refractories Co.*)

nions, which permit it to be tilted from the horizontal to the vertical position. The bottom consists of a refractory containing a large number of holes through which air can be blown under considerable pressure. A typical converter is shown in Fig. 21.8.

In operation, the converter while on its side is charged with molten iron from the mixer or blast furnace. It is then rocked into an upright position after the air blast has been applied, thus forcing air up through the molten metal and oxidizing the carbon and silicon to bring it down to the right composition. Experiments are being made on the use of oxygen-enriched air for the blast, which may well alter the conditions for the refractory.

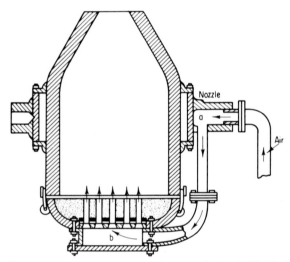

FIG. 21.8 Bessemer converter. (*From "Iron and Steel," by H. M. Boylston.*)

Type of Refractory Used. Most of the early converters were acid-lined with silica brick or natural furnace stone such as sandstone or mica schist. The bottoms of the converter are usually made of fireclay and must be frequently replaced, as 10 to 30 heats is the average life. The walls of the converter, however, may give a life of 200 or 300 heats before extensive replacements are necessary.

Today most converters are basic-lined so that they can handle high-phosphorus ores. In France at the present time about 60 percent of the steel production comes from Thomas basic converters. These vessels usually have a magnesite lining. In the past 10 years one-half or more of the steel made on the Continent is from Thomas converters, while in the United States and Great Britain the open hearth accounts for the larger part of the production.

One of the problems with the converter is the fumes injected into the atmosphere. Today efforts to end air pollution are gradually forcing an end to the practice of open disposal. In many of our steelmaking areas, exhaust control is required, and in West Germany, after 1970, all converters must be equipped with dust separators. This will probably make it advisable to abandon the many small Thomas converters and go over to a single vessel of the LD type.

21.8 The Oxygen Converter. *Description of the Process.* The various methods used to convert iron to steel were suddenly changed when oxygen became available at $20 to $30 a ton. In 1952, the LD process was started in Austria, at first with a silica lining to treat low-phosphorus iron, but soon adapted to other irons with a basic lining. The great advantage of the process was the large production from a small vessel and the low consumption of refractories. The LD converter consists of a vessel, as shown in Fig. 21.9, with a filling and pouring opening at the top. Trunnions allow the vessel to be tilted. A water-cooled oxygen lance can be inserted through the top, after the molten iron and scrap has been put in. Capacities are such as to handle up to 300 tons of metal, and larger ones will be used in the near future. The time of blowing is only 30 min. A

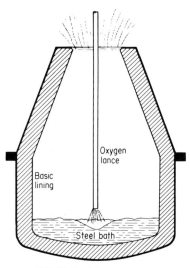

FIG. 21.9 LD converter.

modification of the process is the LDAC, where powdered lime is blown in and the resultant slag tapped off before the final blow, to remove phosphorus.

In 1956, the Kaldo process was announced in Sweden. This process uses a vessel which is revolved on its axis to bring the hot metal and slag in contact with heated walls; and this axis can be tilted for pouring, as shown in Fig. 21.10. The Kaldo converter is thermally efficient, is able to melt up to 50 percent scrap, and gives a good-quality steel, but the wear on the basic brick is severe. Campaigns of 22,500 tons have been made with 13 lb of refractory used per ton of steel.

A third type of converter was developed in Germany and called a rotary, as shown in Fig. 21.11. Here the vessel is elongated on a horizontal axis and is rotated while blowing with one or two oxygen lances.

Most of the new installations being put in now are of the LD type, so that further discussions will be confined to this process. While at first only low-carbon steel was produced, now the process is able to handle a variety of compositions.

LD Linings. Many types of basic linings have been tried in the oxygen converters, such as 98 percent magnesite burned brick, tar-impregnated magnesite brick, tar-bonded magnesia, and tar-bonded dolomite. The latter, because of its low cost and good life, is now used extensively.

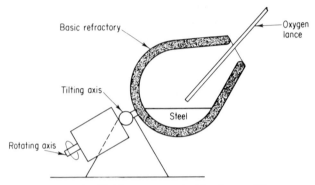

FIG. 21.10   Kaldo converter in blowing position.

Over 400 heats and 50,000 tons of production have been reported per lining, and refractory consumption as low as 6 lb per ton of steel is not unusual.   In oxygen converters the wear rates in Table 21.1 are reported by Lakin.[24]

The lining wear is not uniform; so it is most economical to use more resistant and more costly refractories in the danger spots.   For that reason, fusion-cast basic blocks have been used around the mouth opening, below the keeper plate.   In Kaldo converters the lining wear is more rapid than in the LD operations, so that walls up to 3 ft thick are required for a contemporaneous reline.

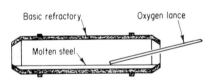

FIG. 21.11   Rotary oxygen converter.

The shelf life of the dolomite brick is about 2 weeks; so many European steel plants make their own linings.   In the United States, steel plants still buy from the refractory manufacturer.

21.9 Electric-arc Furnaces. *Principle of Operation.* Furnaces of this type are used to melt iron and steel by the heat of an arc formed between the slag layer and the overhead electrodes, usually three in number, which can be raised and lowered to accommodate the changes in metal height and loss in electrode length.   There are many variations in types

TABLE 21.1   Rate of Wear of Basic Refractories in BOF Operation

Type of operation	Type of refractory	Wear per heat, mm
LD...............	Tar-bonded dolomite	2.0
LDAC............	Tar-bonded dolomite	3.0
LD...............	Tar-bonded magnesite	1.5
LD...............	Tar-impregnated magnesite	0.7
LDAC............	Tar-impregnated magnesite	1.5
Kaldo.............	Tar-bonded dolomite	15.0

of furnace, but most of them consist of a cylindrical, refractory-lined shell with a domed roof and a charging door in the side. Nearly all of them are on trunnions so that they are poured by tilting.

The acid-type furnace is used mainly for steel castings and iron, whereas the basic furnace is most commonly used for production of steel ingots.

Construction. The bottom is built up much like that of the open hearth with a course of firebrick against the shell, a course of magnesite brick, and then a fettled-in bottom of dolomite, as shown in Fig. 21.12.

The side walls are usually metal-cased, unburned magnesite brick, with fused-cast blocks set into the places where the wear is greatest (such as the slag line, opposite the electrodes, and opposite the mast). An interesting construction used on the Continent consists of casting large reinforced blocks of tar-bonded dolomite, perhaps six for the whole wall. These can be installed with a minimum of labor and shutdown time.

The roof has always been of a dense silica brick, which has given good service. Now, when the power is being stepped up, the furnace builder is turning to high-alumina brick.[16,21] If the lifting equipment were strong enough, fusion-cast basic brick would be promising.

21.10   High-frequency Induction Melting Furnaces.   *Principle of Operation.*   This type of furnace melts metal by inducing high currents in the charge by means of an external coil surrounding the crucible. This furnace differs from all others in that the heat is generated directly in the metal itself and the crucible is not subjected to any other heating. High-frequency induction furnaces are made in small units for experimental melts of a few pounds up to 5-ton steel-melting furnaces. This kind of melting has certain advantages, particularly for alloy steels, in that a thorough stirring of the melt takes place and the analysis of the resultant steel can be accurately controlled.

Refractories. In Fig. 21.13 is shown a cross section of a large high-frequency induction furnace for melting alloy steels. The high-frequency current is carried by the water-cooled coil, inside which is a formed mica sleeve or troweled on sillimanite mix for coil protection. On the inner

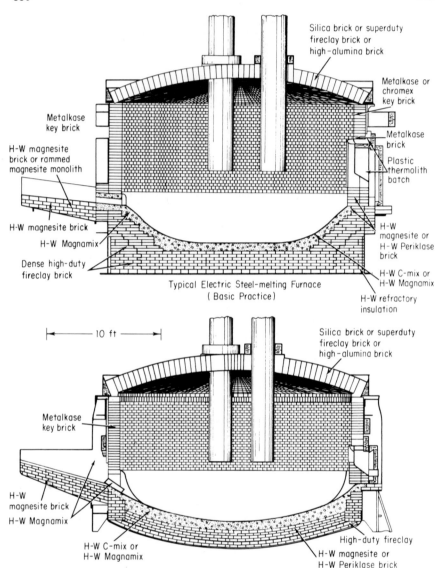

Silica brick or superduty
fireclay brick or
high-alumina brick

Metalkase or
chromex
key brick

Metalkase
key brick

H-W magnesite
brick or rammed
magnesite monolith

Metalkase
brick

Plastic
thermolith
batch

H-W magnesite brick

H-W Magnamix

Dense high-duty
fireclay brick

H-W
magnesite or
H-W Periklase
brick

H-W C-mix or
H-W Magnamix

H-W refractory
insulation

Typical Electric Steel-melting Furnace
(Basic Practice)

|← 10 ft →|

Silica brick or superduty
fireclay brick or
high-alumina brick

Metalkase
key brick

H-W
magnesite brick

H-W Magnamix

H-W C-mix or
H-W Magnamix

High-duty fireclay

H-W magnesite or
H-W Periklase brick

Typical Electric Steel-melting Furnace
(Basic Practice)

FIG. 21.12   Electric-arc steel furnace.   (*Harbison-Walker Refractories Co.*)

side of this is rammed a thin layer of refractory to form the melting chamber. For the highest efficiency, this refractory must be thin, and yet it must not crack during the melting operation. The preferred method of putting in this refractory is to ram it by hand inside a core made in the form of a steel or asbestos cylinder. The core can be withdrawn or melted out while, at the same time, fritting the surface of the refractory together into a strong mass.

A satisfactory refractory may be a rammed ganister bonded with a little clay or silicate of soda for acid furnaces. Electrically sintered magnesite

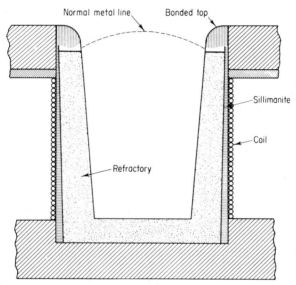

FIG. 21.13 General arrangement of coreless induction furnace. (*Courtesy of B. W. Magalis, Babcock & Wilcox Co.*)

or magnesia-alumina combination with an organic binder is used for basic furnaces, although linings are fritted in place with no bond by melting out an iron sleeve. With patching, 100 to 200 heats can be obtained with a single lining. Fused alumina and zircon have also been tried with good success in special cases.

For the smaller furnaces, prefired crucibles are generally used, magnesite being usually preferred.

21.11 Furnaces for Forming Operations. *Soaking Pits.* The soaking pit as shown in Fig. 21.14 consists of a rectangular or cylindrical furnace generally heated by tangential burners using either gas or oil as fuel. Regeneration is often employed. The removable cover, automatically operated, allows the heavy ingots to be placed or removed with an overhead crane.

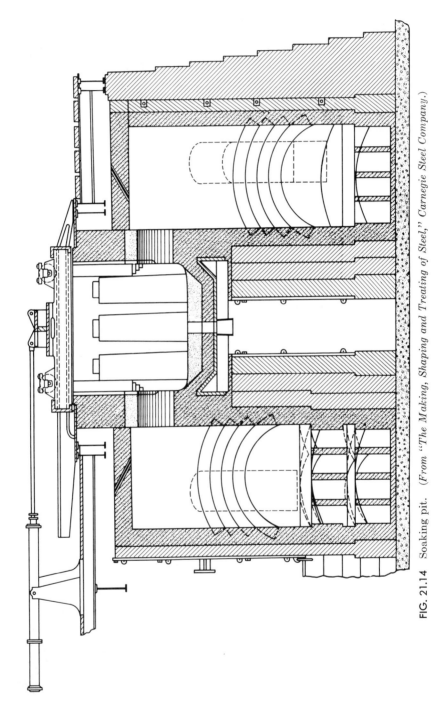

FIG. 21.14 Soaking pit. (From "The Making, Shaping and Treating of Steel," Carnegie Steel Company.)

The bottoms of these furnaces are made of rammed coke and molasses or rammed chrome plastic to withstand the action of mill scale, which forms in comparatively large quantities in furnaces of this type. The side walls are usually made of superduty fireclay or silica brick, but unburned magnesite or chrome brick have been tried with success.

The cover is one of the most important parts of the furnace, as it must withstand considerable mechanical shock due to the frequent opening and closing; and at the same time, it should be as light as possible for the quick operation necessary for efficiency. For this reason, covers have recently been made almost entirely from insulating firebrick, because it not only reduces the weight of the refractory in the cover but also

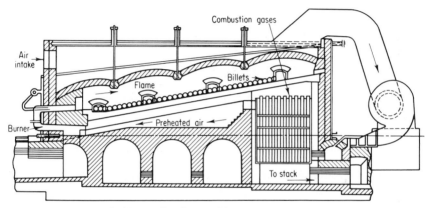

FIG. 21.15   Continuous billet-heating furnace.

permits much lighter steelwork to be employed. These refractories must be carefully selected so that they will stand the spalling conditions of frequent opening and closing of the cover, have a light weight and a good insulation value.

Billet-heating Furnaces. Billet-heating furnaces may be of the continuous type, a typical example of which is shown in Fig. 21.15. Here the billets pass through the furnace in a continuous layer on water-cooled rails and are heated mainly by convection of the hot combustion gases passing over and under them. This type of furnace is a true recuperative type, because of the counterflow action, so that the efficiency is comparatively high. The hotter parts of the furnace are made of superduty or high-alumina[15] fireclay brick or kaolin brick, but the upper structure of the cooler portions is generally made from low-heat-duty brick with good insulation. The bottom of the furnace, which must resist the action of mill scale, is usually constructed of rammed chrome plastic at the hotter end.

Rotary-hearth billet-heating furnaces have many advantages, such as small floor space, ready handling of the billets, and good fuel consumption. They are replacing the horizontal type in many modern mills.

Batch-type billet-heating furnaces are constructed in a large number of designs, but here the bottom is usually made of chrome and the upper structure of insulating firebrick. The value of insulating firebrick is particularly evident in the doors, which are decreased in weight so that they can be operated more rapidly and with less discomfort to the workman.

Forging Furnaces. Forging furnaces are made in many types. The refractory used is generally of fireclay, but the use of lightweight refractories is becoming quite prevalent in modern construction. The floor of the furnace may be of dense fireclay brick or rammed chrome plastic.

21.12 Heat-treating and Stress-relieving Furnaces.
Furnaces of this type are required to withstand only moderate temperatures. Heat-treating furnaces for high-speed steel might require a maximum temperature of 2400°F (about 1300°C), whereas stress-relieving furnaces would reach only 1400°F (about 760°C). For this reason, it has been found that the insulating firebrick can be used very successfully in many constructions. It has the advantage of giving a material saving in fuel, lower cost of burner equipment, and in many cases, increased life.

21.13 Bright Annealing Furnaces.
Furnaces of this type operate with a controlled atmosphere, usually with hydrogen. The temperatures are low; and for this reason, the insulating firebrick has seen extensive use both in the continuous and in the intermittent types. Because of the strong reducing action of the hydrogen, it is necessary to employ a refractory that will be stable under these operating conditions. For that reason, bricks low in iron oxide content are desirable.

21.14 Continuous Casting.
Since the Third Edition of "Refractories" was written, continuous casting of steel has come into widespread use, for it bypasses the steps of ingot casting, soaking pit, and the breakdown mill. These facilities are very costly parts of the steel mill, so that continuous casting allows a small mill to operate efficiently.

The continuous-casting method is illustrated in Fig. 21.16. Molten steel is poured at a controlled rate into a tundish to remove slag and then into a cooled copper mold where solidification continues. The ingot is then drawn down by rollers and bent into a horizontal direction for easy handling and cut to length. Nearly any cross section may be cast, up to 48 in. wide.

The amount of refractory used in this process is small, but it serves an important function. The holding ladle is usually lined with 3 in. of superduty fireclay brick, backed up with a $1\frac{1}{2}$-in. safety lining, which in turn is backed up with a high-temperature insulating layer.   In some installations the ladles are bottom-poured,[17] while in others they are poured from the lip, having a slag bridge with automatic controls to keep a constant rate.   The tundish is lined with $2\frac{1}{2}$ in. of superduty fireclay brick.   The nozzle size depends on the size of the piece cast but usually runs from $\frac{3}{4}$ to 1 in.   Nozzle materials are dense fireclay, high alumina,

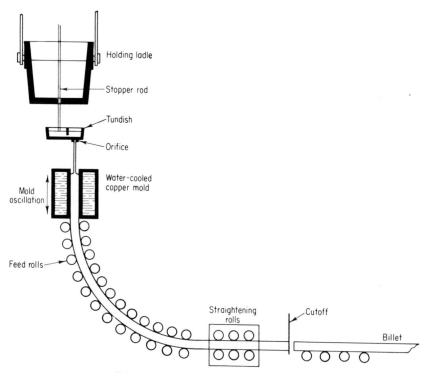

FIG. 21.16   Continuous casting of steel.

magnesite, or zircon.[22]   Fireclay stopper rods are often used to control the flow rate for large sections, while zircon or zirconia is best for small streams.   Johnson et al.[22] give an excellent discussion of the refractories used at Appleby-Frodingham, and Halliday[4] discusses those at Barrow.

It should be recognized that in this method of casting, any slag or dirt in the steel goes into the ingot, as there is no place for it to escape.   Therefore, the refractories must be selected to give minimum erosion.

Casting speeds have increased with reciprocating molds to 50 in. per min, and undoubtedly this will be increased in the future.

BIBLIOGRAPHY

1. Buell, W. C.: "Open Hearth Furnaces," Penton Publishing Company, Cleveland, 1936.
2. Barlow, T. E., and P. D. Humont: Gun-placed Silica Cupola Linings, *Bull. Am. Ceram. Soc.*, 33, 301, 1954.
3. Workman, G. M.: Assessment of Ladle Firebrick Quality and Performance, *Trans. Brit. Ceram. Soc.*, 57, 551, 1958.
4. Halliday, I. M. D.: Continuous Casting at Barrow, *J. Iron Steel Inst.*, 191, 121, 1959.
5. Latour, A.: Lining and Operation of Hot-metal Mixers, *Trans. Brit. Ceram. Soc.*, 59, 432, 1960.
6. Aldred, F. H., and N. W. Hinchliffe: Blast-furnace Refractories, *J. Iron Steel Inst.*, 199, 241, 1961.
7. Storey, C.: Blast-furnace-stove Refractories, *Trans. Brit. Ceram. Soc.*, 60, 783, 1961.
8. Fitchett, K., and H. M. Richardson: The Properties of Magnesite Mixer Blocks and Their Reaction with Mixer Slags, *Trans. Brit. Ceram. Soc.*, 60, 627, 1961.
9. De La Rosa, M.: History of Carbon Linings at Inland Steel Co., *Blast Furnace, Coke Oven, Raw Mater. Comm. Proc.*, p. 209, 1962.
10. Collison, W. H.: Carbon as a Blast Furnace Refractory, *Blast Furnace, Coke Oven, Raw Mater. Comm. Proc.*, p. 203, 1962.
11. Bonnot, M.: Carbon Refractories for Blast Furnaces, *Centre Doc. Sid. Circ. Inf. Tech.*, (10), 2219, 1962.
12. Kerr, W. R.: Carbon Bosh Experience at Kaiser Steel Corporation, *Blast Furnace, Coke Oven, Raw Mater. Comm. Proc.*, p. 227, 1962.
13. Savioli, F.: The Wear of Refractories in Steel Ladles, *Trans. Brit. Ceram. Soc.*, 61, 343, 1962.
14. Bloch, J.: Lining Repairing Equipment, British Patent 907,007, 1962.
15. Workman, G. M.: High-alumina Bricks in the Roofs of Continuous Slab Reheating Furnaces, *Trans. Brit. Ceram. Soc.*, 61, 753, 1962.
16. Halm, L.: Electric Steel-furnaces and Refractories with a High Alumina Content, *Bull. Soc. Franc. Ceram.*, 57, 95, 1963.
17. Gallagher, L. V., and B. S. Old: The Continuous Casting of Steel, *Sci. Am.*, December, 1963.
18. Strelets, V. M., N. V. Pitak, A. I. Kulik, and M. S. Logachev: The Life of Zircon Nozzles in Continuous Steel Casting, *Refractories (U.S.S.R.)*, (3–4), 177, 1963.
19. Richardson, H. M.: Blast-furnace-stove Refractories, *Trans. 9th Intern. Ceram. Congr.*, p. 371, 1964.
20. Petit, D.: Present State of Construction of Cowper Stoves, *J. Iron Steel Inst.*, 202, 305, 1964.
21. Lakin, J. R.: Refractories in Electric Arc Furnaces, *Trans. Brit. Ceram. Soc.*, 63, 221, 1964.
22. Johnson, R., J. W. Middleton, and D. Ford: Continuous Casting at Appleby-Frodingham, *J. Iron Steel Inst.*, 202, pt. 3, 193, 1964.
23. Brown, R. W.: Fused-cast Refractories for Blast-furnace Linings, *Blast Furnace Steel Plant*, 53, 311, 1965.
24. Lakin, J. R.: Progress in Refractories for Steelmaking, *J. Brit. Ceram. Soc.*, 2, 101, 1965.

CHAPTER TWENTY-TWO

Refractories in the Nonferrous Industry

22.1 Introduction. There has been very little change in the nonferrous industry since the Third Edition of "Refractories" was written. Perhaps the more extensive use of basic refractories and the increase in furnace capacities are the important innovations.

22.2 Copper Production. *Outline of the Process.* The common copper ores contain between 1 and 2 percent copper and must be concentrated by ore-dressing methods, which bring the percentage of copper up to between 20 and 30 percent.   If this concentrated ore contains free sulfur or a too high quantity of iron sulfide, it must be roasted to drive off a portion of the sulfur as SO_2.   This roasting takes place at a low temperature in a multiple-hearth furnace (Fig. 22.1); and although no serious refractory problems are encountered here, the hearths of fireclay brick must be abrasion-resistant.   The roasted ore is then smelted, which is carried out at the present time in a reverberatory type of furnace, as the blast furnace is seldom used.   In this operation, the copper sulfide is separated from the gangue and some of the iron as a matte, which is a mixture of copper and iron sulfide.   This matte is then treated in a converter where air is blown through the melted sulfides, thus oxidizing the sulfur and the iron, the latter being taken up in the slag and the former passing out of the furnace as SO_2.   The crude copper, called "blister copper," is then treated in an anode furnace where it is further oxidized, and the base-metal oxides formed are taken up in the slag and removed.   The copper is then reduced by covering with coke and introducing green poles which give off CO and water vapor, thus reducing all but a very small amount of the copper oxide to metallic copper.   The copper from this process is cast in anodes

357

and treated in a cell to form electrolytic copper cathodes; at the same time, the gold and silver can be recovered in the slimes.   The electrolytic copper cathodes are then melted in a fining furnace and cast into commercial shapes.

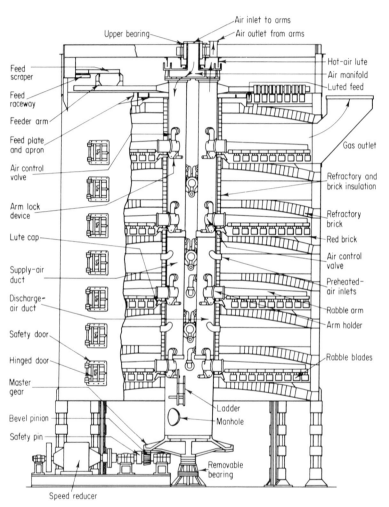

FIG. 22.1   Wedge roaster.   (*Courtesy of Bethlehem Foundry and Machine Company, G. Thorpe.*)

Reverberatory Smelters.   This type of furnace, as shown in Fig. 22.2, very much like a long open-hearth furnace, is fired at one end, usually with gas, oil, or powdered coal.   The charge, which in the earlier furnaces was fed through openings in the center of the roof, is now fed through openings

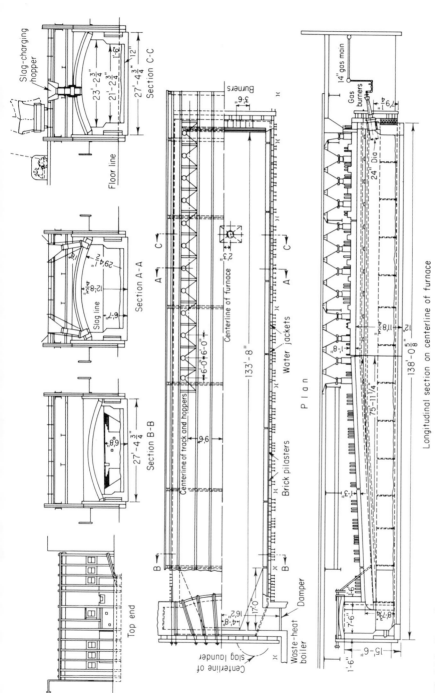

FIG. 22.2 Reverberatory copper-matting furnace. (*Courtesy of American Institute of Mining and Metallurgical Engineers, Paper by Frederick Laist.*)

359

in the roof near the skewbacks.   As the charge melts, it flows down to the cooler end of the furnace where the matte and slag can be tapped off separately.   The size of these furnaces is usually 20 to 25 ft in width and 120 ft in length, capable of handling 500 to 1,000 tons of charge per day, although a few furnaces even larger than this have been constructed.

The bottom of the furnace is made of silica rammed in and fritted in place much in the same way as the acid bottom is put in an open hearth. The bottom must be sufficiently dense to prevent the penetration of the liquid matte and requires frequent repairs.

The side walls of the furnace are generally silica brick, although magnesite brick have been used in the lower side walls.   Water cooling at the burner end has been used in some furnaces; but at present, it is not common.   Silica brick are also used in the end walls, but high-alumina brick have also been successful in some cases.

The roof of the reverberatory furnace is exposed to a great deal of dust, which fuses and produces a severe slag attack.   Silica brick are generally used in the roof in spite of the fluxing effect.   Within the last thirty years, magnesite brick either of the hard-fired, low-iron type or chemically bonded brick have been used quite extensively in the portion of the arch near the skewbacks, especially around the charging holes.   Suspended roofs of chemically bonded magnesite have been tried with very good success in Canada.   The life of the silica roof runs about 100 days; but with the use of magnesite, a somewhat longer life has been reported.   Insulation of these furnaces is beginning to be recognized as good practice, and probably all furnaces of this type will be insulated to some extent in the near future.

Copper Converters.   Two types of converter are generally used at present, known as the "Great Falls" and the "Peirce-Smith."   The converters are mounted on trunnions for pouring the slag and metal, and the capacity runs from 100 to 300 tons per day.   Converters are almost always lined with magnesite brick backed up with some magnesite grain between the lining and shell.   Some unburned magnesite brick have been tried in converters, apparently with considerable success.   The life of these linings is quite long, averaging 5 to 10 years, as a layer of magnesite is built up on the inner surface and protects the lining.   Around the tuyères, however, the cutting action may be quite severe, and repairs have to be made in this zone after about 6 months or less.   In Fig. 22.3 is shown a Peirce-Smith type copper converter.

Copper Anode Refining Furnace.   Most of these furnaces are of the reverberatory type, perhaps 35 ft long and 15 ft wide, with a capacity of 250 to 400 tons of copper.   The fuel is oil, pulverized coal, or gas, firing being from one end.   After melting, air is blown into the bath to saturate with Cu_2O, while a small amount of siliceous flux is sometimes added to

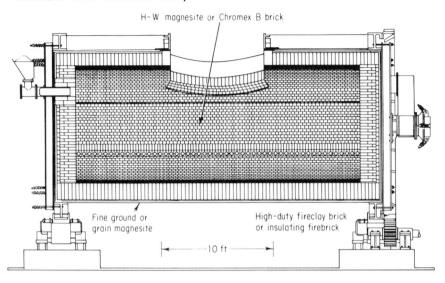

H-W magnesite or Chromex B brick

Fine ground or
grain magnesite

High-duty fireclay brick
or insulating firebrick

|◄———— 10 ft ————►|

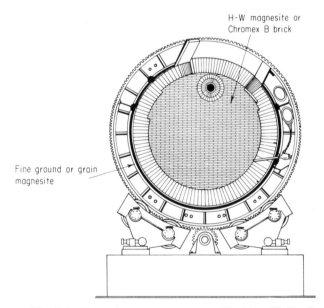

H-W magnesite or
Chromex B brick

Fine ground or grain
magnesite

FIG. 22.3 Typical copper converter. (*Harbison-Walker Refractories Co.*)

slag the base metals. The slag is then skimmed off, and the charge is covered with coke and poled to reduce the Cu_2O. The refining cycle takes somewhere around 24 hr. The bottoms of these furnaces are quite similar to the bottoms of the smelting furnaces, i.e., ganister rammed on top of clay brick. In the melting operation, at the surface of the ganister is formed copper silicate, which acts as a cementing material to give a strong, firm body.   Often bottoms are kept cool by air circulating through pipes running under the bottom.   Though requiring occasional patching, bottoms may last from 3 to 5 years.   In the side walls, magnesite brick are used to slightly above the slag line, beyond which silica brick are generally used.   Above the slag line, chemically bonded and metal-cased magnesite brick have been successfully tried, giving a life of 200 heats.

The roof of the refining furnace is generally of silica brick, but it is actively attacked by the copper oxide slag thrown against it during the poling operation.   Thus the roof has a life of only 100 to 125 heats.

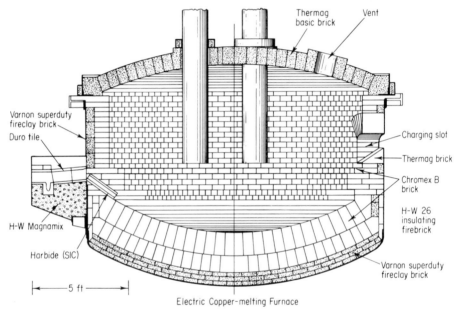

FIG. 22.4   Electric-arc copper-melting furnace.   (*Harbison-Walker Refractories Co.*)

Magnesite and forsterite bricks have been tried in the roofs to some extent, with somewhat increased life.

Now much copper is melted in arc furnaces, an example of which is shown in Fig. 22.4.

Wire-bar Furnaces.   The operations producing wire bars and other commercial shapes from copper cathodes are practically the same as those for producing anodes.   The furnaces are very similar to the anode furnaces previously discussed, except the bottom is made of silica brick instead of silica sand.   Side walls are made of magnesite to slightly above the slag line, and either silica or magnesite brick are used up to the arch.   In the upper parts of the side wall, both unburned and metal-case brick have been successfully tried. The roofs are usually made of silica, although magnesite brick have been used with a somewhat greater life.

22.3 Melting Copper Alloys. *Types of Furnace.* The melting of

brass and other copper alloys is carried out in gas- or oil-fired furnaces with a shallow hearth and an open flame; but for smaller lots or high-zinc alloys, plumbago crucibles are used.

The indirect-arc furnace in a rotating cylindrical shell is much used, but recently, however, the induction-type melting furnace of the Ajax-Wyatt type is favored.

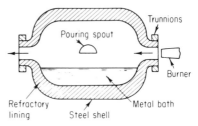

Open-flame Melting Furnaces. In Fig. 22.5 is shown a tilting furnace of this type in which the oil burner is introduced through one end and the charge is put through a door in the

FIG. 22.5 Open-flame brass-melting furnace.

side of the shell. The shell itself can be rotated on trunnions for pouring.

The linings of these furnaces are usually made from dense fireclay brick. For the hearth is used a rammed-in plastic material containing ganister, silicon carbide, or fused alumina with a clay bond. The refractories must

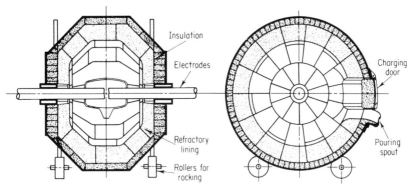

FIG. 22.6 Detroit rocking indirect-arc furnace. (*Redrawn from illustration by Huhlman Electric Company.*)

withstand spalling, and slag attack occurs occasionally. Five hundred to one thousand tons of brass can usually be melted in one lining.

Indirect-arc Melting Furnace. This furnace, as shown in Fig. 22.6, consists of a rotating, refractory-lining shell with an arc placed at the center of the axis so that the radiant heat falls directly on the metal bath and on the refractory walls. By rocking the furnaces, the hot refractory walls can be made to give up their heat to the metal bath and at the same time prevent the walls from being overheated. The lining is of high-alumina shapes or bricks, often with insulation. The furnace capacity runs from 250 to 2,000 lb, and as much as 1,000 tons of brass can be taken from a lining. Failures usually occur as a result of metal penetration in the lining and disintegration due to zinc dust.

The Induction Melting Furnace. This type of furnace, as shown in Fig. 22.7, consists of a crucible for holding the metal, and extending from the bottom of it is a U-shaped channel, which forms a one-turn secondary

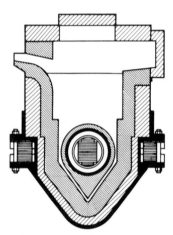

of a transformer by passing around its core. By means of the induced currents, the molten metal circulates through this U-shaped channel and heats the charge in the crucible. Of course, furnaces of this type must be started by a small amount of melted metal, after which pigs can be added.

From the refractory point of view, the greatest difficulty with these furnaces is the U-shaped passage or slot, which has a cross section of only a few square inches and must have thin walls to give a high efficiency. If any cracking occurs and molten metal runs down onto the transformer coil, the furnace is, of course, put out of commission and has to be rebuilt. It ceases to operate after a time because of gradual plugging of the slot by sludge. The refractory that has been found quite successful for these

FIG. 22.7 Section of Ajax-Wyatt furnace. (*J. Am. Ceram. Soc.*)

furnaces melting yellow brass is a plastic mixture of asbestos and fireclay, Adams[19] reporting an 8-year life in one case. Booth[26] reports 12 million pound heats in a fused-alumina rammed lining for yellow brass. For higher-temperature alloys, electrically fused magnesite and silicate of soda, carefully rammed into the furnace shell, have been used. Fused alumina with clay and magnesia-alumina ramming mixtures have also been used. In some cases, prefired shapes of flint clay have proved successful. A life of 600 heats for copper and 50 to 250 heats for nickel and its alloys has been reported.[18,20]

22.4 Zinc. *Outline of the Process.* Zinc ores, which usually contain sulfides but occasionally consist of carbonates or silicates, are generally concentrated by ore-dressing methods to produce a concentrate high in zinc. This concentrate is then desulfurized by roasting or sintering and reduced by carbon at a temperature of 1100 to 1300°C (about 2010 to 2370°F). Because of the fact, however, that metallic zinc is volatile below the temperature at which ZnO is reduced, the process must be carried out in closed vessels with enough excess carbon to prevent oxidation of the vapor. The zinc vapor coming from these closed vessels or retorts is condensed in the liquid form at temperatures around 425 to 500°C (about 800 to 930°F).

Most of the zinc is made in comparatively small retorts of only 2 or 3 cu ft capacity by the batch process. Recently, however, continuous smelters have been in operation.

Batch Zinc Distilling Furnaces. The zinc retort, a cross section of which is shown in Fig. 22.8, is about 8 to 9 in. in inside diameter and 4 to 5 ft long. At the open end is attached the condenser, which tapers down to a diameter of about 3 in. for the escape of the gas. Both the condenser and retort are made of fireclay mixtures, a Missouri fireclay generally being preferred in this country because of its good hot strength. Although the retorts used to be handmade, they are made, at present, entirely by machine, such as a hydraulic press. A good density of structure is necessary to prevent loss of zinc vapor through the walls and also to prevent too much zinc from being lost in the pores of the retort when it is rejected at the end of the run.

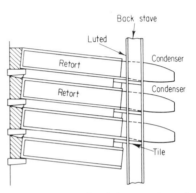

FIG. 22.8 Details of zinc retorts and condensers. (*From J. L. Bray, "The Principles of Metallurgy."*)

Some plants find that retorts containing 75 to 85 percent silicon carbide give enough greater life to pay for this more expensive material.

These retorts with their condensers are set up in banks of 200 or 300 in a long gas-fired furnace, as shown in Fig. 22.9, usually of the Belgian type. A center wall running down the middle of the furnace holds the back end of the retorts, whereas the front end is luted into the wall of the furnace. The firing is usually regenerative, and temperatures of 1250 to 1350°C (about 2280 to 2460°F) are reached inside the retorts, whereas outside temperatures as high as 1450°C (about 2640°F) are sometimes employed. This necessitates a rigid retort to carry the load at these temperatures, and it also means that the center wall of the furnace must be of a high-quality refractory. The side walls of the furnace are usually made of fireclay brick, whereas silica brick would be used for the center wall and roof. Hard-fired, high-alumina brick have been tried in the center walls and should give good results. The life of the retort runs from 30 to 40 days, but the furnace itself, being quite free from slagging action, gives a life of several years without repairs.

The Continuous Zinc Distilled Furnaces. The continuous furnaces, as developed by the New Jersey Zinc Company, use a vertical retort constructed of silicon carbide with a section of about 1 by 5 ft and a height of 20 to 25 ft. The briquetted charge of coal and zinc ore is continuously fed in the top, and the distilled zinc is collected in condensers around the top of the shaft while residues are discharged at the bottom. A single retort has a capacity of about 6 tons of zinc per day, and the life of the

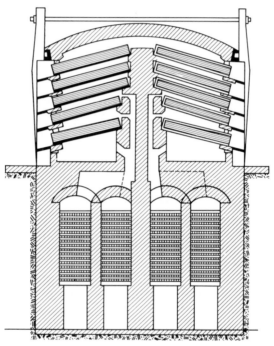

FIG. 22.9 A cross section of a Belgian zinc distillation furnace. (*From C. R. Hayward, "An Outline of Metallurgical Practice," D. Van Nostrand Company, Inc.*)

continuous retorts is estimated at 3 years. The furnaces in which these retorts are placed are fired with gas.

The Electrothermic Zinc Furnace. This type of furnace has been described by MacBride.[33] It is a vertical shaft heated by electrical resistance of the charge itself. The refractories composing the shaft must have a low porosity to prevent zinc penetration, high softening point, and high electrical resistance.

22.5 Lead. *Outline of the Process.* Lead ores are usually sulfides, of which the mineral galena is most common. The ore is concentrated by selective flotation, and the lead concentrate is sintered to a firm mass, usually in the Dwight Lloyd type of continuous machine, which does not employ any refractory. In this process, most of the sulfur is driven off as SO_2 and the lead is converted to an oxide. The process also produces the ore in a lump form, which is more suitable for the smelting operation. The smelting generally takes place in a water-jacketed blast furnace in order to slag off the impurities contained in the ore and to reduce the lead oxide to molten lead. The larger part of the molten lead is removed from the furnace through a "siphon tap," which passes through the crucible wall to the bottom of the crucible. The slag with some occluded lead is tapped from the furnace and allowed to separate in a settler from which the slag overflows to a slag car. The lead is tapped out periodically.

The Lead Blast Furnace. The lead blast furnace, as shown in Fig. 22.10, indicates that very little is used in the way of refractories because of the almost complete water jacket. However, there are refractories, usually high-duty fireclay brick of good density, in the upper part above the water jacket. Abrasion is the most important factor here, as the temperature is comparatively low. Refractories are also used in a crucible below the tuyères, but here again the temperature is not very high, and therefore the refractory requirements are chiefly low porosity and hardness and resistance to the action of molten lead. Dense firebrick are used in the crucible, but it is important that the joints between them be very thin, and often a chrome-base mortar is used for a jointing material. In some cases, plastic refractories of fireclay grog, ash, and coke are rammed into the crucible. Chrome plastics are also used to some extent, especially for the overflow spout.

Lead-refining Furnaces. Molten lead that comes directly from the blast furnace or from the desilvering kettles is treated in a reverberatory type of refining furnace somewhat as shown in Fig. 22.11. Air is blown into the bath to oxidize iron, arsenic, antimony, and tin to dross, and the latter is skimmed off and retreated. High refractoriness is not required, as the maximum temperature is around 1100°C (about 2000°F). Temperature changes are not particularly great; consequently, spalling is not a problem. Slag action is the most serious trouble, as the action of lead oxide is very severe in contact with most types of refractories. Magnesite brick are now generally used in the side walls to somewhat above the slag line, but high-alumina brick have also been tried. The upper parts of the furnace may consist of a medium-duty firebrick with moderate resistance to spalling. The life of a brick in the upper wall depends a great deal on the type of operation and the amount of antimony in the lead.

Other furnaces used in lead refining include dross furnaces, softening furnaces, retort furnaces, cupels, and lead oxide furnaces. While the temperatures are low, the slags are very active, so that basic brick are coming more and more into use.

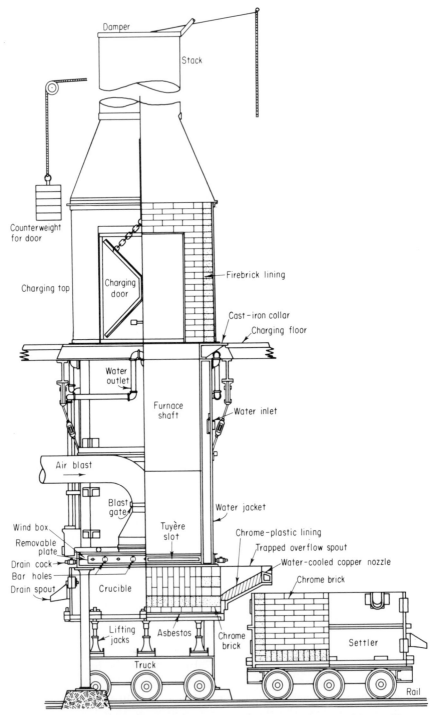

FIG. 22.10 Sectional view of lead-smelting furnace, matting type with outside settling. (*Courtesy of C. H. Mace, The Mace Company, Denver, Colorado.*)

22.6 Tin. There are, at the present time, very few tin smelters in the United States; but in other countries, this is a comparatively important operation. Many types of smelting are used, depending on the kind of ore. The metallurgist is fortunate in having, as his raw material, tin oxide, which can be readily reduced to the metal without great difficulty. The older types of furnace constructed are small blast furnaces much like the ones used for copper and lead. More recently, however, reverberatory furnaces have been used similar to those for copper ores, though on a much smaller scale. As the temperatures are comparatively low and the tin oxide is not an extremely active flux, dense fireclay brick have been found adequate as refractories.

22.7 Aluminum. Little has been published on the refractories used in the aluminum industry; however, M. J. Caprio of the Aluminum Corporation has kindly provided the following information.

The raw material, bauxite or alumina, is calcined in rotary kilns lined in the hot zone with high-alumina block and superduty or high-heat-duty blocks in the cooler areas.

The Hall-Heroult reduction cell is shown in Fig. 22.12. Here the aluminum oxide dissolved in fused cryolite is reduced to metal, which is drawn off, and the liberated oxygen, reacting with the carbon anode, forms CO and CO_2. Along with carbon paste for the lining, nitride-bonded silicon carbide bricks are used in certain places.

Melting furnaces of the reverberatory type have critical areas of the charging ramp, hearth, and lower side walls. Molten aluminum is an active reducing agent which reacts with free silica, producing a silicon impurity in the bath. The properties desired in the refractory are:

1. Resistance to attack by aluminum alloys
2. Chemically inert to gases and solid fluxes
3. Good abrasion resistance
4. Good resistance to thermal and mechanical shock
5. Good resistance to dusting

The refractories used are naturally a compromise between desired properties and cost. Some of the refractories used in the melting furnace are:

1. 85 to 95 percent alumina with ceramic bond
2. 85 to 95 percent alumina with frit or chemical bond
3. High-alumina calcium-aluminate cement castable (lower walls)
4. High alumina, phosphate-bonded ramming mix (lower walls and hearth)
5. Silicon carbide brick bonded with oxide or nitride
6. Zircon brick (lower wall)

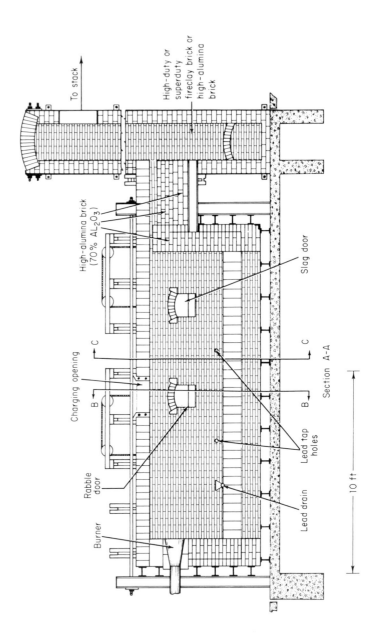

To stack

High-duty or superduty fireclay brick or high-alumina brick

High-alumina brick (70% AL_2O_3)

Charging opening

Rabble door

Burner

Slag door

Lead tap holes

Lead drain

Section A-A

10 ft

C

C

B

B

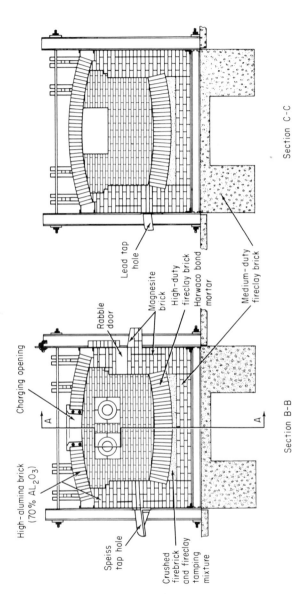

High-alumina brick
(70% AL_2O_3)

Charging opening

Rabble
door

Magnesite
brick

High-duty
fireclay brick

Harwaco bond
mortar

Medium-duty
fireclay brick

Speiss
tap hole

Crushed
firebrick
and fireclay
tamping
mixture

Lead tap
hole

Section B-B

Section C-C

FIG. 22.11 Typical lead-drossing furnace. (*From "Modern Refractory Practice," Harbison-Walker Refractories Company.*)

The refractory selected[39] depends on many factors such as operation, furnace design, life, and cost.

In reverberatory gas- or oil-fired holding furnaces the refractory requirements are not so critical as for the melting furnace, particularly in respect to abrasion resistance and thermal and mechanical shock. The hearth and lower side walls are generally made of high-alumina block, ceramic, frit, or chemically bonded. The noncritical areas of both the melting and holding furnaces such as roof, upper walls, flues, door jambs, arches, and lintels are made of superduty firebrick, or castables of fireclay grog or alumina. Insulating firebrick are used in flues and backup insulation and insulating concrete may be used under the hearth.

Aluminum transfer ladles are generally lined with calcium-aluminate castable or phosphate-bonded ramming mixes.

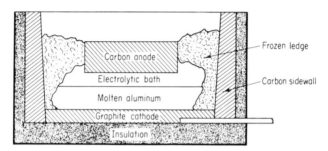

FIG. 22.12 Section of an aluminum reduction cell.

The transfer of molten aluminum in troughs employs linings of various types. The material must have insulating value, be resistant to abrasion and cracking, and have a low density. Such materials are asbestos-bonded concretes, boards reinforced with asbestos or silica-alumina fibers, castables with silica-alumina fibers, foamed silica with wash coats, and alumina-silica fiber paper.

22.8 Nickel. Nickel is smelted in much the same way as copper, the detailed process depending a great deal on the type of ore available. The melting of nickel is often carried out in small open-hearth furnaces that are made almost completely of a dense fireclay brick, although superduty fireclay and 50 percent alumina bricks have been used with considerable success in some of the later furnaces.

BIBLIOGRAPHY

Copper Production

1. Anon.: Decomposition of Refractory Brick in a Copper Reverberatory Furnace, *Eng. Mining J.*, **125,** 244, 1928.

2. Oldright, G. L., and F. W. Schroeder: Suggested Improvements for Smelting Copper in the Reverberatory Furnace, *Trans. Am. Inst. Mining Met. Engrs.*, **76**, 442, 1928.

3. Heuer, R. P.: Refractories (in Copper Industry), *Trans. Am. Inst. Mining Met. Engrs.*, **106**, 278, 1933.

4. Laist, F.: History of Copper Refining Practice, *Trans. Am. Inst. Mining Met. Engrs.*, **106**, 83, 1933.

5. Anon.: Linings for Tin Smelting Furnaces, *Brit. Clayworker*, **44**, 7, 1935.

6. Anon.: Industrial Survey of Conditions Surrounding Refractory Service in the Copper Industry, "Manual of ASTM Standards on Refractory Materials," 1937.

7. Ivanov, B. V.: Service of Dinas Brick in the Arch of a Reverberatory Copper-smelting Furnace, *Tsvetnye Metal.* (9), p. 95, 1938.

8. Honeyman, P. D. I.: Reverberatory Smelting of Raw Concentrates at International Smelting, *Bull. Am. Inst. Mining Met. Engrs.*, 456.

9. Robson, H. C.: Copper Refinery Furnace Firing and Refract., *Bull. Inst. Mining Met.*, 373, 374, 376, p. 25.

10. Heuer, R. P., and A. E. Fitzgerald: Basic Refractories for the Copper Industry, I, *Metals & Alloys*, **19**, 133, 1944.

11. Heuer, R. P., and A. E. Fitzgerald: Basic Refractories for the Copper Industry, II, *Metals & Alloys*, **19**, 1405, 1944.

12. Heuer, R. P., and A. E. Fitzgerald: Basic Refractories for the Copper Industry, III, *Metals & Alloys*, **20**, 68, 1944.

13. Dennis, W. H.: Modern Copper Smelting, *Metal Treatment*, II, **38**, 103, 1944.

14. Kocatopcu, S. S.: Reaction between Copper Reverberatory Slag and Refractories, *J. Am. Ceram. Soc.*, **28**, 65, 1945.

15. Dennis, W. H.: Refractories in the Extraction of Copper, *Metal Treatment*, **12**, 183, 1945.

Melting Copper Alloys

16. Gillett, H. W., and E. L. Mack: The Electric Brass Furnace Refractory Situation, *J. Am. Ceram. Soc.*, **7**, 288, 1924.

17. St. John, H. M.: Refractories for Brass Foundry Furnaces, *Trans. Am. Foundry-men's Assoc.*, **439**, 1926.

18. Campbell, D. F.: Recent Developments in Electric Furnaces, *J. Inst. Metals*, **41**, 37, 1929.

19. Adams, W., Jr.: The Ajax-Wyatt Induction Furnace, *Trans. Am. Electrochem. Soc.*, **57**, 462, 1930.

20. Robiette, A. G.: The Low-frequency Induction Furnace and Its Scope, *Metallurgia*, **3**, 175, 1931.

21. Klyutscharev, J. A., and S. A. Lowenstein: Acid Lining for Melting Nonferrous Metals in High Frequency Furnaces, *Feuerfest*, **9**, 157, 1933.

22. Patrick, W. L.: Fused Magnesia in the Nonferrous Metal Industry, *Metal Ind.* (*London*), **48**, 231, 1936.

23. Letort, Y.: Refractory Furnace Linings for the Melting of Nonferrous Metals, *Bull. Assoc. Tech. Fonderie*, **11**, 458, 1937.

24. Patch, N. K. B.: Nonferrous Melting Requires Excellent Furnace Operation, *Foundry*, **65**, 25, 1937.

25. McDowell, J. S.: Refractories in Nonferrous Metallurgical Furnaces, *Mining Congr. J.*, **25**, 17, 1939.

26. Booth, S. H.: Use of Refractories in Melting Copper and Copper Alloys in an Ajax-Wyatt Induction Furnace, *Bull. Am. Ceram. Soc.*, **19**, 171, 1940.

27. Dennis, W. H.: Refractories in the Extraction of Copper, *Metal Treatment*, **12**, 147, 1945.

Zinc Production

28. Rossmann, W. F.: Refractory Material for Zinc Retorts, U.S. Patent 1,424,120, 1921.
29. Varian, J. P.: Retorts for Zinc Ores, *Eng. Mining J.*, 113, 363, 1922.
30. Endell, K., and W. Steger: Tests on Clays for Making Zinc Retorts, *Metall u. Erz.*, 20, 321, 1923.
31. O'Harra, B. M.: Bibliography on Zinc Retorts and Condensers, *Bull. Mo. School Mines Met.*, 8, 15, 1925.
32. Spencer, G. L., Jr.: High-silica Retorts at the Rose Lake Smelter, *Trans. Am. Inst. Mining Met.*, 96, 119, 1931.
33. MacBride, W. B.: Notes on Refractories for the Electrothermic Zinc Industry, *Bull. Am. Ceram. Soc.*, 14, 389, 1935.
34. Weissmann, L.: Use of Native Clays and Silicon Carbide in the Production of Zinc Muffles, *Hutnické Listy*, 1936, 353.
35. Dennis, W. H.: Zinc Smelting and Refining, *Mine and Quarry Eng.*, 10, 211, 1945.

Lead Production

36. Anon.: Industrial Survey of Conditions Surrounding Refractory Service in the Lead Industry, "Manual of ASTM Standards on Refractory Materials," 1937.

General

37. Searle, A. B.: Refractory Materials for Electric Furnaces, III, Melting Nickel and Its Alloys, *Metal Ind. (London)*, 38, 569, 1931.
38. Stock, D. F., and J. L. Dolph: Refractories for Aluminum Melting, *Bull. Am. Ceram. Soc.*, 38, 356, 1959.
39. Brown, R. W., and C. R. Landback: Applications of Special Refractories in the Aluminum Industry, *Bull. Am. Ceram. Soc.*, 38, 352, 1959.

CHAPTER TWENTY-THREE

Refractories in
Miscellaneous Industries

23.1 Introduction. The use of refractories is so widespread that it is impossible in the space available to cover the whole field.   In this chapter those areas which use a substantial amount of heavy refractories are discussed.

23.2 Kilns. *Periodic Kilns.*   The periodic kiln is used in the ceramic industry mainly for the firing of refractories and heavy clay products, with the use of coal as the principal fuel, although natural gas and oil are also used.   A typical example of a round downdraft kiln of this type is shown in Fig. 13.14.   The maximum temperatures obtained in these kilns run from 1800°F (about 980°C) for some building brick to 2800°F (about 1540°C) for silica brick.

The refractories used in the kiln are always fireclay brick, using low- or intermediate-duty brick for heavy-clay-products kilns and high-heat-duty or superduty brick for the higher-temperature kilns. Because of the fact that the kilns are heated and cooled fairly rapidly, silica brick would not be satisfactory.

Recently, a number of periodic kilns have been constructed with a lining of insulating firebrick.[10]   They have shown distinct advantages over those lined with heavy types of brick.   Kilns of this type have been used for firing common brick and also for firing glass-tank blocks with temperatures of 2600 to 2700°F (1430 to 1480°C).   The main advantage in using the insulating firebrick is the great reduction in the heat stored in the walls and crown of the kiln.   In many cases, this heat storage has been cut down to 25 percent of the value for heavy bricks.   This means, of course, a substantial saving in the fuel required for heating the kiln, a quicker turnover so that a kiln can produce more ware, and perhaps as important, a more even temperature distribution over the kiln itself.

The refractories in periodic kilns generally have a comparatively long life, many kilns in operation having had only minor repairs in the last 15 or 20 years.   The main source of trouble is spalling of the crown brick over the fireboxes, shrinking and softening of the bag walls, and slagging from the coal ash around the furnaces.   With the use of forced draft and forced cooling of the kilns, the spalling tendencies are increased but, even under these conditions, are not particularly severe.

The use of outer insulation on periodic kilns is increasing, especially on the crown.   Although it is difficult to give any exact figures on fuel saving, the insulation does show a definite increase in economy and also enables a more uniform temperature to be maintained in the kiln.[7]   Of course, it may be necessary, when insulation is used on a high-temperature kiln, to employ a better grade of refractory to withstand the load at the higher mean temperature reached by the lining.

Continuous Kilns.   The tunnel kiln has been described in Chap. 13. In general, the requirement for this type of kiln is a refractory lining that will give long, uninterrupted service at the required temperature level. There is little or no tendency to spall because the temperature at any one place is constant.   Usually the wall bricks must have good strength and abrasion resistance in case ware rubs against them during an accident. Therefore, in the modern kiln construction, low-heat-duty fireclay brick is used in the heating and cooling zones and intermediate- or high-heat-duty brick in the hot zones, although for some types of kiln running at high temperatures, superduty fireclay or kaolin bricks are found useful.

The tunnel kiln, of course, must be very thoroughly insulated throughout its length; otherwise, a large amount of heat would be lost from the great amount of surface exposed.   This is accomplished with insulating brick or blocks on the sides and a loose insulation on the top.

Some tunnel kilns have been constructed with a lining of insulating firebrick; but in this case, a low heat storage is not so important as with the periodic kiln and therefore the insulating firebrick does not have so great an advantage.   However, from one point of view, this type of construction shows real economy,[19] i.e., in the kiln that must make a change in temperature to accommodate various types of ware fired.   It is found that with the kiln lined with the insulating firebrick, a change in schedule can be made in a few hours with a minimum loss in ware during the turnover whereas with a heavy brick kiln, this change of schedule may take a day or more and the loss in unproductive firing is much greater.   Also, the lighter-weight construction can use lighter steelwork and foundations.

Some tunnel kilns are indirect-fired and have a muffle lining. This muffle should, of course, have a high thermal conductivity, good hot strength, and stability over long periods of time. Such muffles are usually made of clay-bonded silicon carbide or clay-bonded fused alumina. There are, however, certain temperature ranges in which the silicon carbide tends to oxidize, and care should be taken to follow the recommendations of the manufacturer of the refractory in its use.

Perhaps the portions of the kiln giving the most trouble are the car bottoms, because they are alternately heated and cooled each time that they pass through the kiln and, for this reason, suffer from thermal shock. There also is the difficulty that setting sand used in kilns firing refractories or heavy clay products will fall into the cracks between the car blocks when they are cool; then as the car heats, the blocks are forced apart and there is a gradual growth in the size of the car top until it eventually rubs on the kiln. For this reason, car bottoms are usually made of large shapes, and some manufacturers blow out the sand from the bottom of the car with compressed air each time that it comes out of the kiln. Considerable success has been had lately with car tops made of a lightweight castable mix, which can be put on the car very cheaply by unskilled labor, a construction that seems to give good life in many installations.[5,15]

Kiln Furniture. Since the Third Edition was prepared, saggers have been gradually going out of use except for hard porcelain, some floor tiles, and electronic parts. There has been no real change in the structure of dinnerware saggers, but saggers for special technical ceramic pieces have been developed by the use of high-alumina mixes and higher firing temperatures.

Most ware in the United States is now fired in open settings,[13] usually consisting of a series of shelves spaced with corner posts. The setter material must have good hot strength, good thermal shock resistance, and volume stability. For lower temperature ranges, high-alumina refractories are used; while for higher ranges, frit-bonded silicon carbide gives good results if it has a protective coating to slow down the oxidation. Recently there has been interest in sandwich tiles, where the center layer is silicon carbide and the outer layers aluminous refractory.

In Fig. 23.1 is shown a photograph of a floor-tile sagger setting on a tunnel-kiln car, while in Fig. 23.2 is shown an open setting for glost hotel china, and in Fig. 23.3 some setters and saggers for high-temperature use in firing technical ceramics.

23.3 Glass-melting Refractories. The melting of glass is a very inefficient process, both in capital equipment required and in fuel used. The melting is carried out by applying heat to the top of a layer of batch (sand, limestone, and soda ash), which is floating on another layer of melted glass. The thermal conductivity of the batch is low; so it serves

FIG. 23.1   A setting for vitrified floor tile, using a special sagger.   (*R. C. Remmey Son Co.*)

FIG. 23.2   Glost hotel china plates in horseshoe setters and cups in open saggers. (*Courtesy of Phillip Dressler.*)

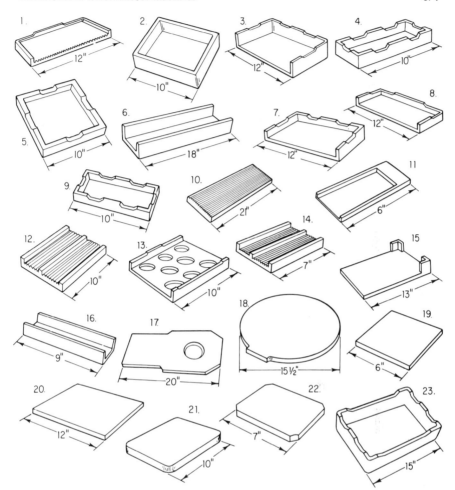

FIG. 23.3 Saggers and setters. 1. Half sagger V-groove bottom. 2. Box sagger. 3. Half sagger. 4. Sagger. 5. Sagger. 6. Open-end sagger. 7. Half sagger. 8. Half sagger. 9. Sagger. 10. V-grooved setter. 11. Tray. 12. Grooved setter. 13. Perforated sagger. 14. Grooved setter. 15. Setter—integral posts. 16. Open-end sagger. 17. Setter plate. 18. Circular setter. 19. Setter. 20. Setter. 21. Pusher tile. 22. Pusher tile. 23. Sagger. (*Babcock & Wilcox Company.*)

as a protective blanket to prevent heat from reaching the pool below. This pool moves sluggishly and requires considerable time to fine, that is, dissolve the last sand grains and remove the gas bubbles. The result is that only 200 or 300 lb of glass can be melted in 24 hr per sq ft of surface. Compare this with a modern steel furnace where 30 tons per sq ft are melted in a day. There must be a better way to melt glass.

Electric melting of glass increases the thermal efficiency of the process, but no large electric melters are used in the United States; and in Europe the recent availability of natural gas has reduced electric melting there. There have been other methods tried on a pilot scale, such as shaft kilns, submerged burners, and rotary furnaces; but none has proved feasible, principally because the melting of the batch must pass through a "sticky" phase.

Pot Melting. The glass made previous to the middle of the nineteenth century was melted in clay pots. Here the glass melt is completely surrounded with heat, so that its temperature and viscosity are uniform. At the present time special compositions, such as optical, art, and colored glasses, are melted in this way. The pots were formerly hand-molded of a mix consisting of highly plastic pot clays and grog, but at present they are usually slip-cast. The furnaces for heating the pots are constructed of high-heat-duty firebrick, sometimes with a silica roof.

Day Tanks. These are small reverberatory furnaces that are filled with batch melted and fired in about 24 hr, then worked out, and filled again. The refractories are similar to those used in the continuous furnaces described later.

Unit Melters. These are small continuous tanks, direct-fired without regenerators. They are not so thermally efficient as the larger continuous tanks but are lower in first cost and are therefore used for lower-capacity operations. The refractory problems are greatly simplified as there are no checkers.

Continuous Tanks for Containers. These tanks now conform closely to the schematic design shown in Fig. 23.4 for a natural-gas-fired unit. If producer gas were used, checkers would be provided for the gas as well as the air. The batch is fed in through the doghouses and spreads out over the melt, where it gradually disappears. The glass then passes through a submerged throat into the refining section and then out through the forehearths to the feeders.

The refractories in the early tanks consisted of clay flux blocks for the bottom and side walls, silica for the roof, and high-duty fireclay for the checker chambers and checkers. Today many types of refractories are used to attain good life with higher melting rates.

The roof is still made of silica refractories, but the parts exposed to glass are often of fused-cast blocks of alumina, or alumina-zirconia-silica compositions while dense, sintered zircon is often used for paving. As mullite tends to break down when exposed to alkali vapor, the burner ports, breast wall, and upper checker chambers are now made of pure alumina or basic refractories such as chemically bonded magnesite, forsterite (magnesium silicate), or fired magnesite. The upper checkers are now generally basic brick, and in some of the newer tanks, all basic checkers are installed. The more recent construction specifies insulation for the checker chambers, port uptakes, and often the roof.

Fabranic[1] gives an excellent description of the refractories problems in the glass tank. Baque[2] describes some of the fused-cast blocks, and Van Dreser[18] and Rochow and Primm[4] treat basic refractory use. Van Dreser and Cook[16] find 89 percent MgO checkers give the best results in their tests. Abbey[6] shows excellent pictures of refractory use, and

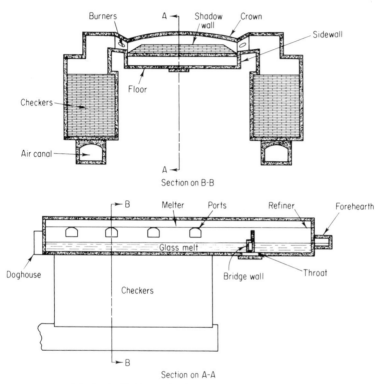

FIG. 23.4 Refractories in a container glass tank.

Knauft[3] discussed refractories for the forehearth and feeder. Poole[22] deals with the influence of refractories on glass quality.

Flat Glass Tanks. These tanks are large, often 30 ft wide and 100 ft long inside, as shown schematically in Fig. 23.5. Unlike the container tank, no throat is used, but often "floaters" of fireclay prevent floating batch from getting into the refining end. The glass is drawn out of the tank in the Colburn or Libby-Owens process through a floating fireclay debiteuse, as shown in Fig. 23.6. In the rolling process the glass passes under a baffle and over a lip to the rolls, as shown in Fig. 23.7.

The tank refractories are much like those described for the container tank, except that water cooling is often used at the level of the glass melt where the attack on the refractories is most severe. The floaters and debiteuse are hand-molded from glass pot clays and grog. When new ones are put in, they are preheated in a kiln and lifted into place while hot.

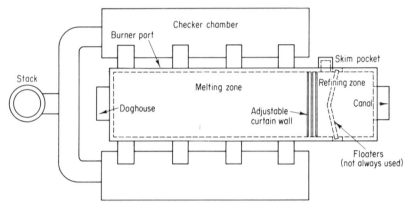

FIG. 23.5   Plan view of a flat glass tank.

There seems to be very little modern literature published on this type of tank.

Continuous Tanks for Ophthalmic and Optical Glass. Crown glass is now commonly made in small tanks with capacities of 100 to 400 lb per hr. The melting is done in a gas-heated fusion-cast tank, often with

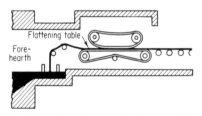

FIG. 23.6   Drawing sheet glass. (*"Elements of Ceramics,"* Addison-Wesley Publishing Co.)

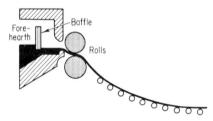

FIG. 23.7   Rolling sheet glass. (*"Elements of Ceramics,"* Addison-Wesley Publishing Co.)

auxiliary electric boosting, using molybdenum electrodes. The fining chamber is of platinum, electrically heated, and a platinum stir chamber homogenizes the glass before it flows out of the platinum exit tube. Very high quality glass can be produced in this way.

23.4   Refractories in the Cement and Lime Industry. *Shaft Kilns for Lime Burning.* Shaft kilns are still used extensively for lime and dolomite burning. There are several types. The mixed-feed kiln is like a cupola in that a mixture of coke and limestone is fed in the top and air is blown in the bottom. Sizes go up to 25 ft in diameter and 100 ft in height. Fuel-fired shaft kilns may use coal or oil burned in Dutch ovens at the base of the kiln or gas in a center burner. These kilns are usually lined with high-duty or superduty firebrick, often backed up with insulating fire-brick. In some cases, silica brick have been used for the lining, and in high-production forced-draft kilns, high-alumina linings have been found necessary.

Rotary kilns are now used for burning lime and dolomite in the larger plants and, in construction, are very similar to kilns for Portland-cement manufacture. The lining of these kilns is usually made of superduty fireclay block or high-alumina block. For the hot zone, magnesite blocks have been successfully used with steel-sheet inserts between the individual blocks.

Rotary Cement Kilns. These kilns are continuous-production units fed with a batch of limestone and shale or slag to give a clinker of

	%
SiO_2	23
Fe_2O_3	3
Al_2O_3	8
CaO	63
MgO	2
SO_3	1

In the United States the charge is usually fed in dry, but in Europe the charge is introduced into the kiln as a slurry. The temperature in the hot zone runs from 1450 to 1600°C (about 2640 to 2910°F), gradually falling off toward the feed end. A sketch of a typical kiln is shown in Fig. 23.8.

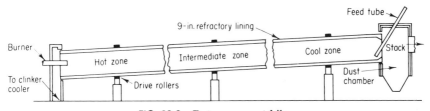

FIG. 23.8   Rotary cement kiln.

The refractories in the cooler end of the kiln are usually 6-in. blocks of dense fireclay brick to withstand the abrasion of the charge. The inter-mediate zone may use dense superduty firebrick. The hot zone is usually lined with 9 in. of high-alumina blocks (60 to 70 percent Al_2O_3). In the last few years, however, basic linings[17] have been used in the hot zone to a considerable extent.   These blocks may be fired magnesia or dolomite compositions, or in some cases chemically bonded.   Often steel plates are put between the blocks.

The main function of the hot-zone lining is to permit building up a coating on the refractory 2 or 3 in. thick that is constantly being renewed. The composition of the coating is similar to that of the clinker but apparently forms because of the comparatively cool face of the refractory. Basic linings are believed to hold the coating better than high-alumina linings.   This may be due to composition or differences in thermal conductivity.   At any rate, it is known that the life of the lining is short if a coating does not build up.

There is little agreement in the industry as to the value of an insulating layer between the lining and the shell.   There is no question about the heat saving, but some feel that the insulation raises the face temperature of the refractory too high to build a good coating.   Insulation may be used in "islands" so that the heavy lining brick may reach to the shell in limited areas.   For the lower-temperature portions, a 2½-in. layer of insulation may be used, but it is essential that it can stand the varying load without disintegration as the kiln revolves.   Budnikov et al.[20] discussed the pros and cons of insulation.   Huggett[11] believes that deformation of the shell and spalling cause some lining failures, while Brisbane and Segnit[12] comment on the fluxing action of alkalies.

23.5 Gas Production.   *By-product Coke Ovens.*   In this type of oven, the coking is carried out in a narrow chamber 16 to 22 in. wide, 10 to 12 ft high, and 30 to 36 ft long.   The bituminous coal is fed in at the top until the chamber is full; then the coking takes place by heat applied to

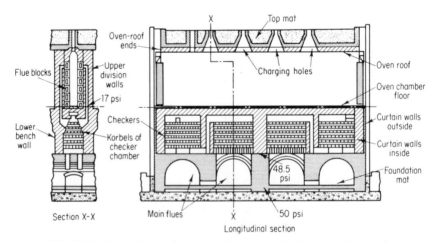

FIG. 23.9   By-product coke oven.   (*American Refractories Institute.*)

the outer walls of this chamber. After the coking has been finished, the charge is pushed out of the chamber laterally and at once quenched. The gases escaping during the coking period (12 to 30 hr) are collected in a large main and treated for removal of the by-products. A section of the Koppers type of by-product coke oven is shown in Fig. 23.9. It will be seen that the heating is regenerative, the gas usually being supplied by the coking process itself.

The walls of the coking chamber are comparatively thin to give low thermal resistance and therefore must have good mechanical strength. The refractory used is silica brick, the pores of which are more or less filled with coke after long operation. The bricks must, of course, be quite resistant to the abrasion of the entering and leaving charge, and there is a very rapid temperature change when the coal is dropped into the red-hot oven chamber. Care must be taken not to allow the temperature to drop to the critical temperature for silica brick. Sometimes fireclay brick are used at the end of the ovens where spalling may occur.

23.6 Incinerators. Incinerators are used extensively for combustion of refuse and require refractories capable of standing sudden temperature changes, but they need never undergo very high temperatures. Therefore, a low-heat-duty brick with good spalling resistance is generally satisfactory.

23.7 Domestic Heating. The increasing use of oil burners for domestic. heating has made a really large field for refractories, even though the number of bricks in any one unit amounts to only a few dozen. The temperatures in this service are low, but the bricks must be quite resistant to spalling because of the intermittent action of the burner. In many types of installation, standard-size brick are used; but in a few cases, special shapes are assembled to reduce the labor cost.

The use of the insulating firebrick in this type of service has been increasing rapidly in the last few years because it permits a high temperature to be built up on the surface of the refractory immediately after the burner has started, so that efficient combustion takes place during the whole operation with less production of soot and lower consumption of oil. As the burner is in operation only for short periods of time, the immediate attainment of efficient combustion is very important. It has also been found that the sound-absorbing qualities of the porous insulating firebrick are very helpful in reducing the noise of the burner flame.

Today much of the refractories in domestic-heating units consists of rather thin, flexible sheets of high-temperature fiber mats. This saves space and is easy to install or replace.

23.8 Steam-power Generation. Boiler refractories at one time comprised a sizable proportion of all refractories used, but each year the amount becomes less because of the greater proportion of waterwalls. The smaller boilers are now made in package units all ready to connect up and use with what little refractory there is all set in place.   The large power-station steam generators have now cut down the refractories used to little more than burner blocks.   Therefore, it is felt advisable to reduce the chapter on "Refractories in the Generation of Steam Power" to a section, in spite of the fact that there are older boiler settings using considerable amounts of refractory.

Tube and refractory walls are made up of waterwall tubes, which are part of the heat-absorbing surface of the boiler, spaced at intervals by

FIG. 23.10   Membrane boiler wall.   (*Babcock & Wilcox Company.*)

vertical fins and backed with 2 to 3 in. of block insulation.   The insulation is sufficiently cooled by the tubes to allow this type of wall to be used on the boiler convection zone for any type of firing, on high-duty furnaces fired by gas or stoker, and on low-duty pulverized-coal-fired furnaces. The limiting factor with pulverized-coal furnaces is excessive slag accumulation rather than deterioration of the refractory.   A typical wall of this type is shown in Fig. 23.10.

Closely spaced tube walls (see Fig. 23.11) made of touching waterwall tubes can be used for any type of firing to form a continuous steel envelope enclosing the furnace of a steam-generating unit.   When its area is properly proportioned, such an enclosure will eliminate trouble from excessive accumulation of slag. It is usually sealed against air leakage with insulating concrete, insulated with block insulation or insulating firebrick, and covered with a metal casing or a plastic insulation finish.

The castable refractory backing is reinforced with expanded metal lath placed close to its outer surface, which makes the castable act as a reinforced-concrete mat outside the tubes. Application of the concrete may be by hand or by cement gun. In either case, the metal lath is

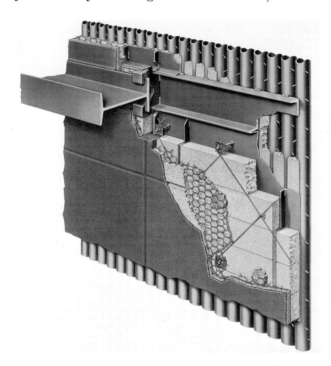

FIG. 23.11 Cross section of boiler wall with close-set tubes. (*Babcock & Wilcox Company.*)

Installed on the wall first and the castable applied through it, so that the lath will act as a form.

This construction can be made with no expansion joints. The drying shrinkage of the castable (about 0.5 percent at 600°F) plus a small difference between the thermal expansion of the boiler steel and the castable (the castable coefficient of expansion is about two-thirds that of steel) results in a small theoretical difference between the width and height of the tube wall and that of the refractory backing after the unit has been brought up to operating pressure. The effect of this slight differential expansion is a series of hairline cracks very well distributed by the reinforcing metal lath. The resultant enclosure is tighter against air leakage than one with expansion joints and requires very little maintenance.

One particular application of stud tube walls is the slag-tap furnace, where it is necessary to keep the coal ash above its fluid temperature in order to remove it as a liquid. This construction is very resistant to the fluid slag found in this type of furnace. A section of this construction is shown in Fig. 23.12.

A complete radiant boiler for slag-tap operation is shown in Fig. 23.13 with various types of waterwall.

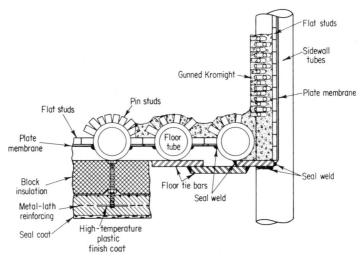

FIG. 23.12 Recovery-furnace section to show cooling. (*Babcock & Wilcox Company.*)

23.9 Paper-mill Refractories. *Sulfite Process.* For sulfur- and pyrite-burning furnaces, high-heat-duty or superduty fireclay brick are generally employed.

Soda and Sulfate Processes. Refractories are required here for smelters, rotary incinerators, rotary and shaft kilns, and sludge-recovery kilns. The slagging action is very severe in the smelting and recovery furnaces, as the soda salts are active fluxes. For the lower parts of these furnaces, soapstone or chrome brick are used, and dense high-heat fireclay brick are often employed in the upper portions.

Black-liquor Recovery Furnaces. The economical operation of a modern pulp mill necessitates the recovery of chemical and heat values in the waste liquor. In the sulfate or kraft process, in which the recovery of chemicals is most generally employed, the wood chips are digested with a hot solution of sodium hydroxide, $NaOH$, and sodium sulfide, Na_2S, to remove the noncellulose portion of the wood. The spent cooking liquor

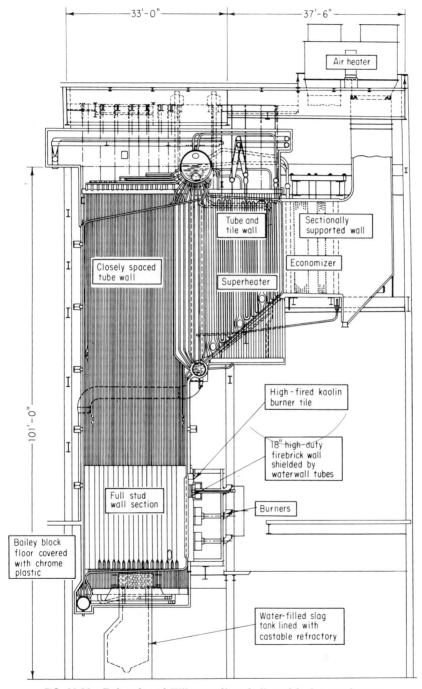

FIG. 23.13 Babcock and Wilcox radiant boiler with slag-tap furnace.

(black liquor) is concentrated by evaporation until it contains from 55 to 65 percent solids. It is then sprayed into a recovery furnace where it drops and falls to the floor as "char." On the floor, the carbonaceous material is burned with a deficiency of air. In the reducing atmosphere thus formed, the sodium salts are smelted out of the char. The resulting smelt containing mostly sodium carbonate, Na_2CO_3; sodium sulfide; a small percentage of sodium sulfate, Na_2SO_4; and traces of other sodium salts are discharged from a spout.

The conditions existing in this furnace are extremely severe as far as action on the refractories is concerned, primarily in view of the aggressive chemical activity of the fused sodium salts. In the earlier forms of recovery units, soapstone blocks were generally used as being the refractory best able to withstand the action of these molten sodium salts, but this material had a maximum life of only 10 to 12 months. In recent years, furnaces of this kind have been entirely water-cooled with a high degree of success. The upper walls are flat stud, while the lower walls and floor where a higher temperature is required are of full stud construction. The plastic chrome ore is quite resistant to the action of soda because it is inert chemically and is aided by the cooling action of the water tubes. This type of furnace requires little maintenance.

23.10 Enameling. *Cast Iron.* Cast-iron enameling is generally carried out in recuperative-type muffle furnaces. The muffle, which is comparatively thin, is made of high-conductivity material, either silicon carbide or bonded alumina. The furnaces are generally underfired, and the refractory requirements of the piers holding up the hearth are rather severe, as the brick are heated all the way through and must carry a heavy load. For this purpose, high-burned kaolin or mullite brick have been found very satisfactory. The walls of the furnace itself are generally made of high-heat-duty firebrick, and insulation is often used outside this. Since the doors of these furnaces must be quick-acting, they are often lined with a lightweight castable mixture or with insulating firebrick that are tied into the door casing. This not only permits a rapid operation of the door but makes conditions much more comfortable for the men working around the mouth of the furnace.

Sheet-steel Enameling. In this type of furnace, a muffle construction is used; but in many of the modern continuous furnaces, very little refractory is required, as the heat is supplied by radiant metal tubes inside which the combustion takes place. The walls of the furnace outside these tubes can readily be constructed of insulating firebrick, since the requirements in the way of temperature, load, and spalling are not at all severe.

Enamel Fritting. Enamel frit is made in rotary oil-fired furnaces or in small tanks. In the former case, the lining is made from circle blocks of fireclay material. It is desirable to have the blocks as free from iron specks as possible. In the tank, the conditions are about the same as for glass melting, and therefore dense flux blocks are generally used.

23.11 Refractories for the Petroleum Industry. The FCC process for cracking crude oil to give a range of products used today is now used extensively in refineries. The principal units are the regenerator, where the catalyst is reactivated by burning off its coating of coke; the reactor, where the feed is cracked by the catalyst; and the fractionator, where the products of the reaction are separated. Then there are connecting lines for transfer of materials. The temperatures are not high, 510 to 620°C (about 950 to 1150°F), but abrasion may be severe. The practice today is to use refractory linings inside the steel shell of castables reinforced by steel mesh. These linings can be placed by gunning at much less cost than bricklaying. Sometimes an insulating castable is placed first and then a hard castable is placed over it. The petroleum industry has worked closely with refractory manufacturers to develop linings with the best possible properties. Fireclay or kaolin grog bonded with pure high-alumina cement is often used, although the more costly alumina grain with phosphate bond has shown excellent abrasion resistance.

There are many references to refractory linings in the literature. Wygant and Bulkley[8] give an excellent picture of the lining materials available. Paul[9] gives an overall discussion of the problems encountered with linings, and Venable[14] gives erosion-resistance values. The overall design problem is ably handled by Wygant and Crowley.[21]

BIBLIOGRAPHY

1. Fabranic, W. L.: Refractories for the Glass Furnace Superstructure, *Bull. Am. Ceram. Soc.,* **29,** 96, 1950.
2. Baque, H. W.: Fused Cast Refractories, *Bull. Am. Ceram. Soc.,* **29,** 9, 1950.
3. Knauft, R. W.: Refractories for the Glass Tank Forehearth, *Bull. Am. Ceram. Soc.,* **29,** 47, 1950.
4. Rochow, W. F., and H. E. Primm: Refractories for Glass Furnaces, *Bull. Am. Ceram. Soc.,* **30,** 287, 1951.
5. Thompson, N. M.: Refractory Concrete for Tunnel Kiln Car Tops, *Bull. Am. Ceram. Soc.,* **32,** 1, 1953.
6. Abbey, R. G.: Refractory Trends in the Glass Industry, *Bull. Am. Ceram. Soc.,* **32,** 110, 1953.
7. Creson, W. F., and J. W. Moffat: Insulating Firebrick in Periodic Kilns, *Bull. Am. Ceram. Soc.,* **33,** 4, 1954.
8. Wygant, J. F., and W. L. Bulkley: Refractory Concrete for Refinery Vessel Linings, *Bull. Am. Ceram. Soc.,* **33,** 233, 1954.
9. Paul, W. B., Jr.: Monolithic Refractories in Fluid Catalytic Cracking Refinery Units, *Bull. Am. Ceram. Soc.,* **33,** 108, 1954.
10. Robinson, R. R., and R. W. Parker: Operating Period Kilns Constructed of Insulating Fire Brick, *Bull. Am. Ceram. Soc.,* **35,** 182, 1956.

11. Huggett, L. G.: Rotary Cement Kiln Linings—Refractory Problems in the Burning-zone, *Trans. Brit. Ceram. Soc.*, **56**, 87, 1957.
12. Brisbane, S. M., and E. R. Segnit: The Attack on Refractories in the Rotary Kiln, *Trans. Brit. Ceram. Soc.*, **56**, 237, 1957.
13. Landback, C. R.: Kiln Car Engineering, *Bull. Am. Ceram. Soc.*, **38**, 48, 1959.
14. Venable, C. R., Jr.: Erosion Resistance of Ceramic Materials for Petroleum Refinery Applications, *Bull. Am. Ceram. Soc.*, **38**, 363, 1959.
15. Brovarone, J. B., and A. O. Downs, Jr.: Insulating Firebrick in Shuttle and Tunnel Kilns, *Bull. Am. Ceram. Soc.*, **40**, 57, 1961.
16. Van Dreser, M. L., and R. H. Cook: Deterioration of Basic Refractories in a Glass Regenerator, *Bull. Am. Ceram. Soc.*, **40**, 68, 1961.
17. Parnham, H.: New Approach to Portland-cement Kiln Refractory Problems, *Refractories J.*, **38**, 284, 1962.
18. Van Dreser, M. L.: Basic Refractories for the Glass Industry, *Glass Ind.*, **43**, 18, 1962.
19. Hawkes, W. H., and A. Moore: The Evaluation of Some High-alumina Materials, *Trans. Brit. Ceram. Soc.*, **62**, 397, 1963.
20. Budnikov, P. P., G. A. Sokhatskaya, and V. I. Shubin: Insulation of Refractory Lining in Sintering Zone of Rotary Cement-kilns, *Refractories, Moscow*, **11**, 546, 1964.
21. Wygant, J. F., and M. S. Crowley: Designing Monolithic Refractory Vessel Linings, *Bull. Am. Ceram. Soc.*, **43**, 173, 1964.
22. Poole, J. P.: Refractory Needs of the Glass Industry, *Bull. Am. Ceram. Soc.*, **44**, 672, 1965.

CHAPTER TWENTY-FOUR

Brickwork Construction

24.1 Introduction. The life of a furnace depends not only on the refractory used but also on the way in which it is installed. Much practical experience is necessary to handle bricklaying in an effective manner; so all that can be done in this chapter is to set up a few guidelines toward good practice.

24.2 Foundations. The foundations for kilns or furnaces are usually made of reinforced concrete to carry the usual engineering loads. These loads include not only the weight of the furnace itself and its charge but the thrust of the buckstay foot. As concrete loses considerable strength at temperatures above 400°C (about 750°F), it will be found necessary to protect the foundation from overheating if it is expected to maintain its full strength.

In some cases, the foundation is poured with a lightweight concrete containing a quartz-free aggregate. A mixture of crushed brickbats and high-alumina cement is generally preferred, a material that can be heated to elevated temperatures without cracking or entirely losing its strength, although it cannot be depended upon to support any concentrated loads. When reinforcing is used in foundations subjected to heat, the expansion characteristics of the high-temperature concrete and the reinforcing rods should be investigated to prevent a differential expansion from cracking the foundation.

In certain types of furnace, the bottom is supported directly on the steelwork. For example, the open-hearth furnace, a section of which is shown in Fig. 24.1, has the bottom on a steel plate which is supported by

a series of H beams, thereby providing ventilation as well as a rigid
support. The glass-tank bottom, as shown in Fig. 24.2, has the flux
blocks supported directly on H beams, which permits the underside of
the blocks to be air-cooled. Some recent tanks have bottoms consisting
of an iron plate on which is placed a layer of insulating bricks, then a
layer of high-fired superduty firebricks, and on top of this the zircon
paving. In some cases container tanks have had bottoms made from a
ramming mix to give a monolithic structure.

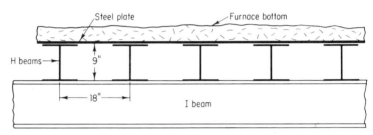

FIG. 24.1   Supports for an open-hearth bottom.

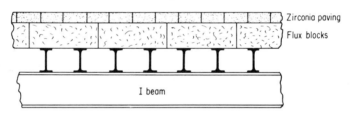

FIG. 24.2   Cross section of glass tank bottom.

24.3   Floors and Hearths. *Insulation.* The floor or hearth of a furnace
is usually insulated from the concrete foundation, both to conserve fuel
and to prevent overheating of the foundation. The insulation is often
accomplished by putting down a layer of hollow tile on top of the concrete foundation. In other cases, discarded steel boiler tubes are set in
the upper surface of the concrete foundation, a few inches apart, to allow
ventilation. Another method uses a layer of cinders tamped over the
concrete foundation, on top of which are laid several courses of firebrick.
In the smaller furnaces, two or three courses of insulating brick may be
laid on the concrete and then heavy brick placed on top of them. Many
furnace designers do not appreciate the large amount of heat that may be
lost through the foundations if the insulation is not well taken care of.

Shapes and Bonds. Most of the hearth structures are laid with standard-size brick. In some cases, however, larger blocks are used in the
hearth to minimize the number of joints and to make larger units for
reducing the tendency to float. In the case of the blast furnace, hearth

blocks 18 × 9 × 4½ in. are generally used. Sometimes these bottoms are keyed in to prevent floating. In the case of glass tanks, the bottom is made of large flux blocks often 24 × 48 in. across the face. Here again, the purpose is to minimize the joints and prevent the bottom from floating.

Expansion Joints. It is just as important to have expansion joints in the hearth as in the walls, because failure to take care of this expansion will force out the base of the furnace and bend the buckstays or cause buckling of the bottom. The expansion joints may be put in the hearth, although expansion joints directly under the walls are not considered good practice by some designers. The amount of expansion to be left is considered more fully under the section on walls.

Monolithic Construction. Many types of furnace hearth are constructed of a monolithic slab. The material may be magnesite grain as in the open-hearth furnace, chrome plastic as in the bottom of a powdered coal-fired boiler, or beach sand as in a malleable-iron furnace. There are two general methods of putting in the monolithic material. The first is to place it in dry, perhaps with a little slag, and sinter it down in layers until a firm hearth is produced. The other method is to ram in the plastic mix while it is wet to a homogeneous structure, then dry it out, and fire it in place. Refractory concrete, made with either ordinary or lightweight aggregate, is used to form monolithic hearths, particularly for car-bottom furnaces and tunnel-kiln cars.

24.4 Walls. The walls of a furnace are primarily for the purpose of retaining the heat in the working chamber. To be satisfactory, however, they must be stable under the severe operating conditions encountered. Experience has shown that certain limiting factors must be understood before satisfactory construction of walls can be accomplished.

Thickness. The thickness of the wall depends a good deal on the conditions encountered. From the point of view of stability, the higher walls must be thicker, as shown in Table 24.1. On the other hand, if the wall is subjected to severe slag attack or spalling, extra thickness must be built into it to give it a reasonable amount of life. Table 24.2 gives

TABLE 24.1 Safe Wall Heights for Unanchored Walls

Wall thickness, in.	Maximum height* of wall without anchors but arch supported on top of wall, ft
4½	3
9	7
13½	12
18	15

* When wall does not support arch, the safe height is decreased 20 percent.

TABLE 24.2   Walls Made of One Kind of Brick

Wall thickness, in.	Number of straight brick per sq ft
2½	3.6
4½	6.4
9	12.8
13½	19.2
18	25.6
22½	32.0
27	38.4
36	51.2

Refractory lining of a composite wall

9, fifth course tie	14
9, fourth course tie	14.4
4½, fourth course tie	8.0
4½, third course tie	8.5
4½, alternate header and stretcher	9.6
Alternate 9- and 13½-in. courses	16.0

Backing-up insulation of a compound wall

9, fifth course tie	11.5
9, fourth course tie	11.2
4½, fourth course tie	4.8
4½, third course tie	4.3
4½, alternate header and stretcher	9.6
Alternate 9- and 13½-in. courses	16.0

the number of brick required for various thicknesses of wall, both solid and composite, data that will be useful in brickwork estimating. Table 24.3 gives the weight of various types of refractory for computing wall weights.

Bonds.   There are four ways of laying a standard brick, as indicated in

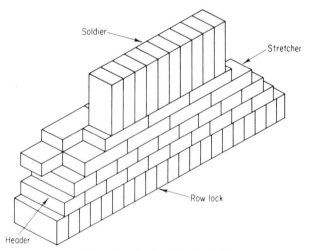

FIG. 24.3   Methods of laying brick.

TABLE 24.3 Weight of Refractories

Material	*Weight per cu ft*
Fireclay brick	120–140
Superduty brick	130–150
High-alumina brick:	
50	125–135
60	130–140
70	140–150
Fused Al_2O_3	153–196
Kaolin brick	130–135
Silica brick	100–110
Magnesite brick:	
Burned	160–175
Chemical bonded	180–185
Chrome brick:	
Burned	180–190
Chemical bonded	180–195
Silicon carbide brick	136–158
Forsterite brick	150–160
Insulating firebrick:	
1600°F	19–30
2000°F	28–40
2300°F	30–45
2600°F	39–55
2800°F	40–60
Insulating brick	20–35
Lightweight, castable	55–80 (45–75 fired)
Fireclay plastic	120–130
Chrome plastic	175–200
Mortar, high-temperature	90–120
Mortar, chrome	125
1 cu ft of firebrick requires	Seventeen 9-in. straights
1 cu ft of red brick requires	21 standard brick
1,000 firebrick closely packed occupy 60 cu ft	

Fig. 24.3. Nine-inch walls of one material are usually laid with all headers or with alternate header and stretcher courses. The rowlock courses are used for changing the height of the course, for sills, and for some floors but are not often used in wall construction with refractories. The soldier courses are not particularly stable and would not be used where they are expected to carry any amount of load. In building the thicker types of wall, it is necessary to bond the different layers together into a strong unit. Typical methods of bonding are shown in Fig. 24.4 for walls of different thickness. It should be noted that at the end of a wall, it is better to break the joint with a 6¾-in. brick, referred to as a large 9-in. straight, rather than with a soap, as this construction gives much greater stability.

With composite walls, e.g., a combination of heavy brick and insulating firebrick, it is necessary to tie one layer into the other.   The general rule is that the tie should be made of the stronger layer, as it is less apt to break off.   In many cases, the tie brick can be made with soaps rather than straights when using an insulating-firebrick lining, which reduces the heat conductivity of the whole.

Many types of furnace construction are carried out with bricks larger than the 9-in. size for more rapid laying of the brick or for more solid structures.   The 3-in. brick series is commonly used; and in some of the steel furnaces, still larger units are found useful.   It should be noted that the dimensions of the standard brick are not ideal for the purpose of efficient laying.   Strictly speaking, the thickness of the brick should be definitely related to the width and length, so that the brick could be tied into the structure no matter in what position it was used.   For example, if we consider the length of the brick 9 in. and use a $\frac{1}{16}$-in. joint, the dimensions would be $9 \times 4^{15}\!/\!_{32} \times 2^{15}\!/\!_{16}$.   Under these conditions, a rowlock course would equal two stretcher courses and one soldier course would correspond exactly to three stretcher courses. It will be seen that the 3-in. brick closely fulfills these conditions.

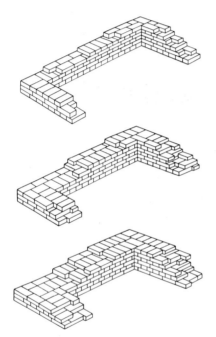

FIG. 24.4   Bonding in walls using brick of standard sizes.   (*Harbison-Walker Refractories Co.*)

Expansion Joints.   All except the smallest refractory structures must have allowance made for expansion on heating.   Considerable experience is necessary to know how many and what size expansion joints are to be used, but it is better to allow a little too much than not enough.   An expansion joint in a wall can be a straight vertical joint, or it can be a broken joint, as shown in Fig. 24.5.   In other types of construction, particularly when using silica and magnesite bricks, the expansion is taken care of by inserting a piece of cardboard in each vertical joint of the wall.   This burns out and allows the individual bricks to expand freely.

When regular expansion joints are used, it is common in high-temperature furnaces to place them not more than 10 or 15 ft apart, usually at the corners of the furnace first and then in intermediate positions if necessary.   In Fig. 24.6 are shown typical expansion joints near a corner.

Figure 24.7 shows an expansion joint used with insulating firebrick inside a steel casing. If it is necessary to keep the expansion joints gastight, loosely packed asbestos rope or high-temperature wool can be used on the cooler face to seal them entirely.

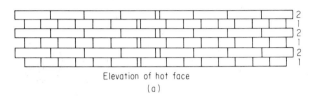

Elevation of hot face
(a)

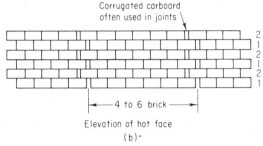

Elevation of hot face
(b)-

FIG. 24.5 Expansion joints in walls. (*a*) 13½-in. wall, alternate header and stretcher construction. (*b*) Built of magnesite brick, 9-in. header construction. (*From "Modern Refractory Practice," Harbison-Walker Refractories Company.*)

Horizontal expansion joints to allow for vertical expansion are sometimes required, as in sectionally supported walls and where the arch is supported independently of the walls.

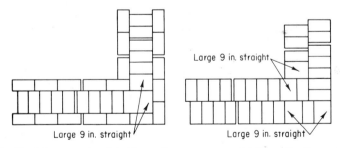

FIG. 24.6 Double expansion joint in wall near corner. 18-in. wall, alternate courses showing header and stretcher construction. (*From "Modern Refractory Practice," Harbison-Walker Refractories Company.*)

In Table 24.4 is given the width of expansion joints in inches per foot of wall for various types of refractory and at various temperatures.

Methods of Anchoring the Wall. When very thin or high walls are used,

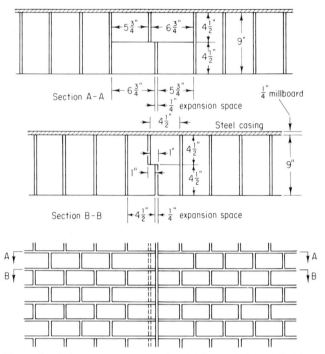

FIG. 24.7 Expansion joint arrangement, 9-in. thick insulating firebrick wall. (*Babcock & Wilcox Company.*)

TABLE 24.4 Width of Expansion Joints in Inches per Foot of Wall Length

Type of refractory	Maximum hot-face temperature			
	1500	2000	2500	3000
Fireclay...............	$\frac{3}{64}$	$\frac{1}{16}$	$\frac{5}{64}$	$\frac{3}{32}$
Magnesite.............	$\frac{3}{32}$	$\frac{1}{8}$	$\frac{3}{16}$	$\frac{1}{4}$
Silica................	$\frac{3}{16}$	$\frac{5}{32}$	$\frac{5}{32}$	$\frac{5}{32}$
Chrome...............	$\frac{1}{16}$	$\frac{5}{64}$	$\frac{3}{32}$	$\frac{5}{32}$
Silicon carbide........	$\frac{1}{32}$	$\frac{3}{64}$	$\frac{1}{16}$	$\frac{5}{64}$
Kaolin...............	$\frac{1}{32}$	$\frac{3}{64}$	$\frac{1}{16}$	$\frac{5}{64}$

they must be anchored at certain intervals to the casing or steelwork to prevent buckling. The design of anchors has been greatly improved in the last few years and has thereby permitted the use of much thinner walls in many types of furnace. This anchoring is usually accomplished by a bolt fitting into a special brick. In Fig. 24.8 is shown one method of anchoring an insulating firebrick wall with a heavy brick tile, while Fig. 24.9 illustrates another method, where alloy through bolts make an excellent construction for low and medium temperatures.

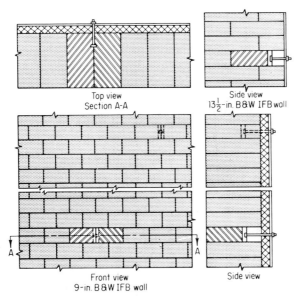

Top view
Section A-A

Side view
13½-in. B&W IFB wall

Front view
9-in. B&W IFB wall

Side view

FIG. 24.8 Tieing IFB walls with firebrick anchors. (*Babcock & Wilcox Company.*)

Figure 24.10 shows a sectionally supported wall of insulating refractories.

High walls, such as those used in boilers, even when 18 to 22½ in. thick must be anchored.   This can be carried out by the use of special anchor tile as furnished by the various manufacturers.

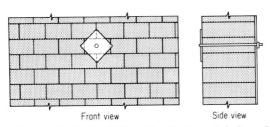

Front view

Side view

FIG. 24.9 Anchoring an IFB wall with through-bolt anchors.

Insulation. When insulation is applied to the outside of the wall, it may be tied in or it may be placed between the refractory and a steel casing in the form of a block.   In some types of construction, the outer face of the wall is made of red brick to protect the insulating material.

This construction is stable, but it increases the heat storage of the wall, which is a disadvantage for intermittent operation, and also requires additional floor space.   Many insulated furnaces simply have the corners protected with angle iron, and the face of the insulation is coated with a hard-face cement and possibly an asphalt coating for waterproofing. Another method of protecting the insulation is to use a casing of approximately 10-gage sheet steel or aluminum.

Monolithic Walls.   In the last 15 years great progress has been made in monolithic walls, either rammed with a plastic mix or gunned with a castable.   The main advance has been the design of anchor blocks to hold

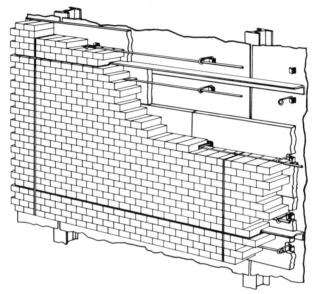

FIG. 24.10   Sectionally supported IFB wall.   (*Babcock & Wilcox Company.*)

the wall in place, as shown in Fig. 7.6.   Walls of this type have given excellent service in many installations.

24.5 Doors, Ports, and Flues. *Door Openings.* The door openings in most furnaces are among the most troublesome points in the structure, both because they mechanically weaken the wall and because the door frame must take considerable abuse from charging and hot gases.   One of the first rules in door construction is to keep the width as small as possible compatible with access to the furnace.   A typical door structure is shown in Fig. 24.11a, where a 4½-in. arch is thrown over the top of the opening and bonded into the wall.   As the front 9-in. bricks are apt to be knocked loose, the sill of the door should be made of a solid construction.   This may be accomplished by using a rowlock course bonded with air-setting mortar or by using larger than 9-in. brick.   Figure 24.11*b* shows a door opening, or port, of somewhat smaller size made up of special shapes, a construction that generally gives excellent service.   In Fig. 24.11*c* is shown a wall opening with a heavy tile across the top.   This construction, though satisfactory for low-temperature work, is to be avoided for high-

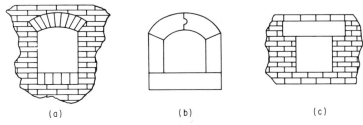

(a)   (b)   (c)

FIG. 24.11   Door openings.

temperature furnaces, as the tile usually cracks in the center and the broken ends sag.

In the low-temperature structures, a steel door jamb is often placed outside the wall to protect the brickwork.   This is welded together from standard angle iron and makes a solid structure, as shown in Fig. 24.12*a*. For higher-temperature furnaces, such as the open hearth, a water-cooled door jamb is generally used.   It holds the brickwork in shape and resists

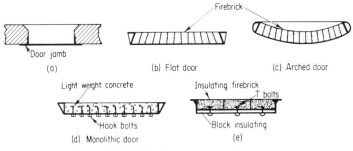

Door jamb
(a)   (b) Flat door   (c) Arched door

Light weight concrete   Insulating firebrick   T bolts
Hook bolts   Block insulating
(d) Monolithic door   (e)

FIG. 24.12   Door construction.

to a considerable extent the abuse to which it is subjected from the charging machine.

Doors.   Doors on furnaces almost always slide in vertical ways with a counterweight to balance them approximately.   The weight of the door should be kept as low as possible in order to reduce the cost of the supports and make the operation quicker.   Two general types of door are the flat door and the arch door, as shown in Fig. 24.12*b* and *c* where the shapes are locked into the door frame to prevent them from bowing in. Such a door will stand considerable mechanical abuse, but it is rather heavy and conducts much heat out of the furnace.

More recently, furnace doors have been lined with lightweight concrete held in place by hook bolts as shown in Fig. 24.12*d*. This construction is very simple to reline, gives a light door, and has a comparatively low heat conductivity. Another construction for higher-temperature use is shown in Fig. 24.12*e*, where the door is lined with insulating firebrick held into the door frame. This gives a lightweight construction and one that is an excellent insulator; the latter is a factor of great importance to the comfort of the men working around the furnace.

Ports and Burner Openings. Ports or openings in the furnace walls, when large, are constructed in the same way as described for the door openings. When the ports are smaller, a special shape often is set into the wall with the proper opening molded into it or a number of shapes are fitted together to form the opening, as would be the case in burner tile as shown in Fig. 24.13. It is often found practical to form the port or opening out of plastic material, which is rammed into place around the form in a rectangular opening in the brickwork. When properly made, such a construction often works out very satisfactorily and avoids the necessity of purchasing special shapes. Such plastic structures, however, are generally not so abrasion resisting as fired shapes and are not so satisfactory for extreme temperature conditions as preburned tile.

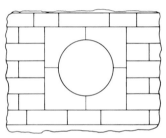

FIG. 24.13 Burner tile set in wall.

When walls are constructed of insulating firebrick, the ports or openings, when small, can often be cut directly in the wall after it has been finished. For larger openings, specially cut shapes can be used to advantage.

Flues and Ducts. Flues and ducts for high-temperature use are lined with a refractory. The smaller sizes are usually rectangular, and the larger ones circular, in which case arch or circle brick are generally employed. Where the moving gases in the duct do not carry a large amount of suspended particles, the ducts can well be lined with insulating firebrick, as a much thinner layer can be used because of its better insulating value. This, in turn, will allow a very much lighter lining, which means lower cost of steelwork and supports. The cost of the duct can often be cut in half by the use of insulating firebrick.

Stacks. Stacks for low-temperature furnaces are often of steel with a brick lining up to a sufficient distance to protect the steelwork. Insulating firebrick and lightweight concrete reinforced with wide-mesh, light-gage, expanded metal lath are frequently used for linings. For really high-temperature work, however, an all-brick stack is desirable such as that used for periodic kilns. Such a stack should have a firebrick lining that is free to expand and contract inside a stable wall such as is shown in Fig. 24.14.

24.6 Sprung Arches. *Stresses and Shapes.* The circular arch, as generally employed, subtends an angle of 60°, which makes the radius of curvature equal to the span. However, there is no standardization in the ratio of span to rise of the arch, as in some cases a high arch is needed for combustion space and in others a very flat arch is desirable to force the flame down on the work. It should be noted that the circular arch in the higher forms becomes relatively unstable because of the fact that the line of thrust of the arch takes the form of an inverted catenary which departs more and more from the circular form as the height

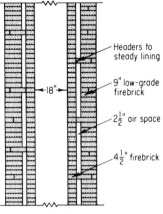

Headers to steady lining

9" low-grade firebrick

$2\frac{1}{2}$" air space

$4\frac{1}{2}$" firebrick

FIG. 24.14 High-temperature stack.

of the arch increases. Therefore, it has been found desirable always to employ the catenary form of arch when the ratio of rise to span is large. The curvature of the circular arch is usually expressed in inches of rise per foot of span.

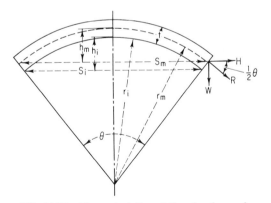

FIG. 24.15 Characteristics of the circular arch.

In Fig. 24.15 is shown a typical circular arch with the principal dimensions given. From this diagram, it is easy to compute the various factors and stresses, which are given in the formulas below. It should be noted that for stress computations, the mean radius r_m must be used, whereas for brickwork estimation, the inner radius r_i is more convenient. The rise, however, is the same in both cases, as $h_m/S_m = h_i/S_i$.

$$r_m = \frac{S_m{}^2}{8h_m} + \frac{h_m}{2}$$

$$h = r_m - \sqrt{r_m{}^2 - \left(\frac{S_m}{2}\right)^2}$$

$$\sin \frac{1}{2}\theta = \frac{S_m}{2r_m}$$

$$\tan \frac{1}{4}\theta = \frac{2h_m}{S_m}$$

$$W = \tfrac{1}{2}ltd$$

$$R = W \operatorname{cosec} \tfrac{1}{2}\theta$$

$$H = W \cot \tfrac{1}{2}\theta$$

$$l = 2\pi r_m \frac{\theta}{360}$$

where θ = the central angle

r_m = the radius at center of arch ring, ft

r_i = the inner radius of the arch ring

S = the span, ft

h = the rise, ft

t = the thickness, ft

W = one-half weight of 1-ft length of arch, lb

H = the horizontal thrust at skewback per foot of length, lb

R = the resultant thrust at skewback per foot of length, lb

l = the length of mean arc

π = 3.1416

d = the density of brick, pcf

In Table 24.5 are given the important factors applying to circular arches of various heights. This table will be found useful for the quick calculation of the characteristics of arches.

The temperature stresses in an arch are difficult to compute, but it may be said that they are greater in the flatter arches because a given expansion of the arch will produce a greater rise and more pinching of the tips of the bricks. To minimize this difficulty, especially with silica brick, tie bolts are often slacked off in a predetermined manner as the arch heats up in order to keep the curvature constant. There is also another temperature stress, due to the temperature gradient through the arch itself, thus causing the tips of the bricks to expand more than the outer face, which puts an additional compression force on the inner surface. This latter effect can be minimized by using insulation on the outside of the arch.

TABLE 24.5

Inches rise per foot span	Central angle	Fraction of a complete circle	Weight factor, W_1	Horizontal-thrust factor, H_1	Resultant-thrust factor, R_1
1.000	37°51′	0.1051	1.02	2.90	3.07
1.250	47 04	0.1307	1.03	2.30	2.51
1.500	56 08	0.1560	1.04	1.87	2.13
1.608	60 00	0.1667	1.05	1.73	2.00
1.750	65 02	0.1807	1.06	1.57	1.86
2.000	73 44	0.2048	1.07	1.30	1.64
2.250	82 13	0.2284	1.09	1.15	1.52
2.500	90 29	0.2513	1.12	1.00	1.44
3.000	106 15	0.2952	1.16	0.76	1.25
6.000*	180 00	0.5000	1.57	0.00	1.00

* Unstable.

To find the weight of the roof multiply together the mean span, length, thickness (in feet), weight per cubic foot, and the factor W_1.

To find the horizontal thrust on one skewback multiply the factor H_1 by one-half the total weight.

To find the resultant thrust on one skewback multiply the factor R_1 by one-half the total weight.

FIG. 24.16 Catenary arch on a large laboratory kiln with front wall removed. (*Babcock & Wilcox Company.*)

The catenary shape has been found particularly satisfactory for some of the higher arches, thus allowing the side walls and crown to be one continuous curve with almost complete elimination of ironwork, as shown in Fig. 24.16. The calculation of the catenary can be carried out as follows.

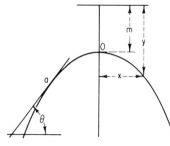

FIG. 24.17 Catenary.

The equation of the catenary as illustrated in Fig. 24.17 is

$$y = \frac{m}{2} \left(e^{x/m} + e^{-x/m} \right)$$

or in hyperbolic functions

$$y = m \cosh \frac{x}{m}$$

The latter form is the most convenient to work with.

The slope angle at any point on the curve is given by

$$\theta = \tan^{-1} \left(\sinh \frac{x}{m} \right)$$

The length of the curve between the apex and any point is

$$l = m \sinh \frac{x}{m}$$

The radius of curvature at any point is

$$r = \frac{y^2}{m}$$

The compression (or tension) at the apex is

$$H = wm$$

where w = the weight per unit length of the curve

Table 24.6 shows the calculation of a typical high-catenary arch.

TABLE 24.6

x	$\dfrac{x}{m}$	$\cosh \dfrac{x}{m}$	y	$y - m$
0	0	1.000	19.0	0
5	0.263	1.035	19.7	7
10	0.527	1.142	21.7	2.6
15	0.789	1.328	25.2	6.2
20	1.054	1.609	30.6	11.6
25	1.317	2.000	38.0	19.0
30	1.579	2.528	48.1	29.1
35	1.841	3.231	61.4	42.4
40	2.108	4.177	79.5	60.5
42	2.210	4.613	88.0	69.0

It will be found, in general, more convenient to lay out the catenary by hanging a sash-weight chain along a vertical surface in such a way as to pass through the three points representing the skewbacks and the apex. The chain thus hanging will form a perfect catenary, and the shape can be traced from this directly on the surface.

Arch Spans and Thicknesses. It is a little difficult to give any hard-and-fast rules regarding the thickness and span of arches, as this ratio depends a great deal on the type of refractory, the temperature of the furnace, and the operating conditions. In general, for heavy brick construction, a 4½-in. thickness would not be used for more than a 5-ft span, a 9-in. arch would be used up to a 12-ft, a 13½-in. arch up to 16 ft, and an 18-in. arch up to 20 ft. However, there have been a number of successful operations at moderate temperatures with much wider arches than these. For example, a firebrick arch of 18-ft span and 9-in. thickness has given an excellent life.

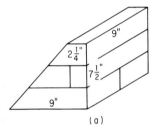 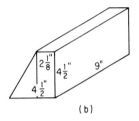

FIG. 24.18 (a) Skewback built with standard 9-in. featheredge brick of the 2½-in. series. (b) Skewback built with standard 9-in. sideskew and endskew brick of the 2½-in. series. (*Courtesy of J. S. McDowell and L. L. Gill, "Steel."*)

In the case of the insulating-firebrick construction, the thickness is seldom over 9 in.; but in moderate-temperature construction, the span often is as great as 15 ft; and in one case, a 22-ft arch has given excellent service.

Skewbacks. The skewback must be made to give the proper angle for the ends of the arch, and it must have sufficient face to accommodate the thickness of the arch ring. Standard skewbacks are made with side skews and featheredge bricks as shown in Fig. 24.18, which accommodate 4½- and 9½-in.-thick arches with rises of respectively 1½ and 2⁵⁄₁₆ in. per ft. However, special skewbacks can be readily obtained from most of the refractory manufacturers, as shown in Figs. 24.19 to 24.21, for an arch of 60° central angle. In the case of insulating firebrick, standard skewbacks can be made up as shown in Fig. 24.22 out of standard shapes for arches having a 60° central angle. In the case of the insulating firebrick, however, it is very easy to grind the skewbacks to any desired slope.

Methods of holding the skewback are shown in Fig. 24.23, but in all cases, the arch thrust should be well supported by the ironwork using either an angle section or a channel. As explained previously, the arch can be supported on top of the wall or supported independently of the

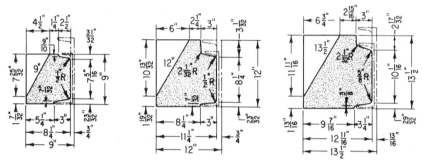

FIG. 24.19 Skewback brick with cutouts for steel channel framework. (*Courtesy of J. S. McDowell and L. L. Gill, "Steel."*)

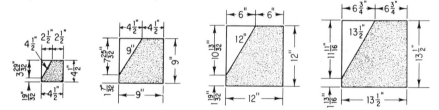

FIG. 24.20 Sixty-degree skewbacks. Shape 60–4½ is 9-in. thick; the others are 4½-in. thick. (*Courtesy of J. S. McDowell and L. L. Gill, "Steel."*)

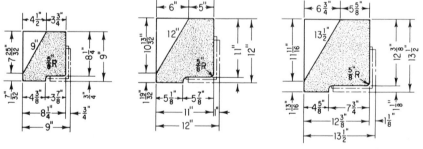

FIG. 24.21 Skewback brick with cutouts for steel angle supporting framework. (*Courtesy of J. S. McDowell and L. L. Gill, "Steel."*)

wall. The latter construction would be used where the life of the side walls is shorter than the arch, such as in a glass tank, in which case a replacement of the walls can be readily made without disturbing the arch. Also, in the case of heavy arches, less load is put on the side walls if the arch is separately supported. On the other hand, wall-supported arches tend to hold the top of the wall from bowing in.

Steelwork. The steelwork for supporting the arch usually consists of an angle or channel running horizontally behind the skewback at the top of the wall; this is held in place by buckstays, the lower ends of which are held in a socket in the concrete foundation and the upper ends tied together with a bolt. The section of iron used in the buckstays depends

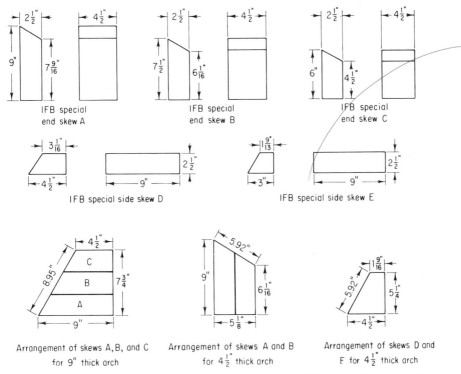

FIG. 24.22 Special IFB skewbacks for sprung arches in which the radius equals the span. Note: Mortar joints $\frac{1}{8}$-in. thick. (*Babcock & Wilcox Company.*)

on the weight of the crown, a crown of insulating firebrick requiring much lighter sections than a heavy brick crown. The strength of these sections can be readily computed from the thrust of the arch, but a considerable factor of safety of at least 5 should be introduced because of the high temperature at which the buckstays go in certain types of construction.

The section of the buckstay has not received so much attention as it should. The I beam in (*A*) of Fig. 24.24 is the usual one because of its high mechanical efficiency. However, from the thermal point of view, it is most inefficient, as the inner flange is at a comparatively high temper-

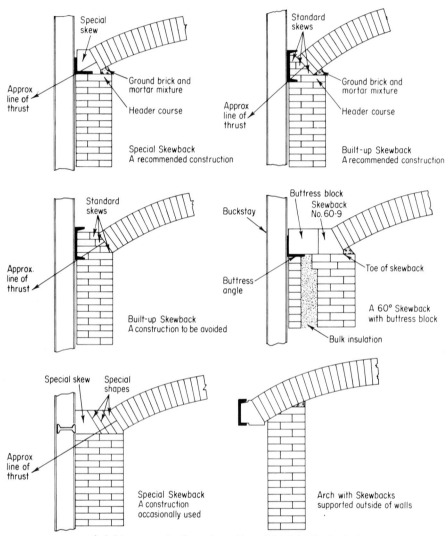

FIG. 24.23 Placement of skewback brick in sprung arches. (*Harbison-Walker Refractories Co.*)

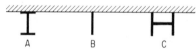

FIG. 24.24 Types of buckstay.

ature whereas the outer flange is at a comparatively low temperature because of the small heat conductivity along the web. This condition produces a maximum of warping, which is often noticed in kiln construction, and thus allows a bowing of the walls. A simple section (*B*) or an H section (*C*) will be found thermally more efficient because it will maintain a much more uniform temperature over the section.

Bonding of the Arch.   Arches can be built up in a bonded construction, such as shown in Fig. 24.25, or in a separate ring construction, such as shown in Fig. 24.26.   The bonded construction has the advantage of giving a more stable structure in case of failure of one or two individual units.   On the other hand, it requires a little more skill in laying in order to get a uniform load on all the bricks.   In the separate ring arch, any one ring can be repaired without disturbing the others, which in some cases is a real advantage.

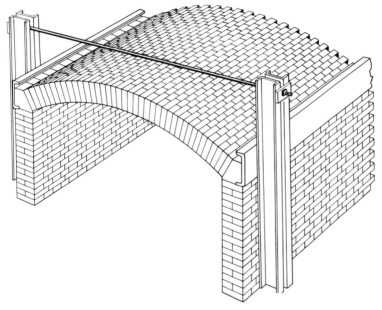

FIG. 24.25   Bonded-arch roof. *(Harbison-Walker Refractories Co.)*

Special Arches.   There are many types of special arch, such as the rib arch, often used in open-hearth-furnace crowns to give stability for long spans.   The jack arch in Fig. 24.27 is made up of special shapes in order to give a flat lower surface.   This type of arch is often used in certain portions of tunnel kilns to conform closely to the charge.

The relieving arch shown in Fig. 24.28 is sometimes used to support the weight of an end wall and to take the load off the main arch.   Relieving arches in modern construction are not used very extensively.

Monolithic Roofs.   Castable roofs with the same type of anchor blocks as shown in Fig. 7.6 are giving good service in many types of furnace.

In Fig. 24.29 is shown a bung arch used in the malleable-iron industry.

These arch rings are wedged tightly into a heavy iron frame so that they can be lifted off as individual units. Brick with good strength and spalling resistance are essential in this type of construction. In Fig. 24.30 is shown a double arch sometimes used in tunnel-kiln construction for preheating of the combustion air. This type of arch should never be

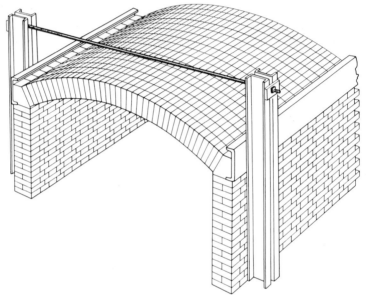

FIG. 24.26 Ring-arch roof. (*Harbison-Walker Refractories Co.*)

FIG. 24.27 Jack arch. This type of arch is formed by special wedge shapes, which may be designed by first laying out a radial sprung arch of proper rise and thickness for the given span, then extending all brick shapes upward to the level of the upper crown surface and downward to the spring line. This arch is designed on a basis of 1½-in. rise per foot of span. (*From "Modern Refractory Practice," Harbison-Walker Refractories Co.*)

used for high-temperature work because of the difficulty in making repairs on the inner arch.

Figure 24.31 shows a perforated arch used in some types of overfired furnace. As this arch is subjected to temperature on both sides, it must be made of material of high load-bearing capacity.

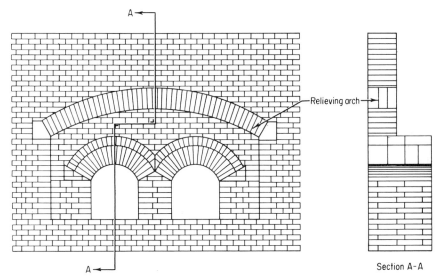

FIG. 24.28 Use of relieving arch to support wall. (*From "Modern Refractory Practice," Harbison-Walker Refractories Co.*)

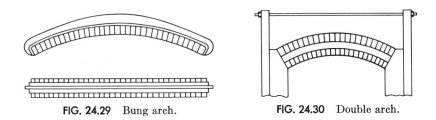

FIG. 24.29 Bung arch.   **FIG. 24.30** Double arch.

Expansion Joints. The longitudinal expansion joints in the crown should, in general, be aligned with the expansion joints in the wall. It is often desirable to recess the end of the crown into the end wall as shown in Fig. 24.32, giving a labyrinth joint. Expansion joints coming in the center of the crown should be covered with a row of splits cemented in place, also shown in Fig. 24.32.

Brickwork Estimation for Sprung Arches. The sprung arch is made up of various combinations of straights, arches, wedges, or keys. The simplest method of finding the number of shapes in a given arch ring consists in deter-

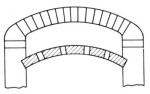

FIG. 24.31 Perforated arch.

mining the count needed for a complete circle and then multiplying the number of each shape by the fraction of the circle covered by the arch. Tables are found in all manufacturers' handbooks.

For example, let us find the number and kind of bricks in a 9-in. arch with an 11-ft span (*S*, of Fig. 24.25) and a rise of 1.608 in. per ft. Here the inner radius is equal to 11 ft; a circle of 22 ft inside diameter requires 91 No. 1 wedges and 264 straights. Multiplying these figures by the fraction of a circle 0.1667, the final count is 15 No. 1 wedges and 44 straights per ring. This calculation allows nothing for joints that may average $\frac{1}{32}$ to $\frac{1}{16}$ in. for the modern dipped joint.

The arch can be computed for any case without the tables by finding the total number of bricks in the ring, which is the outer arc in inches divided by 2.5 or 3.0 in. The number of tapered brick is found by dividing the

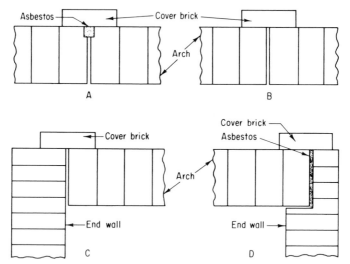

FIG. 24.32 Construction of expansion joints. (*Babcock & Wilcox Company.*)

difference between the inner and outer arcs in inches by the taper of a single brick in inches.

When using insulating firebrick, it is often advisable to have them all ground with such a taper that the joints will all be radial. This construction, of course, makes the best type of arch.

Forms. Steel forms are to be preferred for arch construction that is often rebuilt. On the other hand, wooden forms are less expensive and quite satisfactory for one job. The forms are usually lagged with 2 × 2-in. oak strips.

The form should be accurately placed, with wedges or jacks for lowering. The arch should be built up from both skewbacks simultaneously to meet at the apex. The keys should be carefully fitted and not driven in too hard; otherwise the crown will be strained and the key cracked. Insulating-firebrick keys should never be hammered into place.

24.7 Domes. *Stresses in Domes.* As the stresses in domes are redundant, it is difficult to obtain an exact evaluation. The band stress, however, is approximately given by

$$\frac{\text{Total weight of dome}}{\cos \frac{1}{2}\theta}$$

or

$$\frac{2\pi r_m h_m t d}{\cos \frac{1}{2}\theta}$$

in pounds, where $d =$ the bulk density of the refractory, lb per cu in. The other symbols are shown in Fig. 24.33.

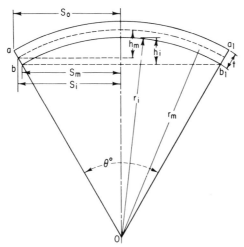

FIG. 24.33 Section of a dome.

It should be noted that in the same way as for the cylindrical roof, the spherical dome is not the most stable form. Fellows[3] and Goodier[4] have shown that this limiting form approaches the surface of revolution formed by an ellipse. Any dome *inside* this surface is stable.

Brickwork Estimating. The dome can be constructed of radial blocks with a double taper to fit the particular radius used, or it can be made out of standard shapes such as straights, wedges, and keys, as is the usual practice in kiln construction. It will be seen, however, that perfect fitting cannot be obtained in either case and mortar must be used to make up the difference. The estimation of brick for the dome is a little complicated, but the following method will probably be found satisfactory for most cases.

With the Use of Special Blocks. The spherical dome can be best constructed of special shapes having a large end of $4\frac{1}{2} \times 2\frac{1}{2}$ in. and a taper such that a projection of their corners will pass through the center of curvature of the dome. This allows strictly radial joints and either 2½- or 4½-in. rings. However, as the rings approach the top of the crown, even these blocks will not fit perfectly unless hand-cut. In practice, such shapes can be economically made by grinding insulating firebrick, but with heavy bricks this is not generally feasible. Table 24.7 gives the dimensions of the small end of the bricks for 9-in.-thick domes having various radii of curvature.

TABLE 24.7   Radial-dome Brick

Size of dome, radius of curvature (inside surface)	Size of inner ends	
	Width, in.	Thickness, in.
4 ft = 48 in.	3.79	2.11
6 ft = 72 in.	4.00	2.22
8 ft = 96 in.	4.12	2.29
10 ft = 120 in.	4.18	2.33
14 ft = 168 in.	4.28	2.37
18 ft = 226 in.	4.33	2.40
24 ft = 288 in.	4.36	2.42
30 ft = 360 in.	4.39	2.44
40 ft = 480 in.	4.42	2.45

With the Use of Standard Shapes. Most domes are built up in 2½-in. rings with combinations of straights and keys alternated with 4½-in. rings of straights and wedges to give the curvature in both directions. A by no means accurate fit can be obtained, and mortar must be used to fill the joints. The estimation of the brickwork has been rather empirical, but the following method will give fairly close results.

Referring to Fig. 24.33, where a cross section of a typical dome is shown, the following relations are evident, all dimensions being in inches. It is assumed, as in the usual construction, that all the rings are 2½ in. wide.

The area of the outer surface of the dome

$$A = 2\pi(r_m + \tfrac{1}{2}t)h_m$$

where h_m = the rise and $r_m = \sqrt{r_m^2 - s_m^2}$

$$\text{Circumference of the segment at } a = 2\pi\left(r_m + \frac{1}{2}t\right)\sin\frac{1}{2}\theta$$

$$\text{Circumference of the segment at } b = 2\pi\left(r_m - \frac{1}{2}t\right)\sin\frac{1}{2}\theta$$

$$\frac{\text{Circumference at } a}{4.5} = \text{number of 9-in. equivalents in first ring}$$

$$\frac{\text{Circumference at } a - \text{circumference at } b}{4.5 - 4.0}$$

$$= \text{number of No. 1 keys in first ring}$$

$$\text{Arc } aa_1 = 2\pi \left(r_m + \frac{1}{2} t \right) \frac{\theta}{360}$$

$$\text{Arc } bb_1 = 2\pi \left(r_m - \frac{1}{2} t \right) \frac{\theta}{360}$$

$$\frac{aa_1}{2.5} = \text{number of rings}$$

$$\frac{aa_1 - bb_1}{2.50 - 1.88} = \text{number of No. 1 wedge rings}$$

Total number of 9-in. equivalents in the dome $= \dfrac{A}{4.5 \times 2.5} = \dfrac{A}{11.25}$

Of these $\dfrac{(aa_1 - bb_1)/(2.50 - 1.88)}{aa_1/2.5}$ are No. 1 wedges.

The remainder have $\dfrac{\dfrac{\text{circumference at } a - \text{circumference at } b}{4.50 - 4.0}}{\dfrac{\text{circumference } a}{4.5}}$ No. 1

keys and the rest are straights.

Example of Dome Calculations

Assume $r_i = 120$ in.

$\qquad\quad t = 9$ in.

$\qquad\quad \theta = 60°$

Circumference of segment at $a = 2\pi(129) \times \frac{1}{2} = 406$ in.

Circumference of segment at $b = 2\pi(120) \times \frac{1}{2} = 377$ in.

$$\frac{406}{4.5} = 90 \text{ nine-inch equivalents in first ring}$$

$$\frac{406 - 377}{4.5 - 4.0} = \frac{29}{0.5} = 58 \text{ No. 1 keys in first ring}$$

$$\frac{aa_1}{2.5} = \frac{129 \times 2\pi \frac{1}{6}}{2.5} = 54 \text{ rings}$$

$$\frac{aa_1 - bb_1}{2.50 - 1.88} = \frac{129 \times 2\pi\frac{1}{6} - 120 \times 2\pi\frac{1}{6}}{0.62} = \frac{135.2 - 125.9}{0.62}$$

$$= \frac{9.3}{0.62} = 15 \text{ No. 1 wedge rings}$$

Total number of brick $= \dfrac{2\pi 129 \times 17}{2\frac{1}{2} \times 4\frac{1}{2}} = 1{,}230$ nine-inch equivalents

In whole dome

$$1{,}230 \times {}^{15}\!/_{54} = 342 \text{ No. 1 wedges}$$
$$(1{,}230 - 242){}^{58}\!/_{90} = 585 \text{ No. 1 keys}$$
$$1{,}230 - 880 = 303 \text{ straights}$$

and oil stills because there is no limit to the span that can be obtained. It also finds application in soaking pit covers where the light weight allows rapid action.

In this type of roof, 9-in. straights can be used, which are easy to drill and assemble. The high insulating value of the insulating firebrick keeps the supporting steelwork comparatively cool. Usually the whole roof, even when as large as 20 by 20 ft, is cemented together with air-set mortar into a monolithic structure with the only expansion joint around the edge. This joint should be sealed for gastightness.

24.9 Bricklaying. It is needless to say that good refractory construction requires the services of an experienced furnace-brick mason. Even under these conditions, complete drawings should be prepared for anything except the simplest structures. Joints should be kept thin, which is possible with modern refractories of uniform size.

When ordering brick, a certain amount in addition to that estimated must be added for losses in cutting, broken bricks, etc. This is called "overage" and amounts to from 2 to 5 percent on large jobs to 10 or 15 percent on small ones. If there is much cutting, the overage must be increased. It also should be remembered that when bricks come in cartons, the number ordered must be an even multiple of the number in the carton, which may make it necessary to purchase more bricks than actually needed.

BIBLIOGRAPHY

 1. Anon.: "Modern Refractory Practice," Harbison-Walker Refractories Co.
 2. Norton, F. H.: The Design of Arches for Kilns and Furnaces, *J. Am. Ceram. Soc.*, **9** (3), 144, 1926.
 3. Fellows, J. A.: Domes for Circular Kilns, *J. Am. Ceram. Soc.*, **15** (9), 508, 1932.
 4. Goodier, J. N.: Stresses in Domes and Crowns of Circular Kilns, *J. Am. Ceram. Soc.*, **16** (5), 220, 1933.
 5. McDowell, J. S., and L. L. Gill: Refractory Arches, I, *Steel*, **102,** 42, 1938.
 6. McDowell, J. S., and L. L. Gill: Refractory Arches, II, *Steel*, **102,** 48, 1938.
 7. McDowell, J. S.: Sprung-arch Roofs of High Temperature Furnaces, I, *Blast Furnace Steel Plant*, **27** (6), 592, 1939.
 8. McDowell, J. S.: Sprung-arch Roofs of High Temperature Furnaces, II, *Blast Furnace Steel Plant*, **27** (9), 947, 1939.
 9. McDowell, J. S.: Sprung-arch Roofs of High Temperature Furnaces, III, *Blast Furnace Steel Plant*, **28** (2), 161, 1940.
10. Anon.: "Ceramic Data Book," Industrial Publications, Inc., Chicago, 1966.

Reference Tables

TABLE A.1 Temperature-conversion Table*

C	0	10	20	30	40	50	60	70	80	90
	F	F	F	F	F	F	F	F	F	F
−200	−328	−346	−364	−382	−400	−418	−436	−454	.....	.....
−100	−148	−166	−184	−202	−220	−238	−256	−274	−292	−310
− 0	+ 32	+ 14	− 4	− 22	− 40	− 58	− 76	− 94	−112	−130
0	32	50	68	86	104	122	140	158	176	194
100	212	230	248	266	284	302	320	338	356	374
200	392	410	428	446	464	482	500	518	536	554
300	572	590	608	626	644	662	680	698	716	734
400	752	770	788	806	824	842	860	878	896	914
500	932	950	968	986	1004	1022	1040	1058	1076	1094
600	1112	1130	1148	1166	1184	1202	1220	1238	1256	1274
700	1292	1310	1328	1346	1364	1382	1400	1418	1436	1454
800	1472	1490	1508	1526	1544	1562	1580	1598	1616	1634
900	1652	1670	1688	1706	1724	1742	1760	1778	1796	1814
1000	1832	1850	1868	1886	1904	1922	1940	1958	1976	1994
1100	2012	2030	2048	2066	2084	2102	2120	2138	2156	2174
1200	2192	2210	2228	2246	2264	2282	2300	2318	2336	2354
1300	2372	2390	2408	2426	2444	2462	2480	2498	2516	2534
1400	2552	2570	2588	2606	2624	2642	2660	2678	2696	2714
1500	2732	2750	2768	2786	2804	2822	2840	2858	2876	2894
1600	2912	2930	2948	2966	2984	3002	3020	3038	3056	3074
1700	3092	3110	3128	3146	3164	3182	3200	3218	3236	3254
1800	3272	3290	3308	3326	3344	3362	3380	3398	3416	3434
1900	3452	3470	3488	3506	3524	3542	3560	3578	3596	3614
2000	3632	3650	3668	3686	3704	3722	3740	3758	3776	3794
2100	3812	3830	3848	3866	3884	3902	3920	3938	3956	3974
2200	3992	4010	4028	4046	4064	4082	4100	4118	4136	4154
2300	4172	4190	4208	4226	4244	4262	4280	4298	4316	4334
2400	4352	4370	4388	4406	4424	4442	4460	4478	4496	4514
2500	4532	4550	4568	4586	4604	4622	4640	4658	4676	4694
2600	4712	4730	4748	4766	4784	4802	4820	4838	4856	4874
2700	4892	4910	4928	4946	4964	4982	5000	5018	5036	5054
2800	5072	5090	5108	5126	5144	5162	5180	5198	5216	5234
2900	5252	5270	5288	5306	5324	5342	5360	5378	5396	5414
3000	5432	5450	5468	5486	5504	5522	5540	5558	5576	5594
3100	5612	5630	5648	5666	5684	5702	5720	5738	5756	5774
3200	5792	5810	5828	5846	5864	5882	5900	5918	5936	5954
3300	5972	5990	6008	6026	6044	6062	6080	6098	6116	6134
3400	6152	6170	6188	6206	6224	6242	6260	6278	6296	6314
3500	6332	6350	6368	6386	6404	6422	6440	6458	6476	6494
3600	6512	6530	6548	6566	6584	6602	6620	6638	6656	6674
3700	6692	6710	6728	6646	6764	6782	6800	6818	6836	6854
3800	6872	6890	6908	6926	6944	6962	6980	6998	7016	7034
3900	7052	7070	7088	7106	7124	7142	7160	7178	7196	7214

°C	°F
1	1.8
2	3.6
3	5.4
4	7.2
5	9.0
6	10.8
7	12.6
8	14.4
9	16.2
10	18.0

°F	°C
1	0.56
2	1.11
3	1.67
4	2.22
5	2.78
6	3.33
7	3.89
8	4.44
9	5.00
10	5.56
11	6.11
12	6.67
13	7.22
14	7.78
15	8.33
16	8.89
17	9.44
18	10.00

* Dr. L. Waldo, in *Metallurgical and Chemical Engineering*, March, 1910.
Examples.   1347°C = 2444°F + 12.6°F = 2456.6°F; 3367°F = 1850°C + 2.78°C = 1852.78°C.

TABLE A.2 Standard Calibration Data for Chromel-Alumel Thermocouples

Emf, mv	Reference junction at 0°C					
	0	10	20	30	40	50
	Temperature, 0°C					
0	0	246	485	720	966	1232
0.2	5	251	490	725	972	1237
0.4	10	256	494	730	977	1243
0.6	15	261	499	735	982	1249
0.8	20	266	504	740	987	1254
1.0	25	271	508	744	992	1260
1.2	30	276	513	749	997	1266
1.4	35	280	518	754	1002	1271
1.6	40	285	523	759	1007	1277
1.8	45	290	527	764	1013	1283
2.0	50	295	532	768	1018	1288
2.2	54	300	537	773	1023	1294
2.4	59	305	541	778	1028	1300
2.6	64	310	546	783	1033	1306
2.8	69	315	551	788	1038	1311
3.0	74	319	555	792	1044	1317
3.2	79	324	560	797	1049	1323
3.4	83	329	565	802	1054	1329
3.6	88	334	570	807	1059	1334
3.8	93	338	574	812	1065	1340
4.0	98	343	579	817	1070	1346
4.2	102	348	584	822	1075	1352
4.4	107	353	588	827	1081	1358
4.6	112	358	593	832	1086	1364
4.8	117	362	598	837	1091	1370
5.0	122	367	602	841	1096	1376
5.2	127	372	607	846	1102	1382
5.4	132	376	612	851	1107	1388
5.6	137	381	616	856	1112	1394
5.8	142	386	621	861	1118	1400
6.0	147	391	626	866	1123	
6.2	152	396	631	871	1128	
6.4	157	400	635	876	1134	
6.6	162	405	640	881	1139	
6.8	167	410	645	886	1144	

TABLE A.2   (Continued)

Emf, mv	Reference junction at 0°C					
	0	10	20	30	40	50
	Temperature, 0°C					
7.0	172	414	649	891	1150	
7.2	177	419	654	896	1155	
7.4	182	424	659	901	1161	
7.6	187	429	664	906	1166	
7.8	192	433	668	911	1171	
8.0	197	438	673	916	1177	
8.2	202	443	678	921	1182	
8.4	207	448	683	926	1188	
8.6	212	452	687	931	1193	
8.8	217	457	692	936	1199	
9.0	222	462	697	941	1204	
9.2	227	466	701	946	1210	
9.4	232	471	706	951	1215	
9.6	237	476	711	956	1221	
9.8	241	480	716	961	1226	
10.0	246	485	720	966	1232	

TABLE A.3 Standard Calibration Data for Chromel-Alumel
Thermocouples

Emf, mv	Reference junction at 32°F					
	0	10	20	30	40	50
	Temperature, °F					
0	32	475	905	1329	1772	2250
0.2	41	484	913	1338	1781	2260
0.4	50	493	922	1346	1790	2270
0.6	59	502	930	1355	1799	2280
0.8	68	510	939	1363	1808	2290
1.0	77	519	947	1372	1818	2300
1.2	86	528	956	1380	1827	2310
1.4	95	537	964	1389	1836	2320
1.6	104	546	973	1398	1845	2331
1.8	113	554	981	1407	1855	2341
2.0	121	563	990	1415	1864	2351
2.2	130	572	998	1424	1873	2362
2.4	139	580	1006	1433	1882	2372
2.6	147	589	1015	1441	1892	2382
2.8	156	598	1023	1450	1901	2393
3.0	165	607	1032	1459	1911	2403
3.2	173	615	1040	1467	1920	2413
3.4	182	624	1049	1476	1930	2424
3.6	190	632	1057	1485	1939	2434
3.8	199	641	1065	1494	1949	2445
4.0	208	650	1074	1503	1958	2455
4.2	217	658	1083	1511	1967	2466
4.4	225	667	1091	1520	1977	2476
4.6	234	675	1099	1529	1986	2487
4.8	243	684	1108	1538	1996	2497
5.0	251	693	1116	1547	2005	
5.2	260	701	1125	1555	2015	
5.4	269	710	1133	1564	2024	
5.6	278	718	1142	1573	2034	
5.8	287	727	1150	1582	2044	
6.0	296	735	1158	1591	2053	
6.2	305	744	1167	1600	2063	
6.4	314	752	1175	1609	2072	
6.6	323	760	1184	1618	2082	
6.8	332	769	1193	1627	2092	

TABLE A.3 *(Continued)*

Emf, mv	Reference junction at 32°F					
	0	10	20	30	40	50
	Temperature, °F					
7.0	341	778	1201	1636	2101	
7.2	350	786	1210	1645	2111	
7.4	359	795	1218	1654	2121	
7.6	368	803	1227	1663	2130	
7.8	377	812	1235	1672	2140	
8.0	386	820	1243	1680	2150	
8.2	395	829	1252	1689	2160	
8.4	404	838	1260	1698	2170	
8.6	413	846	1269	1708	2180	
8.8	422	855	1278	1717	2190	
9.0	431	863	1286	1726	2200	
9.2	440	872	1295	1735	2210	
9.4	449	880	1303	1744	2220	
9.6	457	889	1312	1753	2230	
9.8	466	897	1320	1762	2240	
10.0	475	905	1329	1772	2250	

TABLE A.4 Standard Calibration Data for Thermocouples from Platinum and Platinum Alloyed with 10 Percent Rhodium

Emf, µv	0	1,000	2,000	3,000	4,000	5,000	6,000	7,000	8,000	9,000	10,000	11,000	12,000	13,000	14,000	15,000	16,000	17,000	18,000
											Temperatures, °C								
0	0.0	147.1	265.4	374.3	478.1	578.3	675.3	769.5	861.1	950.4	1037.3	1122.2	1205.9	1289.3	1372.4	1454.8	1537.5	1620.9	1704.3
100	17.8	159.7	276.6	384.9	488.3	588.1	684.8	778.8	870.1	959.2	1045.9	1130.6	1214.2	1297.7	1380.7	1463.0	1545.8	1629.2	1712.6
200	34.5	172.1	287.7	395.4	498.4	597.9	694.3	788.0	879.1	968.0	1054.4	1139.0	1222.6	1306.0	1389.0	1471.2	1554.1	1637.6	1721.0
300	50.3	184.3	298.7	405.9	508.5	607.7	703.8	797.2	888.1	976.7	1062.9	1147.4	1230.3	1314.3	1397.3	1479.4	1562.4	1645.9	1729.3
400	65.4	196.3	309.7	416.3	518.6	617.4	713.3	806.4	897.1	985.4	1071.4	1155.8	1239.3	1322.6	1405.6	1487.7	1570.8	1654.3	1737.7
500	80.0	208.1	320.6	426.7	528.6	627.1	722.7	815.6	906.1	994.1	1079.9	1164.2	1247.6	1330.9	1413.8	1496.0	1579.1	1662.6	1746.0
600	94.1	219.7	331.5	437.1	538.6	636.8	732.1	824.7	915.0	1002.8	1088.4	1172.5	1255.9	1339.2	1422.0	1504.3	1587.5	1670.9	1754.3
700	107.8	231.2	342.3	447.4	548.6	646.5	741.5	833.8	923.9	1011.5	1096.9	1180.9	1264.3	1347.5	1430.2	1512.6	1595.8	1679.3	
800	121.2	242.7	353.0	457.7	558.5	656.1	750.9	842.9	932.8	1020.1	1105.4	1189.2	1272.6	1355.8	1438.4	1520.9	1604.2	1687.6	
900	134.3	254.1	363.7	467.9	568.4	665.7	760.2	852.0	941.6	1028.7	1113.8	1197.9	1281.0	1364.1	1446.6	1529.2	1612.5	1696.0	
1,000	147.1	265.4	374.3	478.1	578.3	675.3	769.5	861.1	950.4	1037.3	1122.2	1205.9	1289.3	1372.4	1454.8	1537.5	1620.9	1704.3	

TABLE A.5 Standard Calibration Data for Thermocouples from Platinum and Platinum Alloyed with 10 Percent Rhodium

Emf, µV	0	1,000	2,000	3,000	4,000	5,000	6,000	7,000	8,000	9,000	10,000	11,000	12,000	13,000	14,000	15,000	16,000	17,000	18,000
									Temperatures, °F										
0	32.0	296.8	509.7	705.7	892.6	1072.9	1247.5	1417.1	1582.0	1742.7	1899.1	2052.0	2202.6	2352.7	2502.3	2650.6	2799.5	2949.6	3099.7
100	42.0	319.5	529.9	724.8	910.9	1090.6	1264.6	1433.8	1598.2	1758.6	1914.6	2067.1	2217.6	2367.9	2517.3	2665.4	2814.4	2964.6	3114.7
200	94.1	341.8	549.9	743.7	929.1	1108.1	1281.7	1450.4	1614.4	1774.4	1929.9	2082.2	2232.7	2382.8	2532.2	2680.2	2829.4	2979.7	3129.8
300	122.5	363.7	569.7	762.6	947.3	1125.9	1298.8	1467.0	1630.6	1790.1	1945.2	2097.3	2247.6	2397.7	2547.1	2694.9	2844.3	2994.6	3144.7
400	149.7	385.5	589.5	781.3	965.5	1143.3	1315.9	1483.5	1646.8	1805.7	1960.5	2112.4	2262.7	2412.7	2562.1	2709.9	2859.4	3009.7	3159.9
500	176.0	406.6	609.1	800.1	983.5	1160.8	1332.9	1500.1	1663.0	1821.4	1975.8	2127.6	2277.7	2427.6	2576.8	2724.8	2874.4	3024.7	3174.8
600	201.4	427.5	628.7	818.8	1001.5	1178.2	1349.8	1516.5	1679.0	1837.0	1991.1	2142.5	2292.6	2442.6	2591.6	2739.7	2889.5	3039.6	3189.7
700	226.0	448.2	648.1	837.3	1019.5	1195.7	1366.7	1532.8	1694.8	1852.7	2006.4	2157.6	2307.7	2457.5	2606.4	2754.7	2904.4	3054.7	
800	250.2	468.9	667.4	855.9	1037.3	1213.0	1383.6	1549.2	1711.0	1868.2	2021.7	2172.6	2322.7	2472.4	2621.1	2769.6	2919.6	3069.7	
900	273.7	489.4	686.7	874.2	1055.1	1230.3	1400.4	1565.6	1726.9	1883.7	2036.8	2187.7	2337.8	2487.4	2635.9	2784.6	2934.8	3084.8	
1,000	296.8	509.7	705.7	892.6	1072.9	1247.5	1417.1	1582.0	1742.7	1899.1	2052.0	2202.6	2352.7	2502.3	2650.6	2799.5	2949.6	3099.7	

429

TABLE A.6　Standard Calibration Data for Copper-Constantan Thermocouple

Emf, μv	0	1,000	2,000	3,000	4,000	5,000	6,000	7,000	8,000	9,000	10,000	11,000	12,000	13,000	14,000	15,000	16,000	17,000	18,000
										Temperatures, °C									
0	0.0	25.3	49.2	72.1	94.1	115.3	135.9	155.9	175.5	194.6	213.4	231.7	249.8	267.6	285.1	302.4	319.5	336.4	353.1
100	2.6	27.7	51.5	74.3	96.2	117.4	137.9	157.0	177.4	196.5	215.2	233.6	251.6	269.4	286.9	304.1	321.2	338.0	
200	5.2	30.2	53.9	76.5	98.4	119.5	140.0	159.9	179.4	198.4	217.2	235.4	253.4	271.1	288.6	305.9	322.9	339.7	
300	7.7	32.6	56.2	78.8	100.5	121.6	142.0	161.9	181.3	200.3	218.9	237.2	255.2	272.9	290.3	307.6	324.6	341.4	
400	10.3	35.0	58.5	81.0	102.7	123.6	144.0	163.8	183.2	202.2	220.8	239.0	257.0	274.6	292.1	309.3	326.3	343.1	
500	12.8	37.4	60.8	83.2	104.8	125.7	146.0	165.8	185.1	204.0	222.6	240.8	258.7	276.4	293.8	311.0	327.9	344.7	
600	15.3	39.8	63.0	85.4	106.9	127.7	148.0	167.7	187.0	205.9	224.4	242.6	260.5	278.2	295.5	312.7	329.6	346.4	
700	17.8	42.2	65.3	87.6	109.0	129.8	150.0	169.7	188.0	207.8	226.3	244.4	262.3	279.9	297.3	314.4	331.3	348.1	
800	20.3	44.5	67.6	89.7	111.1	131.8	152.0	171.6	190.8	209.8	228.1	246.2	264.1	281.6	299.0	316.1	333.0	349.7	
900	22.8	46.9	69.8	91.9	113.2	133.9	154.0	173.6	192.7	211.5	229.9	248.0	265.8	283.4	300.7	317.8	334.7	351.4	
1,000	25.3	49.2	72.1	94.1	115.3	135.9	155.9	175.5	194.6	213.4	231.7	249.8	267.6	285.1	302.4	319.5	336.4	353.1	

TABLE A.7 Standard Calibration Data for Copper-Constantan Thermocouple

Emf, μv	Temperatures, °F																			
	0	1,000	2,000	3,000	4,000	5,000	6,000	7,000	8,000	9,000	10,000	11,000	12,000	13,000	14,000	15,000	16,000	17,000	18,000	19,000
0	32.0	77.5	120.6	161.7	201.3	239.6	276.6	312.7	347.9	382.3	416.0	449.1	481.7	513.7	545.2	576.4	607.1	637.4	667.6	
100	36.7	81.9	124.8	165.7	205.2	243.3	280.3	314.6	351.4	385.7	419.4	452.4	484.9	516.8	548.4	579.4	610.1	640.5		
200	41.4	86.3	128.9	169.8	209.1	247.1	283.9	319.8	354.8	389.1	423.0	455.7	488.1	519.0	551.5	582.5	613.2	643.5		
300	45.9	90.6	133.1	173.9	212.9	250.8	287.6	323.3	358.3	392.5	426.0	459.3	491.3	523.2	554.6	585.6	616.2	646.5		
400	50.5	95.0	137.2	177.7	216.8	254.5	291.2	326.9	361.8	395.9	429.3	462.2	494.5	526.4	557.7	588.7	619.3	649.5		
500	55.0	99.3	141.4	181.7	220.6	258.2	294.8	330.4	365.2	399.3	432.7	465.5	497.7	529.5	560.8	591.8	622.3	652.5		
600	59.5	103.6	145.5	185.7	224.4	261.9	298.4	333.9	368.6	402.6	436.0	468.7	500.9	532.7	564.0	594.8	625.4	655.5		
700	64.0	107.9	149.5	189.6	228.2	265.6	302.0	337.4	370.4	406.0	439.3	472.0	504.1	535.8	567.1	597.9	628.4	658.5		
800	68.5	112.1	153.6	193.5	232.0	269.3	305.6	340.9	375.6	409.3	442.6	475.2	507.3	539.0	570.2	601.0	631.4	661.6		
900	73.0	116.3	157.7	197.4	235.8	273.0	309.1	344.4	378.9	412.7	445.9	478.4	510.5	542.1	573.6	604.0	634.4	664.6		
1,000	77.5	120.6	161.7	201.3	239.6	276.6	312.7	347.9	382.3	416.0	449.1	481.7	513.7	545.2	576.4	607.1	637.4	667.6		

TABLE A.8 Formulas Useful in Calculations

Rectangle:
 Area $= ab$

Triangle:
$$\text{Area} = \frac{ah}{2}$$

Regular Polygon:

Sides	Area
5	$1.720 \times S^2$
6	$2.598 \times S^2$
7	$3.634 \times S^2$
8	$4.828 \times S^2$
9	$6.182 \times S^2$
10	$7.694 \times S^2$
11	$9.366 \times S^2$
12	$11.196 \times S^2$

Circle:
 Circumference $= \pi d = 2\pi r$

$$\text{Diameter} = 2r = \frac{C}{\pi} = 2\sqrt{\frac{\text{area}}{\pi}}$$

$$\text{Radius} = \frac{d}{2} = \frac{C}{2\pi} = \sqrt{\frac{\text{area}}{\pi}}$$

$$\text{Area} = \frac{\pi d^2}{4} = \pi r^2 = 0.7854 d^2$$

 Length of chord, $x = 2\sqrt{a(d-a)^2} = 2r(\sin \tfrac{1}{2}\theta)$

Segment:
$$\text{Area} = \frac{rl - x(r-a)}{2}$$

 Height of arc above chord, $a = r - \sqrt{r^2 - (x/2)^2}$

 Radius, when chord and height are known, $r = \dfrac{(x/2)^2 + h^2}{2h}$

 Length of arc, $l = 2\pi r \dfrac{\theta}{360} = \pi r \dfrac{\theta}{180}$

Sector:
$$\text{Area} = \frac{1}{2} lr = \pi r^2 \frac{\theta}{360}$$

Annulus (Circular Ring):
$$\begin{aligned}
\text{Area} &= \pi(r_2{}^2 - r_1{}^2) \\
&= \pi(r_2 + r_1)(r_2 - r_1) \\
&= 0.7854(d_2{}^2 - d_1{}^2)
\end{aligned}$$

TABLE A.8 (Continued)

Ellipse:

$$\text{Area} = \frac{\pi}{4} Dd = 0.7854Dd$$

$$\text{Perimeter (approx.)} = 2\pi \sqrt{\frac{D^2 + d^2}{8}}$$

Skewbacks:

To find a skew to fit a given arch:
1. Find subtended angle θ
2. Find sine and cosine of $\frac{1}{2}\theta$
3. H = thickness of arch $\times$ sin $\frac{1}{2}\theta$
4. V = thickness of arch $\times$ cos $\frac{1}{2}\theta$
Slope of the skewback from the horizontal = $90° - \frac{1}{2}\theta$

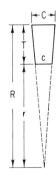

Circle Brick:

Radius of radial brick, $r = \dfrac{cT}{(C - c)}$

Number of radial brick to make a circle:

$$n = \frac{\pi \times \text{diameter of circle}}{\text{thickness of large end}}$$

$$= \frac{2\pi r}{C}$$

Number of straights to fill out a circle of given diameter using given number of radial brick:

$$n = \frac{\pi(\text{given outside diameter}) - (\text{possible diameter with radial brick})}{\text{thickness of straight}}$$

Cylinder:

Volume $= \pi r^2 a$

$\qquad = 0.7854 \, d^2 a$

Cylindrical surface area $= \pi \, da$

Total surface area $\quad = 2\pi r(r + a)$

Regular Triangular Prisms:

Volume $= \frac{1}{2}bha$

Lateral surface area $= (b + m + n)a$

TABLE A.8 (*Continued*)

Pyramid:
$$\text{Volume} = \frac{\text{area of base} \times a}{3}$$
$$\text{Lateral surface area} = \frac{\text{perimeter of base} \times s}{2}$$

Cone:
$$\text{Volume} = \frac{\pi r^2 a}{3}$$
$$= 0.2618 \, d^2 a$$
$$\text{Lateral surface area} = \pi r s$$

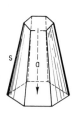

Frustum of Pyramid:
$$\text{Lateral surface area} = \frac{(P + p)s}{2}$$
$$\text{Volume} = \frac{a}{3}(b + B + \sqrt{Bb})$$
where P = perimeter of lower base
$\quad\quad\; p$ = perimeter of upper base
$\quad\quad\; B$ = area of lower base
$\quad\quad\; b$ = area of upper base

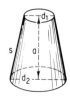

Frustrum of Cone:
$$\text{Volume} = \frac{a}{3}(b + B + \sqrt{Bb}), \text{ or}$$
$$= \frac{\pi}{12} a(d_1^2 + d_1 d_2 + d_2^2)$$
$$\text{Area of conic surface} = \frac{\pi S}{2}(d_1 + d_2)$$

Sphere:
$$\text{Area of surface} = 4\pi r^2 = \pi d^2 = 12.566 r^2$$
$$\text{Volume} = \frac{4\pi r^3}{3} = 4.1888 r^3$$
$$= \frac{\pi d^3}{6} = 0.5236 d^3$$

Segment of Sphere:
$$\text{Area of spherical surface} = 2\pi rh = \frac{\pi}{4}(d^2 + 4h^2)$$
$$\text{Volume} = \pi h^2 \left(r - \frac{h}{3} \right)$$

Name Index

Subject Index

A

Abrasion resistance, 326
Abrasion-resisting metals, 144
Accessory minerals in clay, 68
Adsorbed ions, 70
Aetna Fire Brick Co., 20
Aftershrinkage, 290
(*See also* Reheat shrinkage)
Air furnace, 342, 345
Akron, Ohio, early firebrick making, 19, 21
Alumina, hydrated, 66
Aluminum electrolytic cell, 369, 372
Aluminum holding furnace, 372
Aluminum melting furnace, 369
Aluminum transfer ladle, 372
Amanda Furnace, 21
Amboy, N.J., early history of firebrick making, 16
Amsterdam, early firebrick making, 16
Andalusite, 174, 185
Andreassen pipette, 320
Angstrom unit, 62
Annulus, properties of, 432
Apparatus, for measuring drying shrinkage, 159
  for measuring firing shrinkage, 281, 282
  for measuring particle size, 318
  for measuring viscosity of slips, 116
  for measuring workability of clay, 117
Arch, bonding of, 413
  forms for, 416
  rise of, 405
  span of, 405, 409
  special, bung, 413
    catenary, 407
    calculation of, 408
    jack, 413
    monolithic, 413
    perforated, 414
    relieving, 413
    suspended, 421
  stresses in, 405–407
  thickness of, 409

Auger, 128
  deairing, 132, 133

B

Babcock and Wilcox Co., 22
Back up insulation, 32
  classification of, 33
  properties of, 35
Baddeleyite, 84, 177
Ball clays, 72
Ball mill, 98, 100
  capacity of, 101
  vibratory, 99
Baltimore, early history of firebrick making, 18, 19
Base exchange capacity, 70
Batching, automatic, 150
Bauxite, 75
  deposits of, 77
Bayerite, heat effects on, 168
Bellport Furnace, 21
Bennington, history of early firebrick making, 15, 17–18, 22
Bessemer converter, 344
  refractories for, 346
Bibliographies, 12
Bigelow-Liptak suspended roof, 420
Billet-heating furnace, 353
Black-liquor process, 388
Blast furnace, iron, carbon blocks for, 334
    checkers for, 335
    construction of, 332
    principle of operation, 332
    refractories, life of, 335
    refractories in, 334
    stoves for, 335
  lead, 367
Blister copper, 357
Bloating in overfiring, 185
Boiler, refractories for, 385
  slag tap, 388
  tube-wall, 386
  wall for, 386